CONDUCTION OF ELECTRICITY
THROUGH GASES

IN TWO VOLUMES

VOLUME I

CONDUCTION OF ELECTRICITY THROUGH GASES

by

Sir J. J. THOMSON, O.M., F.R.S.

MASTER OF TRINITY COLLEGE, CAMBRIDGE
PROFESSOR OF EXPERIMENTAL PHYSICS
CAMBRIDGE

and

G. P. THOMSON, M.A.

FELLOW OF CORPUS CHRISTI COLLEGE, CAMBRIDGE
PROFESSOR OF NATURAL PHILOSOPHY IN THE
UNIVERSITY OF ABERDEEN

THIRD EDITION

VOLUME I

General Properties of Ions
Ionisation by Heat and Light

CAMBRIDGE
AT THE UNIVERSITY PRESS
1928

CAMBRIDGE UNIVERSITY PRESS
Cambridge, New York, Melbourne, Madrid, Cape Town,
Singapore, São Paulo, Delhi, Mexico City

Cambridge University Press
The Edinburgh Building, Cambridge CB2 8RU, UK

Published in the United States of America by Cambridge University Press, New York

www.cambridge.org
Information on this title: www.cambridge.org/9781107414273

© Cambridge University Press 1928

This publication is in copyright. Subject to statutory exception
and to the provisions of relevant collective licensing agreements,
no reproduction of any part may take place without the written
permission of Cambridge University Press.

First Edition 1903
Second Edition 1906
Third Edition (in Two Volumes)
Volume I 1928
First published 1928
First paperback edition 2013

A catalogue record for this publication is available from the British Library

ISBN 978-1-107-03883-7 Hardback
ISBN 978-1-107-41427-3 Paperback

Cambridge University Press has no responsibility for the persistence or
accuracy of URLs for external or third-party internet websites referred to in
this publication, and does not guarantee that any content on such websites is,
or will remain, accurate or appropriate.

PREFACE TO THE THIRD EDITION

THE preparation of this Edition was commenced some fifteen years ago and some of it was in type when the War broke out. After the War the pressure of other duties left me with little leisure for the very heavy task of dealing with the vast number of researches on the Discharge of Electricity through Gases which have been made since the publication of the Second Edition. The publication of this Edition is due to my having had the co-operation of my son, Professor G. P. Thomson, who has done most of the work required for its preparation. The growth of the subject has made it impossible to deal with it in one volume of moderate size; it has been necessary to extend the book to two volumes, of which this is the first, and which deals with what may be described as the general properties of ions.

We have adopted a decimal notation for numbering the paragraphs, those that were in the Second Edition are denoted by integers and those dealing with subjects cognate to the original paragraph by this integer, followed by a decimal. Most, though not all, of the original paragraphs have been retained, a few in a shortened form. Otherwise little alteration has been made in them beyond replacing the values of the fundamental constants by the more accurate ones obtained since the publication of the earlier editions. Some of these paragraphs deal with matters which are now chiefly of historic interest and which might have been omitted or curtailed if we had been re-writing the book. There seems, however, to be some advantage in maintaining continuity with the older editions, and perhaps after all Science is more easily digested when it is lightened by something less formidable than mathematical analysis or the precautions which must be taken to get the greatest accuracy in the determination of physical constants. We are indebted for the Name Index to Mrs G. P. Thomson, to whom we offer our sincere thanks.

J. J. THOMSON

TRINITY LODGE
August 1928

CONTENTS OF VOLUME I.

CHAP.		PAGE
I.	Electrical Conductivity of Gases in a Normal State	1
II.	Properties of a Gas when in the Conducting State	13
III.	Mobility of Ions	85
IV.	Mathematical Theory of the Conduction of Electricity through a Gas containing Ions	193
V.	Effect produced by a Magnetic Field on the Motion of the Ions	215
VI.	Determination of the Ratio of the Charge to the Mass of an Ion	229
VII.	Determination of the Charge carried by the Negative Ion	291
VIII.	On some Physical Properties of Gaseous Ions	310
IX.	Ionisation by Incandescent Solids	338
X.	Ionisation in Gases from Flames	399
XI.	Ionisation by Light. Photo-Electric Effects	435
Name Index		483
Subject Index		487

Plate I. Positive Ray Parabolas . . . *to face p.* 272

Plate II. Wilson Tracks . . . . *between pp.* 336–7

CHAPTER I

ELECTRICAL CONDUCTIVITY OF GASES IN A NORMAL STATE

1. A GAS in the normal state conducts electricity to a slight, but only to a very slight, extent, however small the electric force acting on the gas may be. So small however is the conductivity of a gas when in this state, and so difficult is it to eliminate spurious effects, that there have been several changes of opinion among physicists as to the cause of the leakage of electricity which undoubtedly occurs when a charged body is surrounded by gas. It was thought at first that this leakage took place through the gas; later, as the result of further experiments, it was attributed to defective insulation of the rods or threads used to support the body, and to the dust present in the gas; more recently however it has been shown that there is a true leak through the gas which is not due to the dust or moisture the gas may happen to contain.

2. The escape of electricity from an insulated charged body has attracted the attention of many physicists. Coulomb[1], whose experiments were published in 1785, came to the conclusion from his investigations on the loss of electricity from a charged body suspended by insulating strings, that after allowing for the leakage along the strings there was a balance over, which he attributed to a leakage through the air. He explained this by supposing that the molecules of air when they come into contact with a charged body receive a charge of electricity of the same sign as that on the body and are then repelled from it, carrying off some of its charge. We shall see later on that this explanation is not tenable.

Matteucci[2] experimenting on the same subject in 1850 also came to the conclusion that there was a leakage of electricity through the gas; he was the first to prove that the rate at which this leak takes place is less when the pressure of the gas is low

[1] Coulomb, *Mémoires de l'Académie des Sciences*, 1785, p. 612.
[2] Matteucci, *Annales de Chimie et de Physique*, xxviii. p. 390, 1850.

than when it is high. He found also that the rate of leak was the same in air, carbonic acid and hydrogen. On the other hand Warburg[1] found that the rate of leak through hydrogen was only about half of that through air and carbonic acid; he agreed with Matteucci with regard to the equality of the rate of leak through these gases and could detect no difference between the leaks through dry and moist air; he confirmed Matteucci's observations on the effect of pressure on the rate of leak. Warburg seemed inclined to suspect that the leak was due to dust in the gases. The belief in dust being the carrier of the electricity was strengthened by an experiment made by Hittorf[2] in which a small carefully insulated gold-leaf electroscope was placed in a glass vessel filled with filtered gas; the electroscope was found to have retained a charge even after the lapse of four days. We know now from recent experiments that the smallness of the leak observed in this case was due to the smallness of the vessel in which the charged body was placed rather than to the absence of dust.

Further experiments on this subject were made by Nahrwold[3] and by Narr[4] who showed that the rate of leak from a charged hollow sphere was not increased when the temperature of the sphere was raised by filling it with hot water. Boys[5] made an experiment which showed very clearly that, whatever the cause of the leak might be, it was not wholly due to want of insulation in the supports of the charged body; in this experiment he attached the gold leaves of an electroscope first to a short and thick quartz rod and then to a long and thin one, and found that the rate of leak of electricity from the gold leaves was the same in the two cases; if the leak had been along the supports it would have been much greater in the first case than in the second. Boys also confirmed Warburg's observation that the rate of leak was the same in dry as in moist air.

[1] Warburg, *Pogg. Ann.* cxlv. p. 578, 1872.
[2] Hittorf, *Wied. Ann.* vii. p. 595, 1879.
[3] Nahrwold, *Wied. Ann.* v. p. 460, 1878; xxxi. p. 448, 1887.
[4] Narr, *Wied. Ann.* v. p. 145, 1878; viii. p. 266, 1879; xi. p. 155, 1880; xvi. p. 558, 1882; xxii. p. 550, 1884; xliv. p. 133, 1892.
[5] Boys, *Phil Mag.* xxviii. p. 14, 1889.

3. The subject of the electric conduction through air is evidently of considerable importance in relation to Meteorology and Atmospheric Electricity. Experiments especially bearing on this point were made by Linss[1] on the loss of electricity from charged bodies placed in the open air; he found there was an appreciable loss of charge which, as control experiments showed, was not due to leakage along the supports of the charged body.

An extensive series of open air measurements were made by Elster and Geitel[2] in many different localities and in different states of the weather. They found that the rate of leak varied much from time to time and from place to place, that it was very much smaller in mist or fog than when the weather was bright and clear, that it was greater at high altitudes than at low ones, and that on the tops of mountains the rate of escape of negative electricity was much greater than that of positive. This is doubtless due to the negative charge on the earth's surface, a mountain top being analogous to a sharp point on a conductor, and thus a place where the earth's electric force tending to move away any negatively electrified body is much greater than it is on the flat. In plains they found the rate of leak to be the same for plus and minus charges.

4. Further experiments on the rates of leak from a charged body placed in a closed vessel filled with air were made almost simultaneously by Geitel[3] and by C. T. R. Wilson[4]. The apparatus used by Wilson for this purpose is represented in Fig. 1. Since the quantity of electricity which escapes from the charged body is very small it is necessary that the capacity of the instrument used to measure it should be small; this condition makes it advisable to use a small gold-leaf electroscope rather than a quadrant electrometer. To prevent the leakage from the supports of the gold leaves vitiating the experiments, the brass strip which carries the gold leaf is attached to and insulated from a metal

---

[1] Linss, *Meteorol. Zeitschr.* iv. p. 352, 1887; *Elektrotechn. Zeitschr.* i. 11, p. 506, 1890.

[2] Elster and Geitel, *Ann. d. Phys.* ii. p. 425, 1900.

[3] Geitel, *Phys. Zeits.* ii. p. 116, 1900.

[4] C. T. R. Wilson, *Proc. Camb. Phil. Soc.* xi. p. 32, 1900; *Proc. Roy. Soc.* lxviii. p. 151, 1901.

rod A by a piece of sulphur B, A being insulated by a plug of sulphur from the vessel containing the gas under examination, and connected with a condenser C formed of parallel plates of metal imbedded in a block of sulphur. The brass strip and gold leaf are initially charged to the same potential as the rod by making momentary contact between the rod and the strip by means of a moveable wire; the rod being connected with a large capacity remains at almost constant potential, and thus if there is any leakage of electricity along the sulphur supporting the brass strip and gold leaf, it will tend to keep them charged and not to

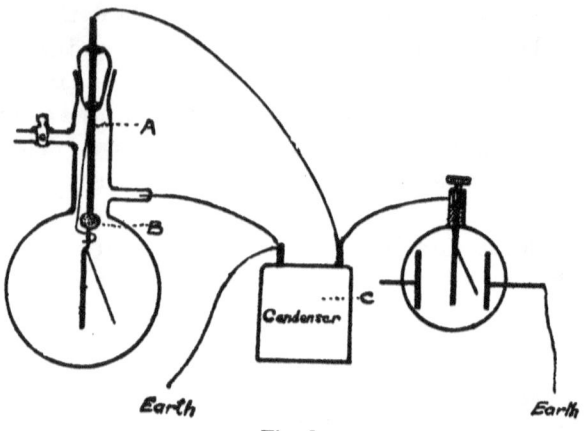

Fig. 1.

discharge them. The position of the gold leaf is read by means of a microscope provided with an eye-piece micrometer scale. The brass strip and gold leaf are used as the charged body and the rate at which the image of the gold leaf moves across the micrometer scale is a measure of the rate of leak through the gas. The following results were obtained by both Geitel and Wilson— the rate of escape of electricity in a closed vessel is much smaller than in the open and the larger the vessel the greater is the rate of leak. The rate of leak does not increase in proportion to the difference of potential between the gold leaves and the walls of the vessel; the rate soon reaches a limit beyond which it does not increase however much the potential difference is increased; provided of course that this is not great enough to cause sparks to pass.

Assuming that the maximum leak is proportional to the volume of the vessel, Wilson's experiments, which were made with vessels less than 1 litre in volume, showed that in dust-free air at atmospheric pressure the maximum quantity of electricity which can escape in one second from a charged body in a closed space whose volume is V cubic centimetres is about $10^{-8}\,V$ electrostatic units. Rutherford and Allen[1] working in Montreal obtained results in close agreement with this.

As the result of a series of experiments made at pressures ranging from 43 to 743 millimetres of mercury, Wilson came to the conclusion that the maximum rate of leak is very approximately proportional to the pressure, thus at low pressures the rate of leak is exceedingly small: this result is illustrated in a striking way by an observation of Crookes[2] that a pair of gold leaves could retain an electric charge for months in a very high vacuum. More recent experiments have shown that it is only in small vessels that the maximum rate of leak is proportional to the volume and to the pressure. With large vessels the rate of leak per unit volume is considerably less than in small vessels. The rate of leak also depends upon the nature of the walls of the vessel. The rate of leak is about the same in the dark as it is in the light, it is thus not due to light, and that it can be wholly due to some invisible form of radiation coming from outside is rendered improbable by the observations of Rutherford and Cooke[3], Cooke[4], McLennan and Burton[5] that though the leak inside a closed vessel can be reduced by about 30 per cent. by surrounding the vessel with thick lead, yet the diminution reaches a limit when the lead is about 2 inches thick, after this no diminution in the leak is produced by increasing the thickness of the lead. The rate of leak in a closed vessel is the same when the vessel is inside a railway tunnel as when it is outside; in the former case any radiation reaching the gas from outside must have travelled through many feet of solid rocks (see however § 6·1).

[1] Rutherford and Allen, *Phys. Zeits.* iii. p. 225, 1902.
[2] Crookes, *Proc. Roy. Soc.* xxviii. p. 347, 1879.
[3] Rutherford and Cooke, *Phys. Rev.* xvi. p. 183, 1903.
[4] Cooke, *Phil. Mag.* vi, 6. p. 403, 1903.
[5] McLennan and Burton, *Phys. Rev.* xvi. p. 184, 1903.

5. Geitel (*loc. cit.*) made the very interesting observation that the rate of leak in a closed vessel increases, after the refilling of the vessel with fresh air, for some days, when it reaches a constant value at which it remains for an indefinitely long time. The most obvious explanation of this result is that it is due to the settling down of the dust, as Elster and Geitel (*loc. cit.*) have shown that the presence of dust, fog, or mist diminishes the rate of leak. This explanation is however rendered untenable by some later experiments[1] made by the same physicists, in which they found that the period required for the gas to attain its maximum conductivity was not appreciably diminished by filtering the dust out of the air by sending it through water, or by extracting the moisture from the gas: thus if the increase in the rate of leak is due to the settling down of some foreign matter from the gas, this matter must be something which cannot be got rid of by filtering the gas through water-traps or plugs of glass-wool.

6. Another aspect of this phenomenon is the very interesting fact discovered by Elster and Geitel[1] that the rate of leak in caves, and cellars where the air is stagnant and only renewed slowly, is very much greater than in the open air: thus in some experiments they made in a cave—the Baumannshöhle in the Harz Mountains—they found that in the cave the electricity escaped at seven times the rate it did in the air outside, even when this was clear and free from mist. They found too that in a cellar whose windows had been shut for eight days the rate of leak was very considerably greater than it was in the air outside. These experiments suggest that something producing abnormally great conductivity slowly diffuses from the walls surrounding the gas, and that this diffusion goes on so slowly that when fresh gas is introduced it takes a considerable time for the substance from the walls to again diffuse through the volume.

This explanation is in accordance with modern knowledge of radioactivity. The minute traces of radium and thorium products in the soil give rise to 'emanations,' radioactive gases which can diffuse out from porous substances. These emanations themselves produce conductivity in any gas with which they are mixed, and by their decay give rise to a series of secondary solid products

[1] Elster and Geitel, *Phys. Zeits.* ii. p. 560, 1901.

which decaying in their turn also give rise to conductivity in the gas near them. As these substances have very different rates of decay and vary in their power of making a gas conducting, the variation of conductivity with time may be very complicated.

The experiments we have described show that the rate of leak of electricity through gas in a normal state is influenced by a great variety of circumstances, such as the pressure of the gas, the volume of gas in the electric field, the thickness of the walls of the vessel containing the gas, and the amount of dust or fog held in suspension by it; all these effects receive a ready explanation on the view to which we are led by the study of the effects shown on a larger scale by gases whose conductivity has been increased by artificial means, namely that the conductivity is due to the presence of charged particles, or "ions." We may at once point out that the increase of the rate of leak with the size of the vessel containing the charged body shows that the conduction is not due, as Coulomb thought, to particles of gas originally uncharged striking against the charged body and receiving a charge which they deliver up to the sides of the vessel; if this were the method by which the electricity escaped the rate of leak would not increase with the size of the vessel. For the sake of completeness we add here an account of recent work on the subject, though this involves reference to the ionic theory to be developed in the next chapter.

6·1. The leakage of electricity through air under normal conditions has attracted a great deal of attention during the last few years and has a very important bearing on theories relating to the constitution of the stars and the transformation of matter into radiant energy.

There are several sources of the conductivity of normal air: part of it may be due to the presence of radioactive substances in the walls of the vessel in which the gas is contained or of a little radioactive emanation in the gas itself. The conductivity we are considering is so small that the presence of a mere trace of a radioactive substance or a trace of radioactive property in ordinary substances would be sufficient to account for it. The metal of which the containing vessel is made has certainly a considerable effect on the conductivity of the gas; this may however be due not to any intrinsic radiation from the metal, but to an excited radiation produced by the passage through the metal of a very penetrating radiation, for the existence of which we shall see that there is very strong evidence. The intrinsic radiation of the metal cannot be the only source of the conductivity of the gas; for, if it were, the conductivity of the gas in a closed vessel would be independent of the surroundings of the vessel. Now McLennan[1] measured the conductivity in a hermetically sealed vessel on land at Toronto in Canada, Cambridge, England, and Bowland, Scotland, and also on board the S.S. *Grampian* while crossing the Atlantic. He found that the rate of leak of electricity through the gas was very much the same at all the land stations, and could be represented by the production in the gas of ions at the rate of 9 ions per c.c. of the gas per second; the leak over the sea was however considerably less than this and could be represented by the production of only 6 ions per c.c. per second in the gas. McLennan and Wright[2] had previously found that the leak over the ice in Toronto Bay was considerably less than that on the shore. Pacini[3] also found a diminution in the rate of leak out at sea compared with that on shore. These experiments indicate that some of the leak is produced by radiation coming from the land and that this radiation is cut off by the ocean. Rocks and soil are known to be radioactive, as is also the water in many rivers and streams. Confirmation of the existence of this terrestrial radiation was obtained by measurements of the leak at different heights above the surface of the earth. McLennan and McCallum[4] measured the rate of leak at the top of a tower 64 metres high, Wulf[5] at the top of the Eiffel Tower 300 metres high, and Bergwitz[6] at the top of a tower 100 metres high, and found a marked diminution as compared with that at the base of the towers; this is what we should expect from the absorption of the radiation by the atmosphere.

[1] McLennan, *Phil. Mag.* vi. 24, p. 520, 1912.
[2] McLennan and Wright, *Phil. Mag.* vi. 17, p. 310, 1909.
[3] Pacini, *Annali dell' Ufficio Centrale Meteor. e Geod. Italiano*, 27, pt. 1, 1910.
[4] McLennan and McCallum, *Phil. Mag.* vi. 22, p. 629, 1911.
[5] Wulf, *Phys. Zeits.* xi. p. 811, 1910.
[6] Bergwitz, *Habilitationschrift*, Brunswick, 1910.

Observations in balloons at greater heights have brought to light the existence of another radiation coming from the sky and not from the earth, which also makes the gas a conductor of electricity.

Goekel[1] found from balloon observations that the diminution of the rate of leak with height got very small at altitudes of from 1 to 2 kilometres, while Hess[2] at similar altitudes found the rate of leak was actually greater than at sea level.

Kolhörster[3] made a very detailed study of the variation in the rate of leak at high altitudes, and observed very decided increases in the rate of leak with the altitude. The results are given below; the figures in the second column represent the difference in the number of ions produced per second per c.c. of the gas at the altitude given and at the surface of the earth.

Height in kilometres	Relative number of ions per c.c. per sec.
1·0	− 1·5
2·0	+ 1·2
3·0	+ 4·0
4·0	+ 8·3
5·0	+ 16·5
6·0	+ 28·7
7·0	+ 44·2
8·0	+ 61·3
9·0	+ 80·4

Millikan and Bowen sent up self-registering instruments in balloons which were afterwards retrieved; these also registered an increase in the rate of leak with the height, though considerably less than that obtained by Kolhörster.

The variation of the leak with the height is what would occur if the leak were produced by radiation of two types: (1) a radiation coming from the ground, and (2) another coming from the sky. The effect due to (1) would diminish as the height increased, while that due to (2) would increase.

Estimates of the leak due to the second type of radiation at the surface of the earth have been made by various experimenters.

[1] Goekel, *Phys. Zeits.* x. p. 845, 1909; xii. p. 595, 1911.
[2] Hess, *Phys. Zeits.* xii. p. 998, 1911.
[3] Kolhörster, *Phys. Zeits.* xiv. pp. 1066, 1153, 1913; *Deutsch. Phys. Gesell. Verh.* xvi. 14, p. 719, 1914.

Their results are given in the following table; the amount of the leak is expressed as the number of ions produced per c.c. per second.

Kolhörster	1·4; 1·05; 1·70; 2·0
Pacini	2·1
v. Schweidler	1·6
Hess	2·7; 2·3
Goekel	1·6; 2·4
Millikan and Bowen	1·4

As the total leak at the earth's surface would generally correspond to the production of between 10 and 20 ions per c.c. per second, the rays of type (2) at sea level are only responsible for a small fraction of the leak. The greater part seems to be due to radiation of type (1), for Bergwitz found that in a cavern in a rock-salt mine where both radiations were presumably cut off the rate of leak sank to ·8 ions per c.c. per second.

Millikan and Bowen[1] give the following numbers as representing the number of ions per c.c. per second produced by radiation of type (2):

1·4 at sea level,
2·6 at 1600 metres,
4·8 at 3600 metres,
5·9 at 4300 metres.

Kolhörster[2] made observations at the top and at some distance below the surface of glaciers on the Jungfrau, and deduced the absorption of these rays by ice. He found values for μ, the coefficient of absorption for ice, ranging from $1\cdot6 \times 10^{-3}$ cm.$^{-1}$ to $2\cdot7 \times 10^{-3}$ cm.$^{-1}$, values which are much smaller than that for the γ radiation from any known radioactive substance. He found a diurnal variation in the amount of this radiation and suggested that it might be connected with the Milky Way. A description of apparatus suitable for these measurements is given by Kolhörster in the *Zeits. Instrumentenk.* xliv. p. 333, 1924.

6·2. A very complete investigation of the absorption of these rays by water has been made by Millikan and Bowen[3], who made

---

[1] Millikan and Bowen, *Phys. Rev.* xxvii. p. 353, 1925.
[2] Kolhörster, *Berlin. Berichte*, 1923, p. 366.
[3] Millikan and Bowen, *Phys. Rev.* xxvii. p. 353, 1925; xxviii. p. 851, 1926.

observations on the rate of leak in vessels submerged at different depths in two mountain lakes on Mount Whitney—Lake Muir at an altitude of 11,800 feet and Arrowhead Lake at an altitude of 5100 feet. These lakes are fed by melted snow, so that the water is not likely to contain radioactive matter.

It appears from these measurements that the radiation is by no means homogeneous, for the greater the depth below the surface the more penetrating was the radiation; the absorption coefficient for water at the top of the lake was 3×10^{-3} cm.$^{-1}$ and at the greatest depth observed $1 \cdot 8 \times 10^{-3}$ cm.$^{-1}$. These are of the same order as those obtained by Kolhörster. Millikan's[1] most recent work has given evidence of radiation of absorption coefficient of 10^{-3} cm.$^{-1}$, corresponding to a wave length of ·00021 A.

Millikan and Bowen, however, did not observe any trace of the diurnal variation described by Kolhörster. Applying an expression due to Compton for the connection between the absorption and the wave length, they found that the greater absorption corresponded to a wave length of ·0063 A., the smaller to one of ·00038 A. or $3 \cdot 8 \times 10^{-12}$ cm. These wave lengths are almost comparable with the dimensions of an electron. The quantum of energy corresponding to the smaller wave length is about 32 million volts. This energy is great enough to separate an electron from a positive charge even if the distance between them were as small as 5×10^{-14} cm. The energy is more than half that of the fastest α particle. This is much greater than the energy associated with the γ radiation from any substance hitherto discovered, and many suggestions have been made as to its origin. Thus Jeans and Eddington have suggested that the rays arise from the destruction of matter by the coalescence of an electron with a positive particle and the conversion of their energy into radiation; this would give even more penetrating radiation than that under consideration, though it might be degraded into a less penetrating kind by passing through matter. C. T. R. Wilson has suggested the possibility of the radiation being due to electrons which have acquired energy amounting to millions of volts under the action of something analogous to thunder-storms; such differences of potential are probable, as the loss of energy by

[1] Millikan, *Nature*, cxxi. p. 20, 1928.

collisions varies inversely as the energy of the electron, the more energy the electron got the less would be the resistance it experienced, so that its energy would not reach a steady value but would continually increase under the action of the field. Millikan[1] has made measurements when violent thunder-storms were in progress but found no difference in the ionisation. At present, however, we do not know nearly enough about these rays to come to any decision as to their nature; it is evident, however, that they raise questions of the greatest interest and importance. It would be one of the romances of science if these obscure and prosaic minute leakages of electricity from well-insulated bodies should be the means by which the most fundamental problems in the evolution of the cosmos had to be investigated.

The radiation of type (1) which comes from the ground and air in the neighbourhood of the body is much less penetrating than that which comes from the sky; it is not at all homogeneous and the softer portions are subject to periodic variations, for when the gas through which the leak is observed is contained within a vessel with walls only a fraction of a millimetre thick, well-marked diurnal variations occur in the rate of leak; these disappear when the walls of the vessel are a few millimetres thick. These diurnal variations are presumably due to changes in the rate of escape of radioactive emanations from the soil, due to changes in temperature and pressure.

A considerable increase in the rate of leak occurred when the earth was passing through the tail of Halley's Comet in 1910. Thus Pacini observed that the rate which was normally represented by the production of 30 ions per c.c. per second, rose during the passage to one represented by 50.

[1] Millikan, *Nature*, cxxi. p. 20, 1928.

CHAPTER II

PROPERTIES OF A GAS WHEN IN THE CONDUCTING STATE

7. THE electrical conductivity of gases in the normal state is so small that, as we have seen, the proof of its existence requires very careful and elaborate experiments. Gases may however in various ways be put into a state in which they conduct electricity with so much facility that the detection and investigation of this property becomes a comparatively easy matter; as the study of the properties of a gas when in this state is of the highest importance from the light which it throws on the general phenomena of electric discharge through gases, we shall find it useful to discuss the subject at some considerable length.

8. There are many ways in which gases may be made to possess considerable conductivity or, as we shall express it, be put into the conducting state. Thus gases drawn from the neighbourhood of flames, electric arcs or glowing metals or carbon, or which have diffused from a space through which an electric discharge is passing or has recently passed, are in this state. A gas is put into the conducting state when X-rays or cathode rays pass through it; the same effect is produced by the rays from radioactive substances, also by ultra-violet light of very short wave length. E. Wiedemann has shown that electric sparks give out rays, called by him Entladungstrahlen, which produce the same effect. Air which has passed over phosphorus or which has bubbled through water is also in this state and remains so for some time after it has left the phosphorus or water. We shall have later on to discuss the action of each of these agents in detail, but we shall begin by studying some of the general properties possessed by a gas when in this state, the experimental methods by which these properties may be investigated, and a theory of this state by which they may be explained.

9. A gas when in the conducting state possesses characteristic properties. In the first place it retains its conductivity for some little time after the agent which made it a conductor has ceased to act; its conductivity however always diminishes, in some cases very rapidly, after the agent is removed, and finally it disappears. The persistence of the conductivity may be shown very simply by exposing the air in one place to X-rays or the rays from radium or polonium, and putting, some way off, a charged electroscope covered with a cage made of wire-gauze so as to screen off the region exposed to the rays from the electrostatic field due to the electroscope. If the air is still, the electroscope will retain its charge even when the rays are in action, but if we blow some of the air traversed by the rays towards the electroscope, the latter will begin to lose its charge, showing that the air has retained its conductivity during the time taken by it to

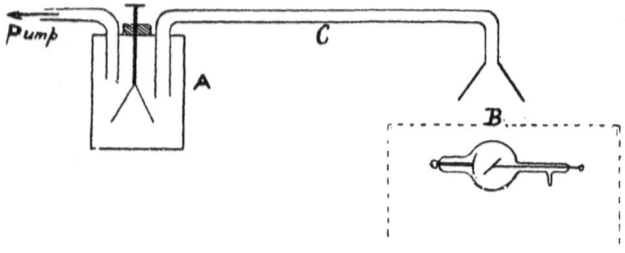

Fig. 2.

travel to the electroscope from the place where it was exposed to the rays. A somewhat more elaborate form of this experiment, which enables us to prove several other interesting properties of the conducting gas, is to place the electroscope in a glass vessel A in which there are two tubes, one leading to a water-pump, while the end of the other C is in the region traversed by the X-rays. The tube used to produce the rays is placed in a box which with the exception of a window at B to let the rays through is covered with lead: this shields the electroscope from the direct action of the rays: if the water-pump be worked slowly so as to make a slow current of air pass from the region traversed by the rays into the vessel A, the electroscope will gradually lose its charge whether this be positive or negative: if the pump be stopped and the current of air ceases, the discharge of the electroscope will cease.

The conducting gas loses its conductivity if it is sucked through a plug of glass-wool or made to bubble through water[1]. This can readily be proved by inserting in the tube C a plug of glass-wool or a water-trap and working the water-pump a little harder so as to make the rate of flow of air through the tube the same as in the previous experiment; it will now be found that the electroscope will retain its charge, the conductivity has thus been taken out of the gas by filtering it through glass-wool or water. The conductivity is very much more easily removed from gases made conducting by the various rays, X-rays, cathode rays, &c., than from the conducting gases derived from flames and arcs; the latter as we shall see require a great deal of filtering to remove their conductivity. If we replace the tube C by a metal tube of fine bore we shall find that the gas loses some of its conductivity when it passes through it, and the finer the bore the more rapidly does the conductivity disappear. The conductivity may also be removed from the gas by making it traverse a strong electric field so that a current of electricity passes through it[2]. To show this, replace the glass tube C by a metal tube of fairly wide bore and fix along the axis of this tube an insulated metal wire; if there is no potential difference between the wire and the tube, then the electroscope in A will leak when a current of air is sucked through the apparatus; if however a considerable difference of potential is established between the wire and the tube, so that a current of electricity passes through the gas during its passage to A, the leak of the electroscope will cease, showing that the conductivity of the gas has been removed by the electric field.

10. The removal of the conductivity by filtering the gas through glass-wool or water and by transmission through narrow metal tubes shows that the conductivity is due to something mixed with the gas, this something being removed from the gas in the one case by filtration, in the other by diffusion to the walls of the tube. Further, the removal of the conductivity by the electric field shows that this something is charged with electricity and moves under the action of the field; since the gas when in the conducting state shows as a whole no charge of electricity, the

[1] J. J. Thomson and E. Rutherford, *Phil. Mag.* xlii. p. 392, 1896.
[2] *Ibid.*

charges removed must be both positive and negative. We are thus led to the conclusion that the conductivity of the gas is due to electrified particles mixed up with the gas, some of these particles having charges of positive electricity, others of negative. We shall call these electrified particles ions, and the process by which a gas is made into a conductor the ionisation of the gas. We shall show later on how the masses and charges of the ions may be determined, when it will appear that the ions in a gas are not identical with those met with in the electrolysis of solutions.

11. The passage of a current of electricity through a conducting gas does not follow Ohm's law unless the electromotive force

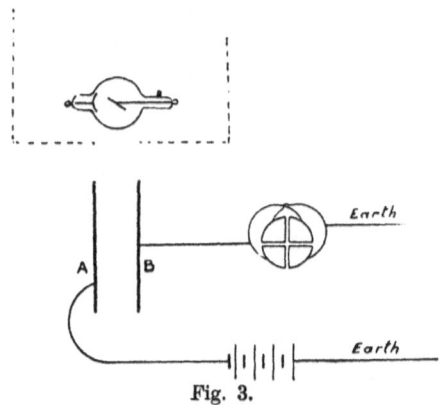

Fig. 3.

acting on the gas is small. We may investigate the relation between the current and potential difference by taking two parallel metal plates A and B (Fig. 3) immersed in a gas, the gas between the plates being exposed to the action of some ionising agent such as X-rays or the radiation from a radioactive substance. One of the plates B is connected with one of the pairs of quadrants of an electrometer, the other pair of quadrants being put to earth. The other plate A is connected with one of the terminals of a battery of several storage cells, the other terminal of the battery being connected with the earth; initially the two pairs of quadrants of the electrometer are connected together, then the connection between the quadrants is broken, as a current of electricity is passing across the air space between A and B, the plate B gets charged up and the needle of the electrometer is deflected; the rate of deflection of the electrometer measures the current passing through the gas. By making a series of observations of this kind we can get the means of drawing a curve such that the ordinates represent the current through the gas and the abscissæ the potential difference between the plates: such a curve is represented in Fig. 4[1]. We see that when the difference of

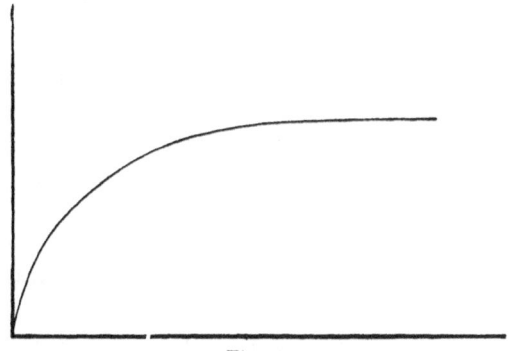

Fig. 4.

potential is small the curve is approximately a straight line, in this stage the conduction obeys Ohm's law; the current however soon begins to increase more slowly than the potential difference and we reach a stage where there is no appreciable increase of

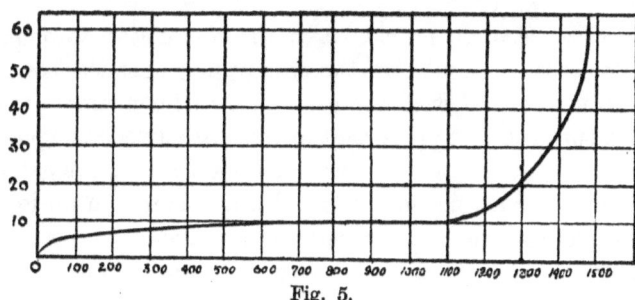

Fig. 5.

current when the potential difference is increased: in this stage the current is said to be saturated. When the potential difference is increased to such an extent that the electric field is strong enough to ionise the gas, another stage is reached in which the current increases very rapidly with the potential difference; curves

[1] J. J. Thomson, *Nature*, April 23, 1896.

showing this effect have been obtained by von Schweidler[1] and by Townsend[2], one of these is shown in Fig. 5. The potential gradient required to reach this stage depends upon the pressure of the gas, it is directly proportional to the pressure; for air at atmospheric pressure it is about 30,000 volts per centimetre, so that in air at a pressure of one millimetre a potential gradient of about 40 volts per centimetre would be sufficient to reach this stage.

12. The saturation current between two parallel plates of given area depends upon the amount of ionisation between the plates; if the ionisation takes place throughout the whole volume of gas between the plates, then the greater the distance between the plates the greater is the saturation current, so that if we use potential differences large enough to produce saturation, then with a constant difference of potential the greater the distance between the plates the larger is the current. Thus the behaviour of the conducting gas is very different from that of a metallic or liquid electrolytic conductor, for if such conductors were substituted for the gas the greater the distance between the plates the smaller would be the current. Under very small potential differences however the three classes of conductors would behave in the same way.

13. The peculiarities shown by the conduction through gases are very easily explained on the assumption that the conduction is due to ions mixed with the gas. Let us for example take the case of saturation. Suppose that in the gas between the plates the ionising agent produces in one second q positive and q negative ions and let e be the magnitude of the electric charge on an ion, then if an electric current i passes between the plates, i/e positive ions are driven against the negative electrode, and the same number of negative ions are driven against the positive electrode in one second; thus in each second i/e positive and negative ions are taken out of the gas by the current. When the gas is in a steady state the number of ions taken out of it in a given time cannot be greater than the number of ions produced in it in the same

[1] von Schweidler, *Wien. Bericht*, cviii. p. 273, 1899.
[2] J. S. Townsend, *Phil. Mag.* vi. 1, p. 198, 1901.

time, hence i/e cannot be greater than q, and thus i cannot be greater than qe: qe is thus the value of the saturation current. If the ions are produced uniformly throughout the gas, and if q_0 is the number of ions produced in one second in unit volume, and V the volume of gas between the plates (Fig. 3), then the number of ions produced in the gas per second is $q_0 V$ and the saturation current $q_0 e V$. Since V is equal to the area of one of the plates multiplied by the distance between the plates, the saturation current is proportional to this distance. This relation between the saturation current and the distance between the plates has been verified by measurements of the saturation currents through gases exposed to X-rays[1].

14. Even when there is no current of electricity passing through the gas and removing some or all of the ions, the number of ions present in the gas does not increase indefinitely with the time which has elapsed since the gas was first exposed to the ionising agent; the number of ions in the gas and therefore its conductivity acquire after a time steady values beyond which they do not increase however long the ionising agent may act. This is due to the recombinations that take place between the positive and negative ions; these ions moving about in the gas sometimes come into collision with each other and in a certain fraction of such cases of collision the positive and negative ions will remain together after the collision, and form an electrically neutral system the constituents of which have ceased to be free ions. The collisions will thus cause the ions to disappear, and the steady state of a gas which is not carrying an electric current will be reached when the number of ions which disappear in one second as the result of the collisions is equal to the number produced in the same time by the ionising agent. Starting from this principle we can investigate the relation between the number of free ions when the gas is in a steady state, the strength of the ionising agent, the rate at which the ions increase on the first exposure to the ionising agent and the rate at which they die away when the ionising agent is cut off.

[1] J. J. Thomson and E. Rutherford, *Phil. Mag.* v. 42, p. 392, 1896.

Growth and Decay of Ionisation in a Gas.
Coefficient of Recombination.

Let us begin with the simplest case, where there is only one kind of positive and one kind of negative ion. Let q be the number of ions (positive or negative) produced in one cubic centimetre of the gas per second by the ionising agent; n_1, n_2 the number of free positive and negative ions respectively per cubic centimetre of the gas. The number of collisions per second between positive and negative ions is proportional to $n_1 n_2$. If a certain fraction of the collisions result in the formation of a neutral system the number of ions which disappear per second in a cubic centimetre will be equal to $\alpha n_1 n_2$, where α is a quantity called the coefficient of recombination which is independent of n_1 and n_2; hence if t is the time which has elapsed since the ionising agent was first applied to the gas we have

$$\left. \begin{array}{l} \dfrac{dn_1}{dt} = q - \alpha n_1 n_2 \\[4pt] \dfrac{dn_2}{dt} = q - \alpha n_1 n_2 \end{array} \right\} \quad \ldots\ldots\ldots\ldots\ldots(1).$$

Thus $n_1 - n_2$ is constant, so that if the gas is uncharged to begin with, n_1 is always equal to n_2. Putting $n_1 = n_2 = n$ the preceding equation becomes

$$\frac{dn}{dt} = q - \alpha n^2 \ldots\ldots\ldots\ldots\ldots(2),$$

the solution of which is, if $k^2 = q/\alpha$, and $n = 0$ when $t = 0$,

$$n = k \frac{(\epsilon^{2k\alpha t} - 1)}{\epsilon^{2k\alpha t} + 1} \ldots\ldots\ldots\ldots\ldots(3);$$

n_0 the value of n when the gas is in a steady state is obtained by making t infinite in equation (3) and is given by the equation

$$n_0 = k = \sqrt{\frac{q}{\alpha}}.$$

We see from equation (3) that the gas will not approximate to a steady state until $2k\alpha t$ is large, *i.e.* until t is large compared with $1/2k\alpha$, that is with $1/2n_0\alpha$ or $1/2\sqrt{q\alpha}$. We may thus take $1/2\sqrt{q\alpha}$ as the measure of the time taken by the gas to reach the steady state under exposure to the ionising agent; as this time varies inversely as $\sqrt{q}$ we see that when the ionisation is feeble it may take a very considerable time for the gas to reach the steady state.

Thus at some distance, say a metre, from an ordinary X-ray bulb it may require an exposure of a minute or two to bring the gas into a steady state.

When the ionisation is produced by X-rays, q is proportional to the pressure; we shall see that when the pressure is not too high α is also proportional to the pressure, so that $\sqrt{q\alpha}$ will be proportional to the pressure, thus the time taken to reach the steady state will be great when the pressure is low. When q and α are proportional to the pressure, n_0 the density of the ions in the steady state will be independent of the pressure. Thus if a large volume of gas at a low pressure is ionised and then suddenly compressed to a small volume, the rate of recombination of the ions in that small volume will be very great. The neutralisation of positively charged atoms by a negative charge is often the source of luminous radiation; hence it would appear not unlikely that if a large volume of ionised gas were suddenly compressed light would be produced. Newall has observed that when ionised oxygen is suddenly compressed it becomes luminous; without further investigation, however, we cannot be sure that this was due to the cause under consideration.

We may use equation (2) to determine the rate at which the number of ions diminishes when the ionising agent is removed; putting $q = 0$ in that equation we have

$$\frac{dn}{dt} = -\alpha n^2 \quad \ldots \ldots \ldots \ldots \ldots \ldots (4),$$

hence
$$n = \frac{n_0}{1 + n_0 \alpha t} \quad \ldots \ldots \ldots \ldots \ldots \ldots (5),$$

where n_0 is the value of n when $t = 0$. Thus the number of ions falls to one-half its initial value in the time $1/n_0\alpha$. We may regard equation (4) as expressing the fact that the rate at which the ions disappear is the same as if the life of each ion were equal to $1/\alpha n$.

14·1. When there are several types of positive and also of negative ions the equations are more complicated, thus if there are n_1 negative ions of one type, n_2 of a second, n_3 of a third and so on, and positive ions $m_1, m_2, \ldots$ of different kinds, we shall have in place of equation (1), the following system of equations:

$$\frac{dn_1}{dt} = q_1 - n_1(\alpha_{11}m_1 + \alpha_{12}m_2 + \alpha_{13}m_3 + \ldots),$$

$$\frac{dn_2}{dt} = q_2 - n_2(\alpha_{21}m_1 + \alpha_{22}m_2 + \alpha_{23}m_3 + \ldots),$$

$$\cdots \cdots \cdots \cdots \cdots \cdots \cdots \cdots \cdots \cdots$$

$$\frac{dm_1}{dt} = p_1 - m_1(\alpha_{11}n_1 + \alpha_{21}n_2 + \ldots),$$

$$\frac{dm_2}{dt} = p_2 - m_2(\alpha_{12}n_1 + \alpha_{22}n_2 + \ldots),$$

where α_{rs} represents the coefficient of recombination between the negative ion of the rth type and the positive ion of the sth type, q_r, p_s represent the rates at which these ions are produced. Except in very special cases such as when the α's are all equal, the m's and n's will no longer be represented by equations of the same forms as equations (2) or (5). Thus as an example take the problem of finding the rate at which the ions disappear when there are two types of negative and one of positive ions; the equations are

$$\frac{dn_1}{dt} = -n_1\alpha_{11}m_1,$$

$$\frac{dn_2}{dt} = -n_2\alpha_{12}m_1,$$

$$\frac{dm_1}{dt} = -m_1(\alpha_{11}n_1 + \alpha_{12}n_2),$$

these are equivalent to

$$n_1^{\alpha_{12}} = Cn_2^{\alpha_{11}},$$

where C is a constant, and

$$\alpha_{11}t = \int \frac{dn_1}{n_1(n_1 + C'n_1^{\alpha_{12}/\alpha_{11}})},$$

an equation which will not be of the same form as (5) unless $\alpha_{12} = \alpha_{11}$. Thus unless the ions were homogeneous or the coefficients of recombination all equal we should not expect the decay of the ions to be represented exactly by equation (5). As a matter of fact however this equation does hold very approximately for the cases which have been tested. This has been shown by Rutherford[1] for gases ionised by X-rays and by uranium, by McClung[2] for gases ionised by X-rays, by McClelland[3] for gases drawn from the neighbourhood of flames and arcs, and by Erikson[4] for gases ionised by α-rays. The analogous case of an insulating liquid ionised by radioactive substances has been shown by van der Bijl's[5] experiments on hexane, carbon tetrachloride, and carbon bisulphide to follow the same law. Sutherland[6] proposed the law

$$\frac{dn}{dt} = -\alpha n^{\frac{4}{3}};$$

there does not however seem any ground for supposing that equation (4) does not hold provided the condition under which it is obtained—that the ions are homogeneous—is fulfilled.

15. *Methods for determining the Coefficient of Recombination.*

I. **Gas Current Method.** In this method which was first used by Rutherford (*loc. cit.*) in 1897 air exposed to X-rays at one end of a long tube is sucked through the tube and the saturation currents measured at different places along the tube. These currents are proportional to the value of n at the place of observation and if the distance of this place from the end of the tube is known as well as the velocity of the air current, the time which has elapsed since the gas was ionised can be calculated. We thus determine the value of n corresponding to a series of values of t, and from these the value of α can be determined by equation (5). Care should be taken to free the gas from dust, as this greatly increases the rate of recombination. The tubes should be so wide that the loss of ions by diffusion to the sides of the tube is small compared with the loss by recombination. This

[1] Rutherford, *Phil. Mag.* v. 44, p. 422, 1897; 47, p. 109, 1899.
[2] McClung, *Phil. Mag.* vi. 3, p. 283, 1902.
[3] McClelland, *Phil. Mag.* v. 46, p. 29, 1898.
[4] Erikson, *Phys. Rev.* xxvii. p. 473, 1908.
[5] van der Bijl, *Ann. der Phys.* xxxix. p. 170, 1912.
[6] Sutherland, *Phil. Mag.* vi. 18, p. 341, 1909.

method was used by McClelland (*loc. cit.*) to determine the coefficient of recombination for gases drawn from flames and by Townsend[1] to determine the coefficients of recombination in air, hydrogen, oxygen and carbonic acid. The method can however only be used when a large quantity of the gas is available.

II. Rutherford's Pendulum Method. Another method also used by Rutherford can be employed even for gases of which only small quantities can be procured. In this method gas confined in a vessel is exposed to the action of an ionising agent such as X-rays. Inside the vessel are two parallel metal plates A and B, between which the ionisation is to be measured (in some of Rutherford's experiments one of these plates was replaced by the case of the vessel, which was made a conductor by lining it with wire-gauze, the other plate was replaced by an insulated wire running down the middle of the vessel). One of these plates A can be connected with an electrometer, the other B with one terminal of a large storage battery, the other terminal of which is kept to earth. A pendulum interrupter is arranged so that as a heavy pendulum swings it strikes against levers, and by this means makes or breaks various connections. While the vessel is under the influence of the rays A and B are connected together and to earth, then A is disconnected from both earth and electrometer and left insulated, and B is disconnected from the earth; the pendulum is then let go: as it falls it first breaks the current going through the primary of the induction coil used to excite the rays, it thus stops the ionisation, then after an interval t (which can easily be varied) it strikes against another lever which has the effect of connecting B with the high potential pole of the battery, thus producing a strong electric field between the plates A and B: this field, if B is charged positively, drives in a very small fraction of a second all the positive ions which exist between A and B against A, so that A receives a positive charge proportional to n; the pendulum in its swing then goes on to disconnect B from the battery and connects it to earth. The plate A is now connected with the electrometer the needle of which is deflected by an amount proportional to the charge on the plate A, *i.e.* to n. By adjusting the apparatus so as to alter the

[1] Townsend, *Phil. Trans.* 193. A, 144, 1899.

time which elapses between cutting off the rays and connecting B with the battery we find a series of corresponding values of n and t; these were found by Rutherford to fit in well with the relation indicated by equation (5).

15·1. III. Rümelin's Method. Another method of determining α which has been used by Rümelin[1] is as follows. The gas is exposed continuously to the ionising agent, the electric field is applied to the gas by means of a rotating sector and is adjusted so as to be zero during a part of the revolution of the sector and to be large enough to produce complete saturation during the remainder of the revolution; the current sent through the gas by this field is measured by an electroscope.

If T is the time of revolution of the sector, t_1 the part of this time during which the gas is free from the electric field, t_2 the part when the saturating field is applied, then with the notation of Art. 14, n_1 the number of ions per unit volume when the field is first applied is given by the equation

$$n_1 = \frac{k(\epsilon^{2kat_1} - 1)}{\epsilon^{2kat_1} + 1},$$

the number of ions produced during the time t_2 is qt_2, hence i the current to the electroscope is given by

$$iT = e\left\{\frac{k(\epsilon^{2kat_1} - 1)}{\epsilon^{2kat_1} + 1} + qt_2\right\} \dots\dots\dots\dots(1).$$

Now let I be the saturation current qe, then putting $2kat_1 = x$; $q = k^2\alpha$, equation (1) may be written

$$iT = \frac{2It_1}{x}\frac{\epsilon^x - 1}{\epsilon^x + 1} + It_2,$$

or if
$$t_2 = \gamma T,$$
$$\frac{i - \gamma I}{(1-\gamma)I} = \frac{2}{x}\frac{\epsilon^x - 1}{\epsilon^x + 1}.$$

From this equation, if we measure i, I and γ, we can determine x; from the value of I we deduce the value of q, and then since $x = 2t_1\sqrt{q\alpha}$ we can determine α.

When the sector rotates so slowly that the ionisation has

[1] Rümelin, *Phys. Zeits.* ix. p. 657, 1908; *Ann. der Phys.* xliii. p. 821, 1914.

time to reach a steady state before the saturating potential is applied $n_1 = \sqrt{\dfrac{q}{\alpha}}$, and hence

$$\sqrt{\dfrac{\alpha}{e}} = \dfrac{I^{\frac{1}{2}}}{T(i - \gamma I)}.$$

Initial Recombination.

The values of α found by Rümelin, who ionised the gases by α-rays in some cases and by β-rays in others, depended to a large extent on the rate of rotation of the sector. The values of α when T was $\tfrac{1}{10}$ of a second were some 17 times those when T was 2 seconds. This indicates that the rate of recombination when the ions are first formed is much greater than when they are older. This effect is also well marked in the case of ionisation by X-rays as is shown in the curves, due to Plimpton[1], which represent the values of α at different intervals after a gas had been ionised by a flash of X-rays.

In these curves $\dfrac{1}{n}$ is plotted against t and the *slope* of the curve gives α.

This effect would be shown by a gas in which the ionisation is not uniform; if the ionisation inside a closed vessel were concentrated in a space occupying only a small fraction of the volume of the vessel, the ions would recombine much more rapidly than if the same number of ions were uniformly distributed throughout the vessel.

From equation (5) we have

$$\dfrac{dn}{dt} = \dfrac{n_0^2 \alpha}{(1 + n_0 \alpha t)^2}.$$

Let N be the number of ions in the vessel at the time t, N_0 the number when $t = 0$, then if V is the volume through which the ions are distributed $N = nV$ and $N_0 = n_0 V$, and the preceding equation becomes

$$\dfrac{dN}{dt} = \dfrac{N_0^2 \dfrac{\alpha}{V}}{\left(1 + \dfrac{N_0 \alpha t}{V}\right)^2};$$

[1] Plimpton, *Phil. Mag.* vi. 25, p. 65, 1913.

in the earlier stages when $N_0 at/V$ is small compared with unity we have
$$\frac{dN}{dt} = N_0^2 \frac{u}{V}.$$

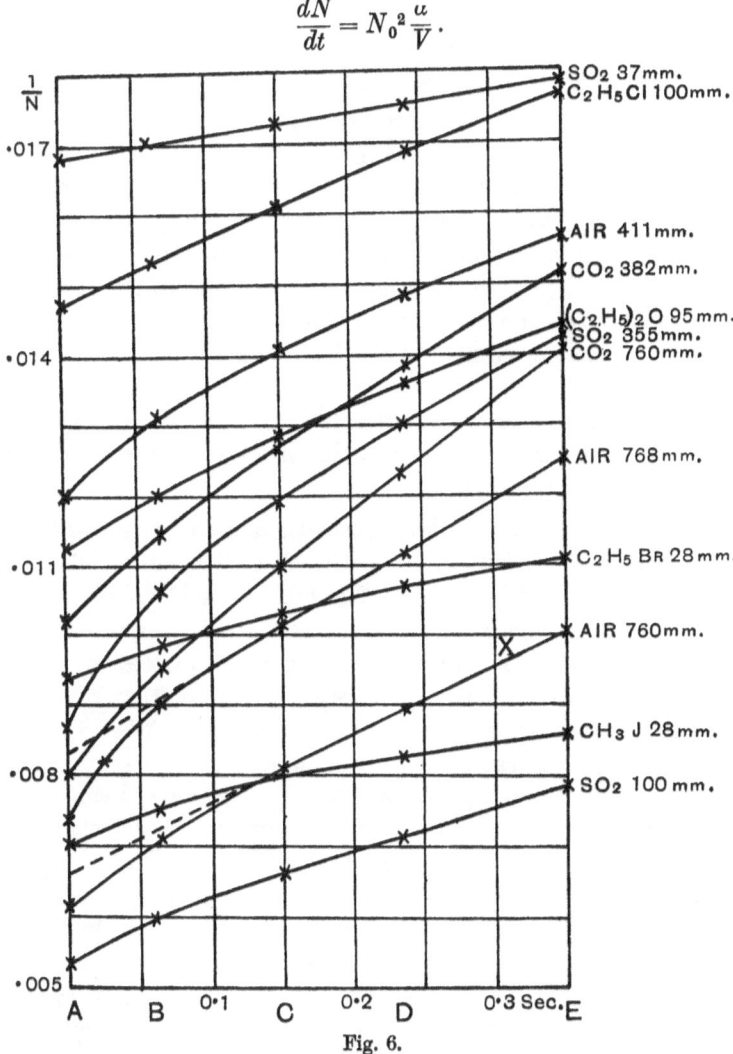

Fig. 6.

Thus for the same amount of ionisation the rate of recombination varies inversely as the volume through which the ions are distributed, so that if initially the ionisation were confined to a small fraction of the volume of the gas the rate of recombination would be much larger than if the gas were uniformly ionised. If the ions are localised to begin with they will gradually diffuse and the distribution will become much more uniform and the rate of recombination will diminish.

When a gas is ionised by α or β rays the ionisation is at first confined to the track of these rays and so occupies only a small fraction of the volume of the gas, and C. T. R. Wilson's photographs show that this is also true when the gas is ionised by X-rays; with these ionisers we should therefore expect the initial rate of recombination to be much greater than that after the ions had had time to diffuse and get uniformly distributed. The greater the concentration of the ions, the greater should be the difference between the initial and final values of α, and it would seem that by the study of this difference for different ionisers, such as α-rays and cathode rays with different velocities, we might get some insight into the intensity of ionisation along the paths of these rays.

The rapid recombination of gases ionised in columns by α or β rays makes them much more difficult to "saturate" than gases which are uniformly ionised throughout their volume. Bragg and Kleeman[1] observed that it was much more difficult to saturate gases ionised by α-rays than those ionised by X-rays. Moulin[2] explained this by the rapid recombination of the ions due to the columnar ionisation; he showed too that the gas is much more easily saturated when the electric force is transverse to the paths of the rays than when it acts along them, a result which evidently would follow from the columnar ionisation. The difference between the transverse and longitudinal saturating electric fields was much more marked at high pressures than at low and with air and carbonic acid gas than with hydrogen—indeed, with this gas it was hardly perceptible. It is also larger when the gas is ionised by α-rays than by β-rays, and disappeared when the gas was ionised by X-rays. This is what we should expect; for although, as Wilson's photographs show, the ionisation when a gas is exposed to X-rays is due to secondary β-rays, they also show that these β-rays are emitted in all directions, so

[1] Bragg and Kleeman, *Phil. Mag.* vi. 11, p. 466, 1906.
[2] Moulin, *Ann. der Phys.* xxii. p. 26, 1911.

15·2. IV. Comparison of the number of ions at any time with the saturation current.

Consider a vessel in which q ions are being produced per unit volume per second: the total number of ions produced per second is $\int q\,dv$, where dv is an element of volume and the integration is extended throughout the vessel; this integral can be determined by measuring the saturation current I through the vessel, since $I = e\int q\,dv$, where e is the charge on an ion.

When the ionisation has got into a steady state, if n is the number of positive or negative ions per unit volume where the ionisation is q and *if there is no loss of ions by diffusion to the sides of the vessel*,

$$\alpha n^2 = q, \quad n = \sqrt{\frac{q}{\alpha}},$$

and the total number of positive or negative ions in the vessel is

$$\frac{1}{\alpha^{\frac{1}{2}}}\int q^{\frac{1}{2}}\,dv.$$

If we apply to the vessel an exceedingly strong electric field for a short time we shall drive to the sides of the vessel all the ions present at that time; if Q is the charge carried by these ions,

$$Q = e\int n\,dv = \frac{e}{\alpha^{\frac{1}{2}}}\int q^{\frac{1}{2}}\,dv.$$

If q is constant throughout the vessel and V the volume of the vessel we see that

$$Q^2 = \frac{e}{\alpha}VI,$$

an equation by which we may determine α.

It is very important to notice that the use of this formula will give too high values for α unless q is constant throughout the volume. For suppose q had the constant value q_1 throughout the volume v_1 and q_2 throughout v_2, where $v_1 + v_2 = V$, then

$$I = e(q_1 v_1 + q_2 v_2),$$

$$Q = \frac{e}{\sqrt{\alpha}}(q_1^{\frac{1}{2}} v_1 + q_2^{\frac{1}{2}} v_2),$$

$$\frac{Q^2}{I} = \frac{e}{\alpha}\frac{(q_1^{\frac{1}{2}} v_1 + q_2^{\frac{1}{2}} v_2)^2}{q_1 v_1 + q_2 v_2},$$

and
$$\frac{Q^2}{VI} = \frac{e}{\alpha} \frac{(q_1^{\frac{1}{2}} v_1 + q_2^{\frac{1}{2}} v_2)^2}{(q_1 v_1 + q_2 v_2)(v_1 + v_2)}$$
$$= \frac{e}{\alpha} \left\{ 1 - \frac{v_1 v_2 (q_1^{\frac{1}{2}} - q_2^{\frac{1}{2}})^2}{(v_1 + v_2)(q_1 v_1 + q_2 v_2)} \right\}.$$

Thus unless $q_1 = q_2$, α is less than the value given by
$$\alpha = \frac{eVI}{Q^2}.$$

This effect of a want of uniformity in the ionisation is of considerable importance in the determination of the coefficient of recombination. For the incidence of the X-rays, or other rays used to ionise the gas, on the sides of the vessel containing the gas gives rise to secondary rays which also ionise the gas, and which being more easily absorbed than the incident rays make the ionisation near the walls of the vessel greater than in the interior. The magnitude of this effect will depend:

1. On the material of which the vessel is made.

2. On the size of the vessel; the importance of the effect is greater for small vessels than for large ones.

3. On the nature and pressure of the gas in the vessel; the effect tends to be greater at a low pressure than at a higher one.

This method has been used by McClung[1], Hendren[2], and Erikson[3].

15·3. V. Gas current from ionisation vessel. Another method which might be used is the following.

Suppose we ionise the gas in a vessel whose volume is V, and through which a stream of the gas is kept flowing, the gas before entering the vessel being carefully screened from ionisation. Let ω cubic centimetres of gas pass through the vessel per second, then if there are n positive and n negative ions in each cubic centimetre of gas in the vessel, the number of either of these ions carried out by the stream per second is ωn; hence if the ionising agent produces q ions per cubic centimetre we have
$$\frac{V dn}{dt} = Vq - V\alpha n^2 - \omega n,$$

[1] McClung, *Phil. Mag.* vi. 3, p. 283, 1902.
[2] Hendren, *Phys. Rev.* xxi. p. 314, 1905.
[3] Erikson, *Phys. Rev.* xxvii. p. 473, 1908.

or when things have reached a steady state
$$Van^2 + \omega n = Vq.$$
If n_1 is the number when the velocity of the stream is ω_1
$$Van_1^2 + \omega_1 n_1 = Vq,$$
hence
$$Va(n_1^2 - n^2) = \omega n - \omega_1 n_1,$$
or
$$a = \frac{\omega n - \omega_1 n_1}{V(n_1^2 - n^2)}.$$

If e is the charge on an ion, Ven_1 and Ven can be determined by applying to the vessel for a short time an electromotive force strong enough to drive in that time all the ions of one sign to the sides of the vessel. Or if the gas on its escape from the vessel passes through a tube connected with an electrometer, the saturation current sent to the electrometer when a large difference of potential is maintained between the tube and a wire along its axis will give ωen and $\omega_1 en_1$.

A modification of this method is to measure the whole number of ions sent through the outlet tube after the ionising agent has been stopped. This number is equal to
$$\int_0^\infty \omega n\, dt,$$
but after the rays are stopped
$$\frac{dn}{dt} = -an^2 - \frac{\omega}{V}n,$$
so that
$$\omega n = \frac{-\omega \frac{dn}{dt}}{an + \frac{\omega}{V}},$$
and therefore
$$\int_0^\infty \omega n\, dt = \frac{\omega}{a} \log\left(\frac{n_0 + \frac{\omega}{aV}}{\frac{\omega}{aV}}\right),$$
where n_0 is the value of n when the ionising agent is stopped, and can be found by measuring the saturation current in the outlet tube while the ionising agent is in action.

Knowing the value of n_0 we can by a graphical method find the value of a.

15·4. VI. Determination of α by measuring the current through an ionised gas. The method by which the first determinations of α were made by Thomson and Rutherford[1] in 1896 requires a knowledge of the velocity of the ions under an electric force; this velocity, as we shall see later on, is proportional to the electric force. Let $k_1 X$, $k_2 X$ be the velocities of the positive and negative ions respectively when the electric force is equal to X; then if there are n positive and n negative ions per unit volume of the gas, i the electric current through unit area is given by the equation

$$i = ne(k_1 + k_2) X \ \ldots\ldots\ldots\ldots\ldots(1),$$

where e is the charge on an ion.

When an electric current is passing through the gas, some of the ions are removed through being driven against the electrode; i/e positive ions are each second driven against each unit area of the negative electrode, and i/e negative ions against that of the positive electrode; hence when we take this into account, we have when i is the current through unit area of the gas

$$V \frac{dn}{dt} = Vq - V\alpha n^2 - \frac{Ai}{e},$$

where V is the volume of gas between the electrodes and A the area of an electrode. When things are in a steady state, $dn/dt = 0$ and therefore

$$q = \alpha n^2 + \frac{A}{V}\frac{i}{e}.$$

If l is the distance between the electrodes $A/V = 1/l$, and eliminating n by means of equation (1) we have

$$q = \frac{\alpha i^2}{e^2 (k_1 + k_2)^2 X^2} + \frac{i}{el} \ \ldots\ldots\ldots\ldots(2);$$

if I is the value of i when the current is saturated $I = qel$ and equation (2) may be written

$$\frac{\alpha}{e} = \frac{(I - i)(k_1 + k_0)^2 X^2}{i^2 l} \ \ldots\ldots\ldots\ldots(3).$$

In this investigation we have assumed that X, the electric force, is constant between the electrodes. In consequence of the

[1] J. J. Thomson and E. Rutherford, *Phil. Mag.* v. 42, p. 382, 1896.

accumulation of the ions arising from their motion in the electric field (see Chapter IV), X will not fulfil this condition except in two cases: (1) when i is small compared with I, when (3) may be written

$$\frac{\alpha}{e} = \frac{I}{i^2}\frac{(k_1+k_2)^2\,X^2}{l},$$

and (2) when i is nearly equal to I, when (3) becomes

$$\frac{\alpha}{e} = \frac{I-i}{I^2}\frac{(k_1+k_2)^2\,X^2}{l}.$$

In this case Riecke[1] and Mie[2] have shown that a closer approximation is given by the equation

$$\frac{\alpha}{e} = \frac{I-i}{i^2}\frac{(k_1+k_2)^2\,X^2}{l}\left\{1-\frac{1}{5}\frac{I-i}{I}\right\}.$$

16·1. VII. Langevin's Method. The principle of this method, which is due to Langevin[3], is as follows. Suppose that A and B are two parallel plates immersed in a gas, let a slab of the gas bounded by the parallel planes a, b be ionised by an instantaneous flash of X-rays. Let there be an electric force between the plates, then all the positive ions produced by the rays would be attracted to the negative plate and all the negative ions to the positive plate, and if the field were infinitely strong they would reach these plates before they had time to recombine, so that each plate would receive N_0 ions, if the flash of X-rays produced N_0 positive and N_0 negative ions. With weaker fields the number of ions received by the plates will be less, as some of these will recombine before they can reach the plates. To find the number in this case we proceed as follows. In consequence of the movements of the ions under the electric field the slab of ionised gas will broaden out and will consist of three

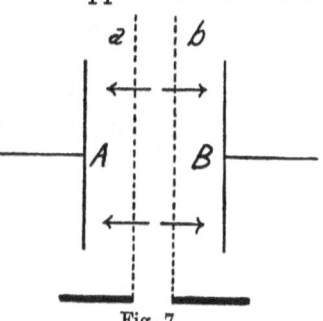

Fig. 7.

[1] Riecke, *Ann. der Phys.* xii. p. 814, 1903.
[2] Mie, *Ann. der Phys.* xiii. p. 857, 1904.
[3] Langevin, *Comptes Rendus*, cxxxiv. p. 533, 1902.

portions, one on the side of the negative plate in which there are nothing but positive ions, a middle portion in which there are both positive and negative ions, and a portion on the side of the positive plate in which there are only negative ions. If n is the number of ions per unit volume in the middle layer at the time t,

$$\frac{dn}{dt} = -an^2 \quad \text{or} \quad n = \frac{n_0}{1 + an_0 t},$$

where n_0 is the value of n when $t = 0$. No recombination occurs in the outer portions, so that all the ions in these portions will reach the plate to which they are nearest.

In the time dt the breadth of each of the outer layers will increase by $X(k_1 + k_2) dt$, where X is the electric force between the plates and k_1 and k_2 the velocities of the positive and negative ions under unit electric force; the number of negative ions added to the layer next the positive plate and of positive ions to the layer next the negative plate in this time is therefore

$$nX(k_1 + k_2) dt = \frac{n_0 X(k_1 + k_2) dt}{1 + n_0 at}.$$

The outer layers will continue to receive fresh ions until the middle layer disappears, which it will do after a time $l/X(k_1 + k_2)$, l being the breadth of the slab ab. Hence N the number of negative and also of positive ions which escape recombination and reach their respective plates is equal to

$$\int_0^{l/X(k_1+k_2)} \frac{n_0 X(k_1 + k_2)}{1 + n_0 at} dt = \frac{X(k_1 + k_2)}{a} \log\left(1 + \frac{n_0 al}{X(k_1 + k_2)}\right).$$

Let $a/4\pi\epsilon(k_1 + k_2) = \epsilon$, and $n_0 l e = Q_0$, Q_0 is the charge on the whole of the positive or negative ions produced by the rays, e being the charge on an ion, then if Q is the charge received by one of the plates we have

$$Q = Ne = \frac{X}{4\pi\epsilon} \log\left(1 + \frac{Q_0 4\pi\epsilon}{X}\right) \quad \ldots\ldots\ldots\ldots(1).$$

If Q_1 is the charge received by either plate when the electric force is X_1

$$Q_1 = \frac{X_1}{4\pi\epsilon} \log\left(1 + \frac{Q_0 4\pi\epsilon}{X_1}\right) \ldots\ldots\ldots\ldots(2).$$

From equations (1) and (2) we can eliminate Q_0 and determine ϵ.

Bloch has shown that similar equations apply to the case when ionised gas is slowly sucked through a long tube and the ions driven to the sides of the tube by a radial electric field.

17·1. On the effect of pressure on the coefficient of recombination. There is a considerable amount of discrepancy in the results obtained by different physicists who have investigated this question. Langevin[1], who was the first to attack this question and who used the method just described, found the following relative values for α at different pressures for air and carbonic acid gas:

Air		CO_2	
Pressure in mm. of Hg	α	Pressure in mm. of Hg	α
152	·05	135	·05
375	·12	357	·27
760	·27	550	·37
1550	·31	758	·51
2320	·26	1560	·47
3800	·18	2380	·33

These results show at low pressures a striking diminution in the value of α with the pressure, and in each case also there is a certain pressure at which α is a maximum, and α seems to diminish pretty rapidly after passing this pressure.

McClung[2], who used Method I, found but little variation in α when the pressure was raised from $\frac{1}{4}$ to 5 atmospheres; it is probable however that his results at low pressures are too large, as with his arrangement there might be considerable loss of ions by diffusion at the lower pressures, at the higher pressures diffusion would not be so important. Plimpton found that α diminished with decreasing pressure below atmospheric.

Hendren[3], who used Method I, found a considerable diminution of α with pressure at pressures less than 1 atmosphere, though

---
[1] Langevin, *Comptes Rendus*, cxxxvii. p. 177, 1902.
[2] McClung, *Phil. Mag.* vi. 3, p. 283, 1902.
[3] Hendren, *Physical Review*, 1905.

this diminution was not nearly so great as in Langevin's experiments; his numbers are given in the following table:

Pressure in cm.	76	45	25	15	10	5	3·5	2	1
a		3·3	2·65	2·07	1·75	1·55	1·31	1·25	1·15	1·00

These numbers are only relative.

A very careful examination of the relation between the coefficient of recombination and the pressure has been made by Thirkill[1] in the Cavendish Laboratory; he ionised the gases by X-rays and measured α by Langevin's method. His results are given in the following table:

Air.

Pressure	743	662	644	462	363	307	197
a/e	3500	3090	3020	2150	1730	1530	920

Carbon dioxide.

Pressure	729	614	498	373	265	175
a/e	3440	3020	2500	1930	1400	1020

Carbon monoxide.

Pressure	757	690	556	409	247
a/e	1820	1600	1310	930	468

Sulphur dioxide.

Pressure	680	504	444	338	200	83·5
a/e	2740	2250	1900	1510	876	548

Nitrous oxide.

Pressure	749	596	430	294	209
a/e	2830	2400	1690	1110	690

Langevin's method involves the mobility of the ions; in deducing the preceding values of α the mobilities used were those given by Wellisch[2].

The results of Thirkill's experiments are represented graphically in Fig. 8. It will be seen that for pressures up to about an atmosphere α is a linear function of the pressure. For these gases, above atmospheric pressure the rate of increase with the pressure according to Langevin's experiments diminishes, α reaches a maximum at a certain pressure and then diminishes as the pressure increases. Erikson[3] found that α for air at 200 was about

---

[1] Thirkill, *Proc. Roy. Soc.* lxxxviii. p. 477, 1913.
[2] Wellisch, *Phil. Trans.* A, 209, p. 249, 1908.
[3] Erikson, *Phys. Rev.* xxvii. p. 473, 1908.

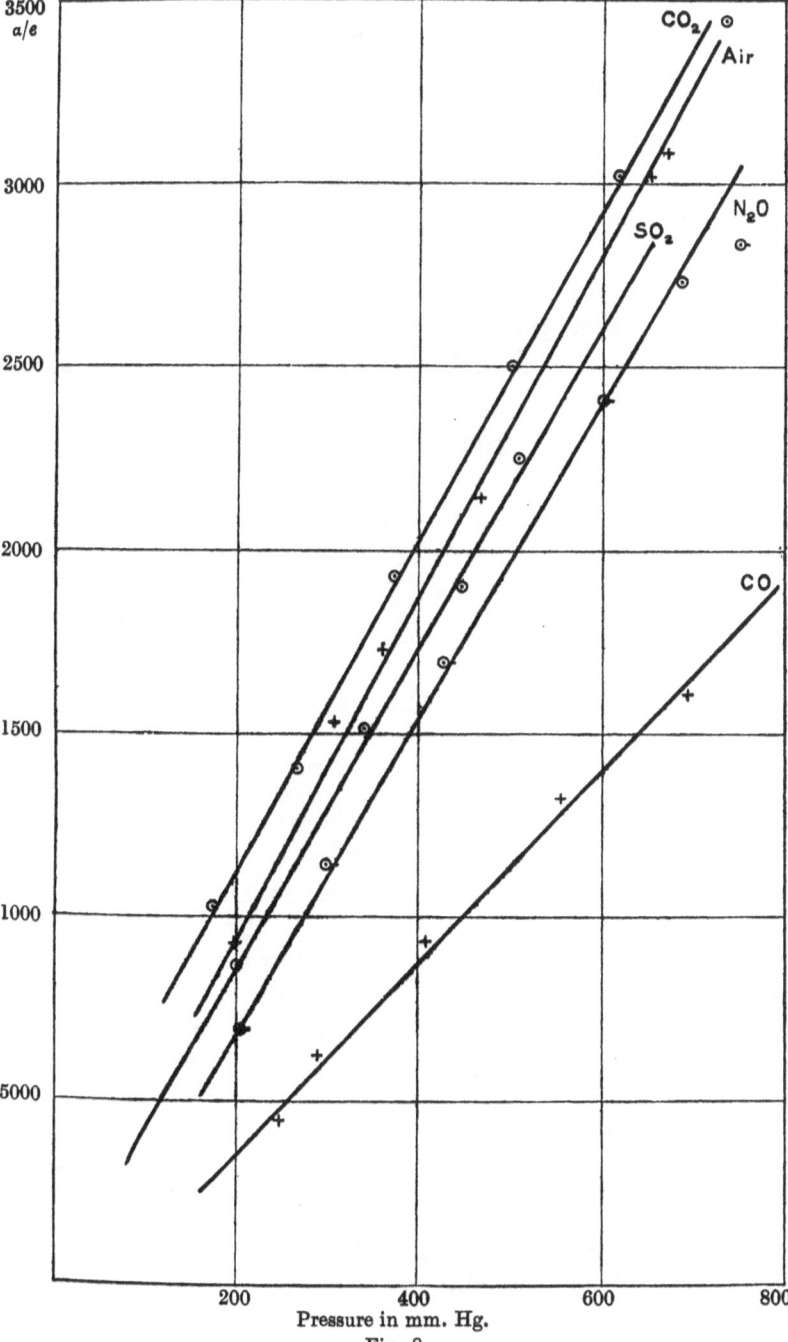
Fig. 8.

30 per cent. greater than α at 400 atmospheres. When liquids are ionised by X-rays or radium rays the coefficient of recombination is very small compared with that for gases at atmospheric pressure, thus for hexane the value of α/e is only 2·26, while for air at atmospheric pressure it is about 3500.

18·1. Effect of temperature on the rate of recombination. The relation between α and the temperature has been investigated by Erikson[1], who used Method I, for air, carbonic acid and hydrogen at constant density when ionised by α

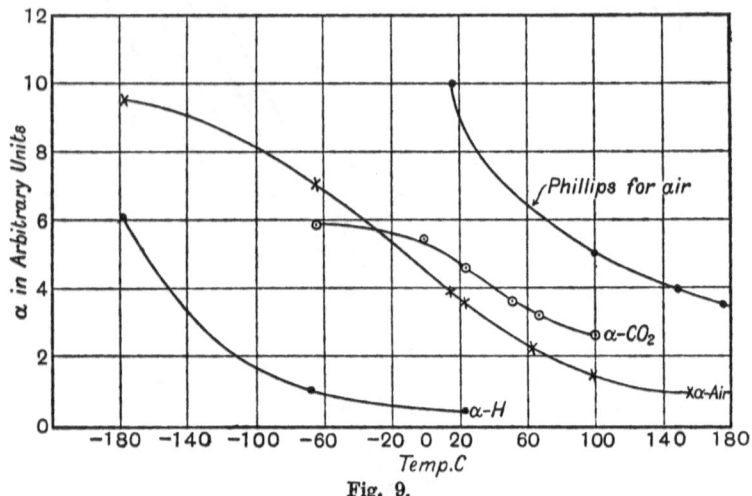

Fig. 9.

rays, and for air at constant pressure ionised by X-rays by Phillips[2], who used Langevin's method. These results are represented graphically in Fig. 9. Fig. 10 is the graph obtained by plotting $\log \alpha$ against $\log \theta$, where θ is the absolute temperature. It will be noticed that for all the gases the value of α when the density is constant is greater at low temperatures than at high. The shape of the curves is different for the different gases, but from Fig. 10 we see that over the whole range of temperatures examined the value of α for hydrogen can be represented by a formula of the type $\alpha = C\theta^{-n}$, where C is a constant, and although

[1] Erikson, *Phil. Mag.* vi. 18, p. 328, 1909; 23, p. 747, 1912.
[2] Phillips, *Proc. Roy. Soc* A, lxxxiii. p. 246, 1910.

this is not true for air and carbonic acid at temperatures much below 0° C., at the higher temperatures an equation of this kind represents with fair accuracy the value of α. The failure of this law at the lower temperatures may possibly be due to the formation of more complex ions at these temperatures. Erikson's experiments make $n = 2\cdot 5,\ 2\cdot 4,\ 2\cdot 2$ for air, carbonic acid and hydrogen

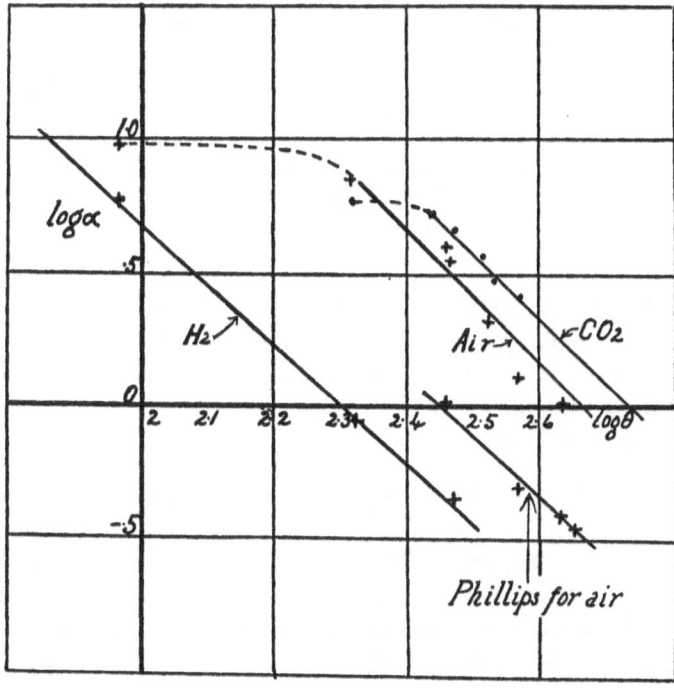

Fig. 10.

respectively; these do not differ much from $\frac{7}{3}$, while Phillips' value of n for air is $2\cdot 2$. Thus according to Erikson α varies as $\theta^{-\frac{7}{3}}$, over a considerable range of temperature for air, carbonic acid and hydrogen, while Phillips' experiments indicate that α varies as $\theta^{-2\cdot 2}$ for air at constant pressure.

18·2. Table of values of α. The following table contains the values of α/e at atmospheric pressure and 0° C. for the gases for which it has been determined:

Values of α/e for various gases at atmospheric pressure and ordinary temperature.

Gas	Townsend. X-rays, Method I	M°Clung. X-rays, Method IV	Langevin. X-rays, Method VII	Thirkill. X-rays, Method VII	Hendren. α-rays, Method IV	Retschinsky. X-rays, Method VI	Rümelin. α-rays, Method III	Rümelin. β-rays, Method III	Erikson. α-rays, Method I
Air	3420	3380	3200	3580	3300	4200	4240	5820	3500
CO_2	3500	3490	3400	3500	—	—	—	—	—
H_2	3020	2940	—	—	—	—	—	—	—
O_2	3380	—	—	—	—	—	—	—	—
SO_2	—	—	—	3000	—	—	—	—	—
N_2O	—	—	—	2960	—	—	—	—	—
CO	—	—	—	1780	—	—	—	—	—

The results ascribed to Thirkill were obtained by extrapolation from his curves, e is 4.77×10^{-10}, so that for air α is about 1.6×10^{-6}

Bloch[2] has shown that α/e for the ions produced when air is passed over phosphorus is between 1 and 6. Jaffé[3] and van der Bijl[4] have obtained values of α/e for the ions produced in insulating liquids by rays from radium; for liquid hexane van der Bijl finds $\alpha/e = 2.26$.

19·1. Theory of recombination of ions. Recombination of ions takes place when a positive and a negative ion get associated so as to form a neutral system, and measurements of the coefficient of recombination give the rate at which these systems are formed. Neutral systems may arise from charged ions in several ways; take the case for example when the positive ion is an atom which has lost an electron and where the negative ions are electrons. The two kinds of ions might form a neutral system:

(1) by the electron going back into the atom;
(2) by the electron describing a closed orbit of small radius round the positive ion.

For the first kind of combination to take place it is necessary for the negative ion to strike against the positive, while for the

---

[1] Retschinsky, *Ann. der Phys.* xvii. p. 518, 1905.
[2] Bloch, *Ann. de Chimie et de Phys.* viii. 4, p. 1, 1905.
[3] Jaffé, *Ann. der Phys.* xxxii. p. 173, 1910.
[4] van der Bijl, *Ann. der Phys.* xxxix. p. 170 1912.

second kind this condition is not necessary but others have to be satisfied. This uncertainty as to the nature of the recombination is one of the difficulties in the way of finding an expression for the rate at which recombination takes place; another difficulty is that the theory depends on the nature of a collision between molecules, and therefore on the nature of the intermolecular forces, and as this is still one of the obscure points in the Kinetic Theory of Gases the treatment of the recombination of ions must be somewhat tentative.

Let us first consider the number of collisions per second which occur between the positive and the negative ions; it is clear that in consequence of the attraction between the electric charges this number will be greater than for uncharged bodies of the same shape and size as the ions—how much greater this must be is shown by the fact that the number of *recombinations* between n positive and n negative ions per c.c. in a second is for air

$$1 \cdot 6 \times 10^{-6} n^2,$$

while the number of collisions in the same time between $2n$ molecules of air is only

$$4 \times 10^{-10} n^2,$$

which is only about $\frac{1}{4000}$ of the number of recombinations between the ions.

We can find an expression for the effect of the attraction between the charges on the chance of a collision in the following way: we shall first take the case when the motion of the ions is not interfered with by collision with the molecules of the gas in which the ions are produced.

Consider two bodies A and B attracting or repelling each other radially in such a way that the attraction between them when they are separated by a distance r is $F(r)$. The centre of gravity G of A and B will move uniformly along a straight line. We shall discuss the motion of A or B round G. If m_1, m_2 are the masses of A and B respectively, and if $AB = r$,

$$BG = \frac{m_1}{m_1 + m_2} r, \quad AG = \frac{m_2}{m_1 + m_2} r.$$

Let us now consider the motion of B round G: the radial

force on $B = F(r) = F\left(\dfrac{m_1 + m_2}{m_1}\rho\right)$ if $\rho = BG$. If v is the velocity of B relative to G when $BG = \rho$ we have by the Conservation of Energy

$$\tfrac{1}{2} m_2 v^2 = \tfrac{1}{2} m_2 V^2 + \int_\rho^\infty F\left(\dfrac{m_1 + m_2}{m_1}\rho\right) d\rho \ \ldots\ldots (1),$$

where V is the value of v when ρ is infinite, the particles being supposed to start at an infinite distance from each other.

If U is the velocity of B relative to A when the particles are at an infinite distance

$$V = \dfrac{m_1}{m_1 + m_2} U.$$

When A and B are at their minimum distance apart, ρ is a minimum and the velocity of B relative to G is at right angles to BG; hence if a is the minimum value of ρ

$$va = Vp,$$

where p is the perpendicular from G on V when ρ is infinite, hence by (1)

$$\tfrac{1}{2} m_2 \dfrac{V^2 p^2}{a^2} = \tfrac{1}{2} m_2 V^2 + \int_a^\infty F\left(\dfrac{m_1 + m_2}{m_1}\rho\right) d\rho \ \ldots\ldots (2),$$

an equation to find a. If σ is the minimum value of AB, and b the perpendicular from A on the velocity of B relative to A when AB is infinite

$$p = \dfrac{m_1}{m_1 + m_2} b, \quad \sigma = \dfrac{m_1 + m_2}{m_1} a.$$

Thus
$$pV = \left(\dfrac{m_1}{m_1 + m_2}\right)^2 bU,$$

so that (2) becomes

$$\dfrac{b^2}{\sigma^2} = 1 + 2\dfrac{(m_1 + m_2)^2}{m_1^2 m_2 U^2} \int_a^\infty F\left(\dfrac{m_1 + m_2}{m_1}\rho\right) d\rho$$

$$= 1 + 2\dfrac{m_1 + m_2}{m_1 m_2 U^2} \int_\sigma^\infty F(r)\, dr.$$

Now suppose we define a collision as the approach of A and B within a distance σ: if there were no forces between the bodies the condition for a collision would be $b = \sigma$, when there are forces the condition is

$$b^2 = \sigma^2 \left(1 + 2\dfrac{m_1 + m_2}{m_1 m_2 U^2} \int_\sigma^\infty F(r)\, dr\right).$$

The number of collisions between two systems A and B is proportional to b^2, hence the effect of the mutual forces is to increase the number of collisions in the proportion of

$$1 + 2\frac{m_1 + m_2}{m_1 m_2 U^2}\int_\sigma^\infty F(r)\,dr \text{ to } 1.$$

If there were no forces between the bodies the number of collisions per c.c. per second between molecules of the types A and B is by the Kinetic Theory of Gases equal to

$$4\sqrt{\pi}n_1 n_2 \sigma^2 (u^2 + v^2)^{\frac{1}{2}},$$

where n_1, n_2 are the numbers of molecules of A and B respectively in unit volume, and

$$\tfrac{1}{2}m_1 u^2 = \tfrac{1}{2}m_2 v^2 = 2 \cdot 02 \times 10^{-16} \times \theta,$$

where θ is the absolute temperature.

It is convenient to express this energy as that acquired by the fall of the charge e on an ion through a certain difference of potential; at 0° C. this difference is $\frac{1}{29}$ of a volt.

If $F(r)$ represents the force between two ions due to their charges

$$F(r) = \frac{e^2}{r^2} \text{ and } b^2 = \sigma^2 \left(1 + 2\frac{m_1 + m_2}{m_1 m_2 U^2}\frac{e^2}{\sigma}\right).$$

Thus the number of collisions between oppositely charged ions is in this case

$$4\sqrt{\pi}n_1 n_2 \sigma^2 (u^2 + v^2)^{\frac{1}{2}}\left(1 + 2\frac{m_1 + m_2}{m_1 m_2 (u^2 + v^2)}\frac{e^2}{\sigma}\right) \quad \ldots(3),$$

since the mean value of U^2 is $u^2 + v^2$.

If each collision results in recombination

$$\alpha = 4\sqrt{\pi}\sigma^2 (u^2 + v^2)^{\frac{1}{2}}\left(1 + 2\frac{m_1 + m_2}{m_1 m_2 (u^2 + v^2)}\frac{e^2}{\sigma}\right) \quad \ldots(4).$$

Now e^2/σ is the work required to separate the ions; and putting $e = 4 \cdot 77 \times 10^{-10}$ and $\sigma = 2 \cdot 95 \times 10^{-8}$, the diameter of a molecule of nitrogen, $e^2/\sigma = e\,(4 \cdot 8 \text{ volts})$. If $m_1 = m_2$

$$\frac{m_1 + m_2}{m_1 m_2 (u^2 + v^2)} = \frac{1}{mu^2} = \frac{1}{e\,(\cdot 069 \text{ volts})} \text{ at } 0° \text{ C.}$$

Thus the term $2\dfrac{m_1 + m_2}{m_1 m_2 (u^2 + v^2)}\dfrac{e^2}{\sigma}$ is in this case about 140, hence the effect of the charge is to increase the number of collisions about 140 times.

For the nitrogen molecules $(u^2 + v^2)^{\frac{1}{2}}$ is $6{\cdot}9 \times 10^4$ at $0°\,C.$; substituting this value we find the number of collisions to be

$$n_1 n_2 \times 6 \times 10^{-8}$$

approximately, so that the coefficient of recombination assuming that each collision results in combination is 6×10^{-8}; this is not nearly so large as the value $1{\cdot}6 \times 10^{-6}$ found by experiment.

We notice from equation (4) that if the nature of the ions does not change with the temperature, the coefficient of recombination would on this theory vary inversely as the square root of the temperature. Again on the assumption that the nature of the ion does not vary with the pressure of the gas, we see that this expression leads to the conclusion that the coefficient of recombination is independent of the pressure. We have seen however that when the pressure is comparable with the atmospheric pressure the coefficient of recombination varies rapidly with the pressure, hence if the preceding theory is true the size of the ion must at these pressures vary rapidly with the pressure, a result which is rendered very improbable by the fact that the determinations of the mobilities of the ions at such pressures lead, as we shall see, to the conclusion that at these pressures there is no appreciable change in the radius of the ion as the pressure changes.

We are thus led to the conclusion that the observed recombination cannot be accounted for by the actual collision of the ions under their mutual attraction, at least as long as we neglect, as has been done in the above investigation, the effect of previous collisions between the ions and the molecules of the gas in which they move.

20·1. Effect on α of collisions with the gas molecules. The collision theory of recombination is however unsatisfactory on other grounds; we have no reason to suppose that recombination should necessarily accompany collision, if the ions were elastic spheres they might rebound and separate with great velocity; thus the condition that a collision should occur is not sufficient for recombination, neither is it necessary, for we might suppose that one ion described an orbit round the other so that the two formed a neutral system without coming into collision.

Let us now find the condition that an ion B should, after making a collision with one of the molecules of the gas, proceed to describe a closed orbit round an ion A of the opposite sign. We shall suppose as before that just after the collision the kinetic energy of the B ion is equal to the mean kinetic energy of a molecule at the temperature of the gas. If M_1, M_2 are the masses of an A and a B ion respectively, then if B makes a collision at a distance r from A, the ions will describe closed orbits round each other if r is less than the value b given by the equation

$$\frac{1}{2}\frac{M_1 M_2}{M_1 + M_2} V^2 = \frac{e^2}{b} \quad \ldots\ldots\ldots\ldots\ldots (1),$$

V being the relative velocity of the ions and the force between them being supposed to be due entirely to their electric charges.

The mean value of V^2 is $u^2 + v^2$, where u and v are respectively the mean velocities of the A and B ions; since

$$\tfrac{1}{2}M_1 u^2 = \tfrac{1}{2}M_2 v^2 = \beta\theta,$$

where
$$\beta = 2\cdot 02 \times 10^{-16},$$

$$V^2 = 2\alpha\theta \left\{\frac{M_1 + M_2}{M_1 M_2}\right\},$$

so that equation (1) becomes

$$\frac{e^2}{b} = \beta\theta;$$

thus b is inversely proportional to the absolute temperature and at $0°$ C. $= 4\cdot 1 \times 10^{-6}$ cm.

To find an expression for the number of recombinations between an A ion and a B ion we proceed as follows.

We first find the chance that a B ion which approaches within a distance b from A should make a collision with a molecule of the gas within this distance.

The chance that the B ion should be at a distance between ρ and $\rho + d\rho$ from a line drawn through A parallel to the relative velocity of the two ions is $2\rho d\rho/b^2$. The length of path described by the ion inside the sphere of radius b is $2\sqrt{b^2 - \rho^2}$, and the chance that it should describe this distance without a collision is

$$\epsilon^{-\frac{2\sqrt{b^2-\rho^2}}{\lambda_2}},$$

where λ_2 is the free path of the ion B through the molecules of the gas. Hence the chance that the ion B should pass without making a collision is

$$\frac{2}{b^2}\int_0^b \epsilon^{-\frac{2\sqrt{b^2-\rho^2}}{\lambda_2}} \rho\, d\rho$$

$$= -2\left(\frac{\epsilon^{-g_2}}{g_2^2} + \frac{\epsilon^{-g_2}}{g_2} - \frac{1}{g_2^2}\right),$$

if $g_2 = 2b/\lambda_2$; hence ω_2 the chance that it should make a collision is given by the equation

$$\omega_2 = 1 + 2\left(\frac{\epsilon^{-g_2}}{g_2^2} + \frac{\epsilon^{-g_2}}{g_2} - \frac{1}{g_2^2}\right).$$

Similarly the chance that the ion A should make a collision with a molecule of the gas within a distance b of the ion B is

$$\omega_1 = 1 + 2\left(\frac{\epsilon^{-g_1}}{g_1^2} + \frac{\epsilon^{-g_1}}{g_1} - \frac{1}{g_1^2}\right),$$

where $g_1 = 2b/\lambda_1$, where λ_1 is the mean free path of the ion A through the molecules of the gas; hence the chance that when the two ions come within a distance b one or other of them should make a collision within this distance is

$$\omega_1 + \omega_2 - \omega_1\omega_2;$$

$\omega_1\omega_2$ is the chance that they should both make collisions within this distance, and this has been included both in ω_1 and ω_2.

If n_1, n_2 are respectively the number of A and B ions in unit volume, U the relative velocity of the ions, the number of times the ions come within a distance b in time δt is

$$\pi b^2 n_1 n_2 U \cdot \delta t,$$

so that the number of collisions which result in recombination is

$$\pi b^2 n_1 n_2 U (\omega_1 + \omega_2 - \omega_1\omega_2)\, \delta t.$$

Hence α, the coefficient of recombination, is equal to

$$\pi b^2 U (\omega_1 + \omega_2 - \omega_1\omega_2).$$

If U_1, U_2 are the average velocities of the ions A and B respectively
$$U = \sqrt{U_1^2 + U_2^2}.$$
Hence
$$\alpha = \pi b^2 \sqrt{U_1^2 + U_2^2}\,(\omega_1 + \omega_2 - \omega_1 \omega_2) \quad \ldots\ldots(2).$$

When the pressure is low enough to make λ_1 and λ_2 large compared with b, g_1 and g_2 are small, and then
$$\omega_1 = \tfrac{2}{3} g_1, \quad \omega_2 = \tfrac{2}{3} g_2,$$
so that
$$\alpha = \frac{4\pi b^3}{3}\left(\frac{1}{\lambda_1} + \frac{1}{\lambda_2}\right) \sqrt{U_1^2 + U_2^2}\ \text{approximately} \quad \ldots(3).$$

Since both $1/\lambda_1$ and $1/\lambda_2$ are proportional to the pressure, this shows that at low pressures α is proportional to the pressure, as Thirkill found. As the pressure increases, g_1 and g_2 increase, and when these are large
$$\omega_1 = \omega_2 = 1\ \text{and}\ \alpha = \pi b^2 \sqrt{U_1^2 + U_2^2} \quad \ldots\ldots(4).$$

Thus at high pressures α becomes independent of the pressure. The experiments of Langevin and McClung show that at pressures higher than one atmosphere the rapid increase of α with the pressure slows down, α seems to attain a maximum at a pressure of a few atmospheres and then decreases as the pressure increases. Equation (2) indicates that α would approach asymptotically a constant value as the pressure increases, but would not decrease unless the ions became more complex and their relative velocity less.

Langevin[1] has deduced an expression for the coefficient of recombination which at high pressures agrees well with observation. He supposes that the ions, like bodies moving through a viscous liquid, move with a velocity proportional to the force acting upon them (we shall see that in the case of the ions this assumption is not true when the electric force is very large); the forces acting on the ions are supposed to be those due to the charges on neighbouring ions. Thus if a positive ion is at A and a negative one at B and there are no other ions within a distance comparable with AB, A and B will move straight towards each other with a velocity proportional to the attraction between them. Suppose that $R_1 F$, $R_2 F$ are respectively the velocities of

[1] Langevin, *Ann. de Chim. et de Phys.* xxviii. p. 287, 1903.

the positive and negative ions when the electric force is F; in this case $F = e/AB^2$, where e is the charge on an ion. The relative velocity of A and B along AB is thus $e(R_1 + R_2)/AB^2$. If n is the density of the B ions, the number of these that in one second would cross the surface of a sphere, centre A and radius AB, is

$$n \times 4\pi AB^2 \times e(R_1 + R_2)/AB^2 = 4\pi en(R_1 + R_2).$$

Thus on the average one would cross the surface every

$$\frac{1}{4\pi en(R_1 + R_2)}$$

of a second, and if it did not get deflected by other ions would move up to A and neutralise it. Thus the average life of an A ion would be

$$\frac{1}{4\pi en(R_1 + R_2)}$$

of a second and if there are m A ions per unit volume, the number that would recombine in one second is $4\pi emn(R_1 + R_2)$, so that the coefficient of recombination α is given by the equation

$$\alpha = 4\pi e(R_1 + R_2).$$

As R_1 and R_2 are inversely proportional to the pressure of the gas, the coefficient of recombination ought to vary inversely as the pressure of the gas and thus be smaller at high pressures than at low. At high pressures this expression gives good agreement with observation. We should expect that it would be a better representation of α at high pressures than at low ones; for its validity depends among other things on being able to find a distance AB large enough to make the electric attraction between the ions so small that the velocity is proportional to the force (the limiting force will increase with the pressure), and yet not so large that there is any danger of A or B coming within a short distance of other ions in their journey towards each other.

20·15. The effect of a charge of electricity on the free path of an ion. In order to compare the experimental values of α with those found on our theory, it is necessary to find expressions for λ_1 and λ_2. This can be done as follows.

If M_1, M_2 are the masses of two colliding particles, the transfer of kinetic energy from one to the other produced by a collision is, omitting terms which average out when a mean is taken,

$$\frac{4M_1M_2}{(M_1+M_2)^2} T \sin^2 \theta,$$

where 2θ is the angle through which the relative velocity of the particles is turned by the collision, $2T$ is the product of the momentum of the system, the relative velocity of the particles, and the cosine of the angle between these vectors. If either of the particles is at rest, T is thus the kinetic energy of the other; if the masses are equal, T is the difference between the kinetic energies of the two particles.

If one of the particles carries an electric charge and the other is an uncharged molecule, the deflection 2θ will be due to the attraction exerted by the charge on the molecule. This will be similar in character to that between an electric charge and an uncharged metal sphere, which varies inversely as the fifth power of the distance between the charge and the sphere, provided this distance is considerable in comparison with the radius of the sphere. We shall proceed to calculate the value of θ on the supposition that the law of force between the charged ion and the uncharged molecule is expressed by Ku^5, where $u = 1/AB$, AB being the distance between the ion and the molecule. With this law of force the equation of the orbit of the molecule relative to the ion is

$$\frac{d^2u}{d\theta^2} + u = \frac{Ku^3}{h^2},$$

where
$$h^2 = V^2 p^2 \frac{M_1 M_2}{M_1 + M_2},$$

where V is the initial velocity of B relative to A, and p the perpendicular from A on V.

Integrating this equation, we have

$$h^2 \left(\left(\frac{du}{d\theta}\right)^2 + u^2 \right) = \frac{Ku^4}{2} + MV^2,$$

where M is written for $M_1 M_2/(M_1 + M_2)$,

$$\left(\frac{du}{d\theta}\right)^2 = \frac{Ku^4}{2h^2} - u^2 + \frac{MV^2}{h^2} \quad \ldots \ldots \ldots \ldots (5).$$

The apsidal distances are found by equating the right-hand side of this expression to zero; they are the roots of the equation

$$u^4 - \frac{2MV^2p^2}{K}u^2 + \frac{2MV^2}{K} = 0 \dots\dots\dots(6);$$

for these roots to be real

$$p^4 > 2K/MV^2.$$

For values of p less than this there will be no apsidal distances; the bodies will come together and collide. If the collisions are like direct collisions between elastic spheres, the relative velocity will be reversed by the collision and the bodies will retrace their paths. In this case the relative velocity will be unaltered in magnitude and reversed in direction by the collision, so that θ will be $\pi/2$ and $\sin\theta$ unity.

The energy lost by an ion at these collisions is therefore

$$\frac{4M_1M_2}{(M_1+M_2)^2}T.$$

If there are N molecules per unit volume, the number of collisions made by an ion per unit length of path for which p is between p and $p+dp$ is $2\pi Np\,dp$, so that the loss of energy due to the collisions where the ion is drawn into the molecule is

$$\frac{4M_1M_2}{(M_1+M_2)^2}T\int_0^{(2K/MV^2)^{\frac{1}{4}}} 2\pi Np\,dp$$
$$= N\pi T\frac{4M_1M_2}{(M_1+M_2)^2}\left(\frac{2K}{MV^2}\right)^{\frac{1}{2}}.$$

There is also a certain amount of energy lost in the collisions where p is greater than $(2K/MV^2)^{\frac{1}{4}}$; in this case the ion is deflected without being drawn into the molecule. It can be[1] shown that this loss amounts to about 10 per cent. of the loss we have just considered, so that the total loss of energy per unit path is

$$4\cdot 4 N\pi T\frac{M_1M_2}{(M_1+M_2)^2}\left(\frac{2K}{MV^2}\right)^{\frac{1}{2}}.$$

We may regard the reciprocal of the coefficient of T as a measure of the energy free path of the ion; for calling this quantity λ, we have, in the distance δx,

$$\delta T = T\frac{\delta x}{\lambda}.$$

[1] J. J. Thomson, *Phil. Mag.* xlvii. p. 337, 1924.

If the ions were uncharged and if the collisions between them and the molecules were like those between elastic spheres, then if s is the sum of the radii of two spheres, p the perpendicular from the centre of one on the velocity relative to it of the other,

$$\sin^2 \theta = 1 - \frac{p^2}{s^2}$$

and

$$\int_0^s \sin^2 \theta \, N 2\pi p \, dp = \tfrac{1}{2} N \pi s^2;$$

so that if λ' is the energy free path for the uncharged ions

$$\frac{1}{\lambda'} = 2\pi N \frac{M_1 M_2 \, s^2}{(M_1 + M_2)^2},$$

so that

$$\frac{\lambda}{\lambda'} = \frac{1}{2 \cdot 2 \, (2K/s^4 M V^2)^{\frac{1}{2}}}.$$

Hence if the free path of the charged ion is to be much less than that of the uncharged one,

$$K/s^4 . M V^2$$

must be considerable.

Now $K/4s^4$ is the work required to separate a charged from an uncharged molecule with which it is in contact, and $MV^2/2$ the mean energy due to thermal agitation; thus for the free path for the charged ion to be much less than that for the uncharged one, the energy required to separate a charged from an uncharged molecule must be considerable, compared with the energy of a molecule due to thermal agitation. But from the Thermodynamics of Chemical Combination this is just the condition that aggregates of charged and uncharged molecules should be formed. Thus the two effects are inseparable; if the free path is shortened by the charges, then aggregates of charged and uncharged molecules must be formed, and when aggregates are formed the path must be shortened.

20·2. Variation of α with temperature. The expression for α at low pressures is

$$\alpha = \frac{4\pi b^3}{3} \left(\frac{1}{\lambda_1} + \frac{1}{\lambda_2} \right) \sqrt{U_1^2 + U_2^2},$$

the quantities λ_1, λ_2 being given, as we see from the above, by expressions of the form
$$\frac{1}{\lambda} = 4\cdot 4 N\pi \frac{M_1 M_2}{(M_1 + M_2)^2}\left(\frac{2K}{MV^2}\right)^{\frac{1}{2}}.$$
We can thus easily find the variation of α with temperature if we are prepared to assume that the complexity of the ions remains unchanged. In this case λ varies as V, i.e. as $\theta^{\frac{1}{2}}$, b^3 varies as θ^{-3}, and $\sqrt{U_1^2 + U_2^2}$ varies as $\theta^{\frac{1}{2}}$. Thus α varies as θ^{-3} at low pressures.

At high pressures $\alpha = \pi b^2 \sqrt{U_1^2 + U_2^2}$, which varies as $\theta^{-\frac{3}{2}}$.

The experimental results at constant density are $\alpha \propto \theta^{-\frac{7}{5}}$ (Erikson); the pressures were of the order of an atmosphere which is in between the regions for which the above formulae apply. While no exact agreement can be claimed, the theory is at least not contradicted by the experiments.

20·3. Value of K. We can, following Langevin, form some estimate of K from measurements of the refractive index of the gas.

If the molecule of a gas when exposed to an electrostatic force X develops an electrostatic moment equal to ωX, the specific inductive capacity of the gas is equal to
$$1 + 4\pi\omega N,$$
where N is the number of the molecules of the gas in a cubic centimetre. In non-polar gases, the specific inductive capacity is equal to μ^2, where μ is the refractive index of the gas; hence
$$\omega = \frac{\mu^2 - 1}{4\pi N}.$$
The electric force to which the molecule is exposed when it is separated by a distance r from an ion is e/r^2; the electrostatic moment of the molecule is thus $\omega e/r^2$, and the attraction between it and the ion is
$$(2\omega e/r^2)(e/r^3) = 2e^2\omega/r^5,$$
hence
$$K = 2e^2\omega = \frac{2e^2(\mu^2 - 1)}{4\pi N}.$$
Now $\frac{1}{2}MV^2 = \beta\theta$, where θ is the absolute temperature; thus
$$\frac{2K}{MV^2} = \frac{2e^2(\mu^2 - 1)}{4\pi N \beta \theta}.$$

WHEN IN THE CONDUCTING STATE

At 0° C. and 760 mm. pressure

$$N\beta\theta = 14\cdot 9 \times 10^5 \text{ and } e = 4\cdot 77 \times 10^{-10};$$

hence at this temperature and pressure

$$\frac{2K}{MV^2} = 2\cdot 4 \times 10^{-26} (\mu^2 - 1).$$

The following table contains the values of $(\mu^2 - 1)$ at 0° C. and 760 mm. for some of the commoner gases:

Gas	$\mu^2 - 1$	$1\cdot 1 \sqrt{2K/MV^2}$
H_2	278×10^{-6}	$2\cdot 9 \times 10^{-15}$
O_2	544×10^{-6}	$4\cdot 0 \times 10^{-15}$
Arg.	566×10^{-6}	$4\cdot 1 \times 10^{-15}$
CO	668×10^{-6}	$4\cdot 4 \times 10^{-15}$
CO_2	898×10^{-6}	$5\cdot 1 \times 10^{-15}$
SO_2	1320×10^{-6}	$6\cdot 2 \times 10^{-15}$

If the free path is determined by the charge on the ion

$$\frac{1}{\lambda} = 4\cdot 4 N\pi \frac{M_1 M_2}{(M_1 + M_2)^2} \left(\frac{2K}{MV^2}\right)^{\frac{1}{2}}.$$

20·4. Numerical value of α. Since the quantities λ_1, λ_2 refer to the frequency of collisions in which the excess energy of the ion is supposed destroyed, it is proper to use for their values the corresponding values of the energy free path as calculated above.

Thus to apply this to oxygen; if the ion has the same mass as the molecule, *i.e.* if there is no aggregation $M_1 = M_2$, and

$$\frac{1}{\lambda} = N\pi \frac{4M_1 M_2}{(M_1 + M_2)^2} 1\cdot 1 \sqrt{2K/MV^2}$$

$$= 3\cdot 4 \times 10^5,$$

$$\frac{2b}{\lambda} = 2\cdot 8;$$

hence $\omega_1 = \omega_2 = \cdot 80$.

This when substituted in equation (2), putting $U_1 = 4\cdot 25 \times 10^4$, gives

$$\alpha = 3\cdot 0 \times 10^{-6}, \text{ which is too large.}$$

If the ion has a mass equal to twice that of a molecule, then
$$\frac{4M_1M_2}{(M_1+M_2)^2} = \frac{8}{9} \text{ and } \frac{2b}{\lambda} = 2\cdot 5;$$
for this value of g_1
$$\omega_1 = \omega_2 = \cdot 77.$$
As the ions have twice the mass their velocity will be $1/\sqrt{2}$ of the previous value, so that
$$\alpha = 2\cdot 1 \times 10^{-6},$$
and is still larger than Thirkill's value.

If the ion has three times the molecular mass,
$$\frac{4M_1M_2}{(M_1+M_2)^2} = \frac{3}{4} \text{ and } \frac{2b}{\lambda} = 2\cdot 1;$$
for this value $\omega_1 = \omega_2 = \cdot 72.$

As the ions have three times the mass of the molecule their velocity will be but $1/\sqrt{3}$ of the average of that of the molecule, so that
$$\alpha = 1\cdot 7 \times 10^{-6},$$
which is the value found by Thirkill.

If the ion in hydrogen is a molecule of hydrogen we find by applying the preceding method that
$$\alpha = 1\cdot 2 \times 10^{-5},$$
which is far higher than the observed value. Pure hydrogen is however so little ionised by X-rays that it is probable that a large proportion of the ions were molecules of air. Since the mass of the oxygen molecule is sixteen times that of the hydrogen one,
$$\frac{4M_1M_2}{(M_1+M_2)^2} = \cdot 25 \text{ approximately,}$$
so that $1/\lambda$ for an oxygen molecule through hydrogen is only a quarter of that of a hydrogen molecule through hydrogen; if all the ions were oxygen molecules this would make $\alpha = 1\cdot 5 \times 10^{-6}$, which is about right.

The preceding investigation is based on the assumption that the force between the ion and the molecule varies inversely as the fifth power of the distance between them; this however will not be true unless this distance is large compared with the radii of the molecule and ion. Thus though the law may be true when the distances are large and the forces weak, it will fail when the forces become large and the effects of the collision most pronounced. To find an accurate numerical value of λ would require much greater knowledge of the forces between ions and molecules than we possess. We can however without more detailed knowledge of this law get some relation between different electrical properties of gases.

20·5. Relation between recombination and mobility of ions. Thus we know that when there is no loss of energy in a collision between two masses M_1, M_2, the momentum parallel to x transferred from one to the other by the collision is, omitting some terms which average out when a mean is taken,

$$\frac{2M_1 M_2}{M_1 + M_2}(u_1 - u_2)\sin^2\theta.$$

Here θ has the same meaning as before and u_1, u_2 are the velocities of A and B respectively parallel to the axis of x.

Hence if there are N molecules per unit volume, an ion in passing through a distance δx will lose an amount of momentum equal to

$$\delta x \times \frac{2NM_1 M_2}{M_1 + M_2}(u_1 - u_2)\int 2\pi p \sin^2\theta \, dp.$$

But from § 20·1 the loss of energy by the ion under the same conditions is

$$\delta x \times \frac{4M_1 M_2 N}{(M_1 + M_2)^2} T \int 2\pi p \sin^2\theta \, dp.$$

We have denoted this by $\delta x \cdot T/\lambda$; hence with this definition, whatever may be the law of force between the ion and the molecule, the loss of momentum by the ion in a distance δx is

$$\frac{1}{2}\frac{(M_1 + M_2)}{\lambda}(u_1 - u_2)\delta x.$$

If the ion under the action of an electric force X has settled down so as to move with a constant velocity,. this force must supply to the ion the momentum it loses by the collision; hence if δt is the time taken by the ion to describe the distance δx and e is the charge on the ion

$$Xe\delta t = \frac{1}{2}\frac{(M_1 + M_2)}{\lambda}(u_1 - u_2)\delta x.$$

Let u_0 be the mean value of dx/dt: then we have

$$u_1 - u_2 = \frac{2e}{M_1 + M_2} \frac{\lambda}{u_0} X \quad \ldots\ldots\ldots\ldots (7).$$

$u_1 - u_2$ is the average velocity of translation of the ion through the gas, and has been the subject of a great many experiments. If $(u_1 - u_2) = kX$, k is called the mobility of the ion: its value for many gases is given on p. 123; hence from (7)

$$k = \frac{2e}{M_1 + M_2} \frac{\lambda}{u_0} \quad \ldots\ldots\ldots\ldots\ldots (8),$$

so that if we know k and $M_1 + M_2$ we can determine the value of λ; knowing the value of λ we can deduce the values of the g's which occur in the expression for the coefficient of recombination, and thus from the mobility deduce the coefficient of recombination. This relation would be true whatever might be the law of force between the ion and the molecule.

At low pressures we saw (p. 47)

$$\alpha = \frac{4\pi b^3}{3} \left(\frac{1}{\lambda_1} + \frac{1}{\lambda_2}\right) \sqrt{U_1^2 + U_2^2}$$

$$= \frac{4\pi b^3}{3} 2e \left\{\frac{1}{M + M_1} \frac{1}{u_0 k_1} + \frac{1}{M + M_2} \frac{1}{u_0' k_2}\right\} \sqrt{U_1^2 + U_2^2},$$

where M is the mass of a gas molecule, M_1 and M_2 are the masses of the negative and positive ions, k_1 and k_2 their respective mobilities, u_0 and u_0' their mean velocities in a given direction. When the positive and negative ions have the same masses and mobilities, this becomes

$$\alpha = \frac{8\pi \cdot eb^3}{3} \left(\frac{2}{M + M_1} \frac{1}{k_1}\right) \frac{\sqrt{U_1^2 + U_2^2}}{u_0}.$$

But
$$u_0 = \frac{1}{\sqrt{3}} U_1 = \frac{1}{\sqrt{3}} U_2;$$

hence
$$\alpha = \frac{16\sqrt{2}\pi eb^3}{\sqrt{3}\,(M + M_1)\,k_1}.$$

It is interesting to see if the value of λ determined by equation (8) approximates to the value we have found on the assumption of a force varying inversely as the fifth power of the distance.

The mobility of the oxygen ion is such that under the force of a volt per cm. the mean velocity of the positive and negative ions at atmospheric pressure is about 1·6 cm./sec.; as a volt per cm. is represented by 10^8 on the electromagnetic system of units

$$k = \frac{1\cdot 6}{10^8}.$$

If there is no aggregation $M_1 = M$, and

$$\frac{e}{M + M_1} = \frac{e}{2M} = \frac{10^4}{64},$$

$u_0 = 2\cdot 7 \times 10^4$, if the ion is monomolecular.

Substituting these values we find

$$\lambda = 1\cdot 4 \times 10^{-6}.$$

This is of the right order, but is too small; if M_1 were greater than M, the value of λ determined from equation (8) would be greater and give a better value of α.

The most direct way of attacking the question of the state of aggregation of the ions would be to measure the value of α at high pressures; since this is equal to

$$\pi b^2 \sqrt{U_1{}^2 + U_2{}^2},$$

it gives us, without any assumption as to the laws of force, the relative velocity of the two ions, and when this is known we can determine whether or not there is any aggregation.

20·6. The free path of an electron. When the moving electrified particle is an electron the free path has to be found by a special investigation, since the mass of the electron is exceedingly small compared with that of the molecules with which it collides. From the expression for the transference of energy given on p. 49 we see that it is very small when the colliding bodies have very different masses; thus there will be very little transference of energy in a collision between an electron and a mass as great as that of a molecule. Thus if the gas through which the electrons are moving is hydrogen, the energy transferred by a collision between an electron and a molecule will be only 2/1850 of that which would be transferred if the collision had been not with an electron but with a hydrogen molecule moving with the same energy. It follows from this that the main loss of energy by the electrons is not by collision with the molecules as a whole but with the individual electrons in the molecule. These electrons have definite periods of vibration and the amount of energy transferred to them by collision with an electron of definite speed will vary very greatly with the frequency of these vibrations.

The energy the moving electron gives up at a collision with an electron oscillating about a fixed position depends upon the relation between the time of vibration of the fixed electron and the time taken by the moving electron to pass over a space equal to the minimum distance between the electrons.

The complete solution of the problem would be very lengthy and difficult: the following very rough solution may be expected to bring out the main features.

Let an electron A approach an electron B which can oscillate about a position of equilibrium O with a frequency n.

Let x be the displacement of B along the line joining B to the apse of the orbit of A. Then the most important part of the collision will be when B and A are in the neighbourhood of the apse, and then we have approximately

$$m\left(\frac{d^2x}{dt^2} + n^2x\right) = -\frac{e^2}{c^2 + p^2t^2}.$$

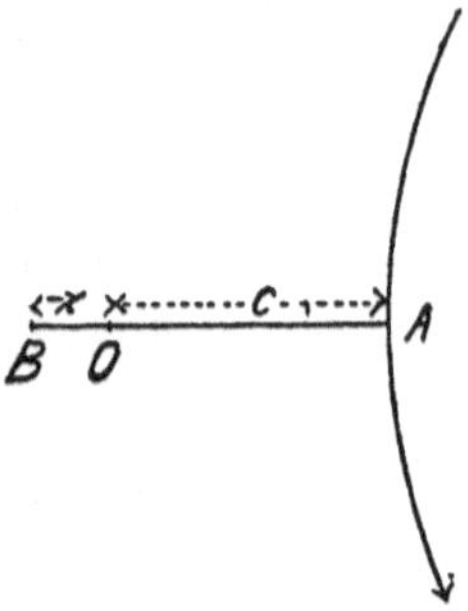

Fig. 11.

The time is measured from the time of passing through an apse, c is the apsidal distance, p a constant which we shall determine later, m is the mass, and e the charge of the electron.

The solution of this equation is

$$x = -\frac{e^2}{mn}\sin nt \int_{-\infty}^{t} \frac{\cos nt'\,dt'}{c^2 + p^2t'^2}$$
$$+ \frac{e^2}{mn}\cos nt \int_{-\infty}^{t} \frac{\sin nt'\,dt'}{c^2 + p^2t'^2}.$$

We may, since the force between the electrons only produces measurable effects when the electrons are near together, take $-\infty$ and $+\infty$ as the limits of integration, so that when the collision is over

$$x = -\frac{e^2}{mn}\sin nt \int_{-\infty}^{+\infty} \frac{\cos nt'}{c^2 + p^2t'^2}\,dt'$$
$$= -\frac{e^2}{mn}\frac{\pi}{cp}\epsilon^{-\frac{nc}{p}}\sin nt.$$

The energy corresponding to this displacement is
$$\frac{1}{2}\frac{e^4\pi^2}{mc^2p^2}\epsilon^{-\frac{2cn}{p}}.$$
This vanishes, both when $p=0$ and when $p=\infty$, and has a maximum when $p=cn$.

To find p we notice that when $t=0$,
$$r\frac{d^2r}{dt^2}=p^2,$$
where r is the distance between A and B, and $r^2=c^2+p^2t^2$ close to the apse: but by the equation of motion of A,
$$m\left(\frac{d^2r}{dt^2}-\frac{v^2}{r}\right)=\frac{e^2}{c^2},$$
regarding B as approximately fixed, where v is the velocity of the electron A at the apse.

Since $r=c$ when $t=0$, and
$$\frac{e^2}{c}=\tfrac{1}{2}m(V^2-v^2),$$
where V is the velocity of projection of A, we find
$$p^2=\tfrac{1}{2}(V^2+v^2).$$
If b is the perpendicular from B on the direction of projection of A,
$$bV=cv.$$

Put
$$\frac{e^2}{b}\tan\theta=mV^2;$$
then we see
$$c=b\cot\frac{\theta}{2};$$
$$v=V\tan\frac{\theta}{2};$$
$$p=\frac{V}{\sqrt{2}}\sec\frac{\theta}{2}.$$

When mV^2 is large compared with e^2/b, θ is nearly $\pi/2$, and we have approximately
$$c=b,\quad v=V,\quad p=V,$$
and the transference of energy is
$$\frac{1}{2}\frac{e^4\pi^2}{mb^2V^2}\epsilon^{-\frac{2nb}{V}},$$

so that when V is large the loss of energy varies as $1/V^2$, and will vanish when V is infinite. We see too that when V is small, $\epsilon^{-\frac{2cn}{p}}$ will vanish, so that the loss of energy vanishes for small velocities as well as great; thus except in special conditions the loss of energy at each collision will be small, and the free path large.

In the case of the ion the free path increases continuously with the energy; in the case of the electron when the velocity is small the free path is long but diminishes until the velocity reaches a critical value; when the velocity is greater than this value the free path again increases with the velocity.

20·7. Rate of recombination of electrons and positive ions. In the calculation of the rate of recombination of positive and negative ions we have assumed that the collision of either a negative or positive ion with a molecule of the gas through which they were moving, resulted in an appreciable loss of the kinetic energy acquired by the ions as they approach under their mutual attraction. This is true when the masses of the two ions are not very unequal, for then the increase in the kinetic energy will be shared between the ions, so that the energy of each will be increased by a finite amount which it will lose on collision. When however the masses differ as much as electrons from molecules, the increase in the kinetic energy when they approach is practically confined to the body of small mass, the electron, so that the positive ion has none to lose by collision. In this case the only collisions that give rise to loss of energy are those of the electrons with other electrons in the molecules, so that from (2), p. 47, α the coefficient of recombination will be given by the equation

$$\alpha = \pi b^2 \sqrt{U_1^2 + U_2^2} \left(1 + 2\left(\frac{\epsilon^{-g_1}}{g_1^2} + \frac{\epsilon^{-g_1}}{g_1} - \frac{1}{g_1^2}\right)\right),$$

where $g_1 = 2b/\lambda$,

and λ is the energy free path of the electron. From what we have seen λ, except in very special cases, will be greater than the geometrical free path or than that for the negative ion; this will tend to make the coefficient of recombination less for the electron than for the ion. On the other hand U_1, the average velocity, will be much greater for the electron than for the ion; this will tend to make the coefficient greater. At very high pressures, when g_1 is very large, the expression for α is

$$\alpha = \pi b^2 \sqrt{U_1{}^2 + U_2{}^2},$$

and this would be greater for the electron than for the ion.

There do not seem to have been any determinations made of the rate of recombination of electrons with positive ions, and for most gases these would be very difficult owing to the electron uniting with an uncharged molecule and becoming for a time a negative ion. There are however some gases such as argon, helium and nitrogen, where the electron remains free, and measurements of the rates of recombination in these would be of great interest.

The loss of energy by electrons through collisions has been measured by Hertz[1] and by Townsend and Bailey[2]. Hertz finds losses for electrons moving through hydrogen, amounting to as much as 6 per cent., and Townsend and Bailey losses in air of, in some cases, 2 or 3 per cent.; the loss depends to a great extent upon the velocity. These losses are much greater than those which occur in collisions between spheres of masses differing as much as those of an electron and a molecule.

20·8. Formation of negative ions and complex ions. The method used in § 20·1 to calculate the rate of recombination of positive and negative ions can be applied to find the rate of combination of any two attracting systems. The expression (2) will always hold if b is defined to be the distance between the systems when the work required to separate them to an infinite distance starting from b, is equal to the energy due to thermal agitation of a monatomic molecule at the temperature of the system.

We shall see however that this is not in all cases the most important method by which the combination can occur. Thus the following calculation on these lines gives only a lower limit to the rate of combination.

When an electron or a charged ion combines with an uncharged molecule, the result is in one case a negative ion, in the other

[1] Hertz, *Deuts. Phys. Gesell.* xix. p. 268, 1917.
[2] Townsend and Bailey, *Phil. Mag.* xlii. p. 889, 1921.

a complex one. If the attraction between the charge and the uncharged molecule is equal to K/r^5, where r is the distance between them, then the work required to separate them by an infinite distance starting from r is $K/4r^4$; hence D, the value of b in the expression for the rate of recombination, is given by the equation

$$\frac{K}{4D^4} = \beta\theta = \frac{e^2}{4\cdot 2 \times 10^{-6}}.$$

We saw, p. 52, that

$$K = \frac{(\mu^2 - 1)\, e^2}{2\pi N},$$

where μ is the refractive index of the gas at atmospheric pressure and N the number of molecules per c.c. at that pressure; hence

$$D^4 = \frac{(\mu^2 - 1)}{8\pi N} \times 4\cdot 2 \times 10^{-6}.$$

For oxygen at atmospheric pressure

$$\mu^2 - 1 = 5\cdot 4 \times 10^{-4}, \quad N = 2\cdot 7 \times 10^{19},$$

hence
$$D^4 = \frac{1\cdot 05}{\pi}\cdot 10^{-29},$$

or
$$D = 4\cdot 2 \times 10^{-8}.$$

This is only 1/100 of the value of b for two oppositely charged ions. The rate of combination of the electrons and ions with uncharged molecules at low pressures is thus

$$\frac{4\pi}{3} \rho\sigma D^3 \,(u_i^2 + u_m^2)^{\frac{1}{2}} \left(\frac{1}{\lambda_i} + \frac{1}{\lambda_m}\right) \ldots\ldots\ldots\ldots(9),$$

where ρ is the density of the ions, σ that of the uncharged molecules, u_i and u_m the average velocities of the ion and molecule respectively, and λ_i and λ_m their free paths. The rate of combination between ions of opposite signs is

$$\frac{4\pi}{3} \rho\rho' d^3 \,(u_i^2 + u_{i'}^2)^{\frac{1}{2}} \left(\frac{1}{\lambda_i} + \frac{1}{\lambda_{i'}}\right) \ldots\ldots\ldots\ldots(10),$$

where ρ, ρ' are the densities of the ions, and d at 0° C. is $4\cdot 1 \times 10^{-6}$.

Now, though d is very much greater than D (in the case we are considering $d = 100D$), yet σ the density of the uncharged molecules is in general so much greater than ρ', the density of the ions, that σD^3 is greater than $\rho' d^3$. At atmospheric pressure

σ is $2{\cdot}7 \times 10^{19}$, so that unless ρ' is greater than $2{\cdot}7 \times 10^{13}$, more ions will begin by uniting with uncharged molecules to form complex ions than with ions of the opposite sign to form neutral systems. It requires very intense ionisation to produce 10^{13} ions per c.c. at atmospheric pressure, far more intense than that produced by X-rays of anything like normal intensity, though not more intense than that which may occur in arcs and luminous electrical discharges; and unless the ionisation has this high value an ion will unite with an uncharged molecule to form a complex ion before it ends its career by combining with an ion of opposite sign. This will be even more certainly the case if there is any other way in which the complex ion can be formed.

In the preceding investigation the neutral molecules are not supposed to exert any electrostatic force except that due to the induced electrification due to the presence in their neighbourhood of charged bodies. There are however some molecules, called polar molecules, which have an intrinsic electrostatic moment[1]; the molecules of water vapour, of ammonia, of hydrochloric acid, are of this type. These produce an electrostatic field stronger than that due to an ordinary molecule and varying with the distance according to a different law. Thus if M be the electrostatic moment of the molecule, the force at a point along the axis of the moment and at a distance r from its centre is $2M/r^3$, and the work required to carry a charge e away to an infinite distance is Me/r^2; thus for this type of molecule the distance b will be defined by the equation

$$\frac{Me}{b^2} = \frac{e^2}{4{\cdot}2 \times 10^{-6}},$$

or
$$b^2 = M \cdot 4{\cdot}2 \times 10^{-6}/e.$$

From Baedeker's experiments on the variation of specific inductive capacity with temperature[2] it follows that for water vapour
$$M = 2 \times 10^{-18} \text{ approximately,}$$
since
$$e = 4{\cdot}8 \times 10^{-10},$$
$$b = 1{\cdot}3 \times 10^{-7}.$$

---

[1] J. J. Thomson, *Phil. Mag.* xxvii. p. 756.
[2] *Zeits. für Physik. Chemie*, xxxvi. p. 305.

Thus b is considerably greater than for the normal uncharged molecule, and thus charged particles will combine more rapidly with polar molecules than with normal ones. We shall meet, when considering the mobilities of ions, many examples of the effect of these molecules in loading up the ions.

The value of b for a collision between a charged and uncharged molecule is as we have seen only 1/100 part of the value for oppositely charged ions and is, except for very high pressures, much less than the distance between two molecules of the gas. Thus when a sphere of radius D is drawn round a molecule A the probability that another molecule will be inside this sphere will be expressed by a small fraction. Thus the chance that the ion makes a collision with the other molecules as it travels through this sphere, is very little greater than that of its hitting the molecule itself. Thus for collisions between electrons or ions and neutral molecules the possibility of their union taking place without the intervention of another molecule becomes much more important than for the recombination of oppositely charged ions, and is probably the most important means by which the combination occurs. For combination between an electron and a neutral molecule, the electron in going through the molecule might collide with the electrons in the molecule and lose so much energy in this way that it is not able to separate from the neutral molecule. We have seen that the loss of energy by a free electron colliding with a bound electron in the molecule is very small, except for special values of the velocity of the electron. There is a complex electrostatic field inside the molecule, and the electron journeying through it might acquire such a velocity from the attraction of the positive charge that it could communicate energy to electrons near its path. The loss of this energy might prevent the electron escaping from the molecule so that it would produce a negative ion. It might, for example, give so much energy to an electron in the molecule as to put that electron in a metastable state; when the electron went back from this to the normal state radiation might be given out.

On this view if V is the relative velocity of an electron and a molecule, k the chance that the requisite loss of energy occurs at a collision, since the number of collisions in a time δt is

$$\rho \sigma V \pi b^2 \delta t,$$

where b is the effective radius of the molecule, *i.e.* when the effect of the attraction between the electron and the molecule is allowed for (see § 19·1), the rate of formation of negative ions is

$$\rho \sigma V \pi b^2 k \ \dots\dots\dots\dots\dots\dots(11).$$

The rate of formation of negative ions on this view is proportional to the pressure of the gas, the rate indicated by the expression (9) is proportional to the square of the pressure, since $1/\lambda$ is proportional to the pressure. Loeb[1] has shown that an electron makes on the average some 27,000 collisions with an oxygen molecule before forming a negative ion, and that this number does not depend upon the pressure; according to the expression (11) the number of collisions is $1/k$ and is thus independent of the pressure. From the expression (9), on the other hand, it would follow that the number of collisions would vary inversely as the pressure.

In collisions between an ion and a molecule there may be a transference of the energy due to the relative motion of the centres of mass of the ion and molecule into energy due to internal motions in the ion or molecule, if these are not monatomic; if this internal energy did not at once flow back again into the energy due to the relative motion of the centre of mass, and it seems unlikely that the whole of it should, the ion and the molecule need not separate, and a complex ion would be formed; the expression for the rate of formation would be

$$\rho \sigma V' \pi b^2 k',$$

where b is the 'modified' sum of the radii of the ion and molecule, and k' the chance that the collision should result in such a transference of energy into the internal energy of the ion and molecule that separation would not take place. The rate of formation given by this expression is proportional to the pressure, whereas that given by (9) is proportional to the *square* of the pressure of the gas.

21. Limit to the size of a gaseous ion. After the charged molecule has increased in size by attracting another molecule, the work required to drag an additional uncharged molecule away from it, starting from a given distance, will be less than the work

[1] Loeb, *Phil. Mag.* xliii. p. 229, 1922.

required to drag an uncharged molecule away from a single molecule. The more complex the ion becomes, the less the work required to drag an uncharged molecule away from it, and at a certain stage of complexity the work required to drag an uncharged molecule from an ion with which it is in contact is less than the kinetic energy the system possesses in virtue of its temperature; when this stage is reached the ion will cease to grow.

The work required to separate an uncharged sphere of radius a from a charged sphere of radius b, the spheres being infinitely nearly in contact, is (see Maxwell's *Electricity and Magnetism*, vol. I. p. 275)

$$\frac{1}{2} \frac{e^2}{a+b - \frac{ab}{a+b}\left\{2\gamma + \psi\left(\frac{b}{a+b}\right) + \psi\left(\frac{a}{a+b}\right)\right\}} - \frac{1}{2}\frac{e^2}{b},$$

where
$$\gamma = \cdot 57712,$$
$$\psi(x) = \frac{d}{dx} \log \Gamma(1+x).$$

Tables by which we can calculate $\psi(x)$ are given in De Morgan's *Differential and Integral Calculus*, p. 587.

From these tables I find the following expressions for the work w:

$$a = b. \qquad w = \cdot 14 \frac{e^2}{a}.$$

$$2a = b. \qquad w = \frac{1}{40} \frac{e^2}{a}.$$

$$3a = b. \qquad w = \frac{1}{156} \frac{e^2}{a}.$$

When b is very large compared with a

$$w = \frac{\pi^2}{12} \cdot \frac{a^3}{b^3} \cdot \frac{e^2}{a}.$$

Now the process of aggregation will stop at about the stage where w becomes less than the kinetic energy of the system which we have seen at 0° C. $= \frac{e^2}{r}$, where $r = 4\cdot 2 \times 10^{-6}$.

Thus if $a = 10^{-8}$, the work required to separate a molecule from an ion whose radius is $3a$ is greater than e^2/r, but the work required to separate a molecule from an ion whose radius is $4a$

is less than e^2/r, hence in this case the radius of the ion cannot exceed three times the radius of the molecule.

If $a = 10^{-7}$, the work required to separate a molecule from an ion of radius $2a$ will be less than e^2/r, hence in this case the radius of the ion cannot exceed twice the radius of the molecule, and we see that the larger the molecule the smaller will be the ratio of the size of the ion to the size of the molecule; with very large molecules it is probable that the ion and the molecule are identical. The ions in different gases will thus not differ so much in size as the molecules of the gases. Since the kinetic energy is greater at a high temperature than at a low one the process of aggregation of molecules will stop at an earlier stage the higher the temperature, so that the ions will be simpler at high temperatures than at low ones. See also § 40·1.

Diffusion of Ions.

22. In addition to the loss of ions arising from the recombination of the positive and negative ions there will be a further loss due to the diffusion of ions to the sides of the vessel. Thus suppose the ionised gas is contained in a metal vessel, then when the ions come in contact with the sides of the vessel their charges are neutralised by the opposite charge induced on the metal and they thus cease to act like ions: the layer of gas next the sides of the vessel is thus denuded of ions, which exist in finite numbers in the gas in the interior; a gradient in the concentration is thus established and the ions diffuse from the interior to the boundary. The problem is closely analogous to that of the absorption of water vapour in a vessel whose sides are wet with sulphuric acid. We shall begin by considering the theory of a very simple case, that of ionised gas contained between two parallel metal plates at right angles to the axis of x. Let n be the number of positive ions per cubic centimetre, q the number of ions produced by the ionising agent per second in a cubic centimetre of the gas, D the coefficient of diffusion of the positive ions through the gas, m the number of negative ions per cubic centimetre, then we see that in consequence of diffusion the rate of increase in the number of positive ions per cubic centimetre is equal to $D\dfrac{d^2n}{dx^2}$, assuming that the surfaces of equal density of the ions are planes at right angles to the axis of x. Thus taking recombination and external ionisation into account as well as diffusion we have

$$\frac{dn}{dt} = q + D\frac{d^2n}{dx^2} - anm,$$

and when things are in a steady state,

$$q + D\frac{d^2n}{dx^2} - anm = 0.$$

Let us consider the special case when the plates are so near together that the loss of ions from diffusion far exceeds that from recombination, then we have

$$q + D\frac{d^2n}{dx^2} = 0 \quad \dots \dots \dots \dots \dots \dots (1).$$

If we take the plane midway between the metal plates as the plane $x = 0$, and if $2l$ is the distance between the plates, then the conditions to be satisfied by n are $n = 0$ when $x = \pm l$; the solution of equation (1) with these conditions is

$$n = \frac{1}{2}\frac{q}{D}(l^2 - x^2) \quad \dots \dots \dots \dots \dots (2).$$

The total number of free positive ions between the plates is equal to

$$\int_{-l}^{+l} n\,dx,$$

and this by equation (2) is equal to

$$\frac{2}{3}\frac{q}{D}l^3.$$

We see from this result how we can measure D. For, if we cut off the rays and apply a strong electric field between the plates, we shall drive all the positive ions against the plate at the lower potential, so that this plate will receive a charge of electricity equal to $\frac{2}{3}\frac{q}{D}l^3 e$, where e is the charge on an ion: if this plate is connected with an electrometer we can measure its charge, which will be proportional to the deflection δ_1 of the electrometer. If the rays are kept on and the field is intense enough to produce the saturation current, the charge received by the plate in one second is equal to $2qle$, hence if δ_2 is the deflection of the electrometer in one second in this case, we see that

$$\delta_1 = \frac{1}{3}\frac{l^2}{D}\delta_2,$$

an equation which enables us to determine D.

23. We have in this investigation neglected the effect of recombination; it is necessary to find the condition that the plates should be sufficiently close together to make this justifiable. An easy way of doing this is as follows: the total number of ions on the hypothesis that the only source of loss of ions is recombination is equal to $2l\sqrt{q/\alpha}$ (see p. 20); the number on the assumption that the loss is entirely due to diffusion is, as we have just seen, $\frac{2}{3}\frac{q}{D}l^3$; hence if $\frac{2}{3}\frac{q}{D}l^3$ is small compared with $2l\sqrt{\frac{q}{\alpha}}$, i.e. l^2 small compared with $D/\sqrt{q\alpha}$, the loss of ions from diffusion will be large compared with the loss by recombination, and we shall be justified in neglecting the latter.

A correction may also be necessary, due to the forces between the charges on the ions, if unequal numbers of positive and negative ions are present in any region. Like recombination this effect depends on the square of n and so can be neglected for weak ionisation.

24. The coefficients of diffusion of the ions in air, oxygen, hydrogen, and carbonic acid gas have been determined by Townsend[1] by a different method; ionised air being sucked through very narrow tubes and the loss of ions suffered in passing through

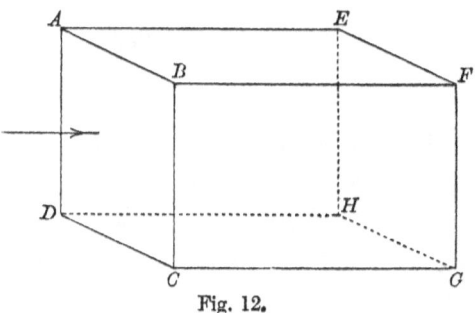

Fig. 12.

[1] Townsend, *Phil. Trans.* A, 193, p. 129, 1900.

a known length of tubing determined. The theory of the method is as follows: ionised gas is sent through a metal tube the axis of which is taken as the axis of z, the gas moving parallel to z and being free from the action of any ionising agent in its course through the tube. Consider the state of things in a small volume $ABCDEFGH$: this volume loses ions by diffusion, and gains them by the gas entering the volume through the face $ABCD$ being richer in ions than that leaving it through the face $EFGH$; when the gas is in a steady state the rates of loss and gain of ions must be equal. If n is the number of ions per cubic centimetre, D the coefficient of diffusion of the ions through the gas, the rate of loss of ions from diffusion is equal to

$$-D\left(\frac{d^2n}{dx^2} + \frac{d^2n}{dy^2} + \frac{d^2n}{dz^2}\right).$$

If v is the velocity of the gas, the rate of gain of ions from the second cause is equal to

$$-\frac{d}{dz}(vn),$$

or, since v does not depend upon z, to

$$-v\frac{dn}{dz}.$$

Hence equating the loss and the gain we get

$$D\left(\frac{d^2n}{dx^2} + \frac{d^2n}{dy^2} + \frac{d^2n}{dz^2}\right) = v\frac{dn}{dz}\dots\dots\dots\dots(1).$$

In the experiments the term $D\frac{d^2n}{dz^2}$ was very small compared with $v\frac{dn}{dz}$; $\frac{1}{n}\frac{dn}{dz}$ being of the order $1/20$, v of the order 100 and D about ·03, so that vn was about 70,000 times $D\frac{dn}{dz}$. For the case of a cylindrical tube of radius a symmetrical about the axis

$$\frac{d^2n}{dx^2} + \frac{d^2n}{dy^2} = \frac{d^2n}{dr^2} + \frac{1}{r}\frac{dn}{dr},$$

where r is the distance of a point from the axis of the tube. Now $v = \frac{2V}{a^2}(a^2 - r^2)$, where $V\pi a^2$ is the volume of gas passing per second through each cross-section of the tube; substituting these values in equation (1) and neglecting d^2n/dz^2 we get

$$\frac{d^2n}{dr^2} + \frac{1}{r}\frac{dn}{dr} - \frac{2V}{Da^2}(a^2 - r^2)\frac{dn}{dz} = 0 \quad \ldots \ldots \ldots \ldots (2).$$

The conditions to be satisfied by n are that $n = 0$ when $r = a$ for all values of z, and that if the ionised gas enters the tube at $z = 0$, $n = n_0$ a constant, when $z = 0$, for all values of r.

To solve this equation put $n = \phi \epsilon^{-\frac{\theta D a^2 z}{2V}}$, where ϕ depends only upon r and θ is a constant to be subsequently determined; substituting this value of n in equation (2) we get

$$\frac{d^2\phi}{dr^2} + \frac{1}{r}\frac{d\phi}{dr} + \theta(a^2 - r^2)\phi = 0 \quad \ldots \ldots \ldots \ldots (3).$$

Put $\quad \phi = 1 + B_1 r^2 + B_2 r^4 + B_3 r^6 + \ldots,$

and we get from (3)

$$4B_1 + \theta a^2 = 0,$$
$$16B_2 + \theta a^2 B_1 - \theta = 0,$$
$$36B_3 + \theta a^2 B_2 - \theta B_1 = 0,$$
$$\ldots \ldots \ldots \ldots \ldots \ldots \ldots$$

thus the first three terms in ϕ are

$$1 - \frac{\theta a^2}{4} r^2 + \frac{1}{16}\left(\theta + \frac{\theta^2 a^4}{4}\right) r^4 \ldots \ldots \ldots \ldots (4).$$

We have to choose such values of θ that $\phi = 0$ when $r = a$; let these values be $\theta_1, \theta_2, \ldots$ and let ϕ_1, ϕ_2 be the values of ϕ when these values of θ are substituted in equation (4); then we may write

$$n = c_1\phi_1 \epsilon^{-\frac{\theta_1 D a^2 z}{2V}} + c_2\phi_2 \epsilon^{-\frac{\theta_2 D a^2 z}{2V}} + c_3\phi_3 \epsilon^{-\frac{\theta_3 D a^2 z}{2V}} + \ldots. (5).$$

To find the values of $c_1, c_2, c_3, \ldots$ we have the condition $n = n_0$ a constant when $z = 0$. Hence

$$n_0 = c_1\phi_1 + c_2\phi_2 + c_3\phi_3 + \ldots \ldots \ldots \ldots \ldots (6).$$

Now from the differential equation (3) we can easily prove the following relations:

$$\int_0^a \phi_n \phi_m (a^2 - r^2) r\, dr = 0 \text{ when } n \text{ and } m \text{ are different} \ldots (7),$$

$$\int_0^a \phi_n^2 (a^2 - r^2) r\, dr = a\left[\frac{d\phi_n}{d\theta_n}\frac{d\phi_n}{dr}\right]^{r=a} \ldots \ldots \ldots (8),$$

$$\int_0^a \phi_n (a^2 - r^2) r\, dr = -\frac{a}{\theta_n}\left[\frac{d\phi_n}{dr}\right]^{r=a} \ldots \ldots \ldots (9).$$

Multiplying both sides of equation (6) by $\phi_n (a^2 - r^2) r$ and integrating from $r = 0$ to $r = a$ we obtain by the aid of equations (7), (8) and (9)

$$c_n = - \frac{n_0}{\theta_n \left[\dfrac{d\phi_n}{d\theta_n}\right]^{r=a}},$$

hence

$$n = - n_0 \left\{ \frac{\phi_1}{\theta_1 \left[\dfrac{d\phi_1}{d\theta_1}\right]^{r=a}} \epsilon^{-\frac{\theta_1 D a^2 z}{2V}} + \frac{\phi_2}{\theta_2 \left[\dfrac{d\phi_2}{d\theta_2}\right]^{r=a}} \epsilon^{-\frac{\theta_2 D a^2 z}{2V}} + \ldots \right\} \quad (10).$$

The number of ions which pass across the section of the tube when $z = 0$ is $n_0 \pi a^2 V$; the quantity which pass across a section of the tube at a distance z from the origin is equal to

$$2\pi \int_0^a n \frac{2V}{a^2} (a^2 - r^2) \, r \, dr;$$

this by equations (7), (9) and (10) is equal to

$$\frac{4\pi V n_0}{a} \left\{ \frac{1}{\theta_1^2} \frac{\left[\dfrac{d\phi_1}{dr}\right]^{r=a}}{\dfrac{d\phi_1}{d\theta_1}} \epsilon^{-\frac{\theta_1 D a^2 z}{2V}} + \frac{1}{\theta_2^2} \frac{\left[\dfrac{d\phi_2}{dr}\right]^{r=a}}{\dfrac{d\phi_2}{d\theta_2}} \epsilon^{-\frac{\theta_2 D a^2 z}{2V}} + \ldots \right\}.$$

The numerical values of θ_1, θ_2 and the quantities in square brackets can be calculated, only the first two terms being important.

If c_1 is the saturation current through the gas after leaving a tube of length l_1, c_2 that after leaving a tube of length l_2, then since the saturation currents are proportional to the numbers of ions given to the gas per second, we have

$$\frac{c_1}{c_2} = \frac{f\left(\dfrac{Dl_1}{a^2 V}\right)}{f\left(\dfrac{Dl_2}{a^2 V}\right)} \quad \ldots\ldots\ldots\ldots\ldots\ldots (11),$$

where f is a known function. Now c_1/c_2 can be determined by experiment, and hence from equation (11) the value of D can be determined by a graphical method.

The apparatus used to measure the value of c_1/c_2 is represented in Fig. 13. *A* is a brass tube 50 cm. long, 3·2 cm. in diameter, provided with an aluminium window *W* through which the X-rays which ionise the gas pass. *C* is another brass tube 17 cm. long fitting accurately into *A* and able to slide along it. *E* is an electrode which is connected to a metal rod *F* passing through an ebonite plug. A series of fine wires were soldered parallel to one another and 2 mm. apart across the end of the tube *C*. The gas entered the apparatus through the glass tube *G* and then before reaching the electrode passed through the tubes *T*. These

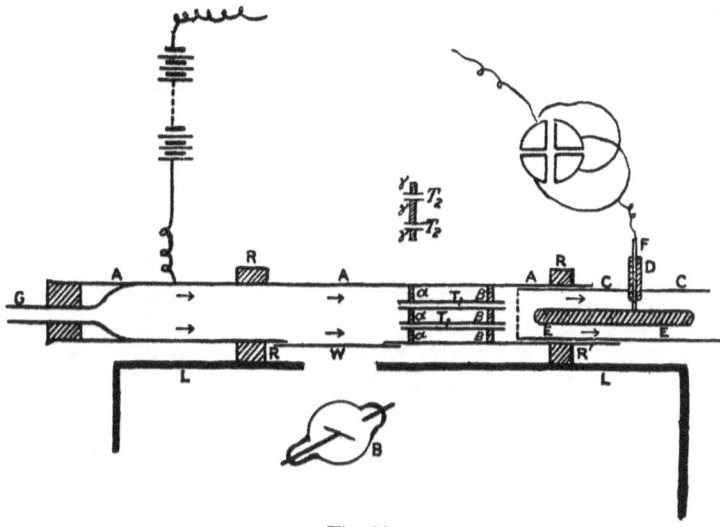

Fig. 13.

were twelve tubes 10 cm. long and ·3 cm. in diameter, arranged at equal intervals and all at the same distance from the axis of the tube *A*; they were soldered into holes bored into two brass discs α and β which fitted so closely into *A* that gas could not pass between the disc and the tube. Another set of twelve tubes only 1 cm. long and ·3 cm. in diameter were fused into another disc γ. The tube *A* was insulated by the two ebonite rings *R*, *R'*. The potential of the tube was raised to 80 volts by connecting it with one of the terminals of a battery of small storage cells, the other terminal of which was connected with the earth. The electrode *E* was connected with one pair of quadrants of an electrometer, the other pair of quadrants being kept to earth. A uniform and measurable stream of gas was supplied by a gasometer, this gas was ionised by the rays as it passed through the tube; some of the ions were lost by diffusion to the sides, all the positive ones which escaped were driven against the electrode E; thus the charge on the electrometer measured the number of positive ions which got through the tubes. By charging the tube A up negatively, the negative ions could be driven against the electrode, and the number of those which get through the tubes determined. After a series of measurements had been made with the long tubes, these were replaced by the short ones, and a similar series of measurements gone through. These measurements, as was explained in the preceding theory, give us the data for calculating the coefficient of diffusion of the ions. For gases other than air, a somewhat different form of apparatus was used, for a description of which we must refer to the original paper.

The loss of ions even in the narrow tubes is not entirely due to diffusion to the sides of the tube, a part, though only a small part, of the loss will be due to the recombination of the ions. To estimate how much was due to this effect, the small tubes T were removed and the deflection of the electrometer observed when the tube C was placed at different distances from the place where the gas is ionised; in a wide tube such as A the loss from diffusion to the sides is negligible, and the smaller deflection of the electrometer when the electrode E is moved away from the place of ionisation is due to the loss of ions by recombination. By making measurements at different distances and knowing the velocity of the gas we can measure in this way the amount of recombination taking place in a given time and hence determine the value of α, the constant of recombination. It was in this way that the values of α given on p. 40 were determined. Knowing α it is easy to calculate the loss of ions from recombination in their passage through the narrow tubes, and then to apply a correction to the observations so as to get the loss due to diffusion alone.

The following tables give the values of the coefficients of diffusion on the C.G.S. system of units as deduced by Townsend from his observations.

TABLE I. COEFFICIENTS OF DIFFUSION IN DRY GASES.

Gas	D for $+$ions	D for $-$ ions	Mean value of D	Ratio of D for $-$ to D for $+$ions
Air	·028	·043	·0347	1·54
Oxygen	·025	·0396	·0323	1·58
Carbonic acid	·023	·026	·0245	1·13
Hydrogen	·123	·190	·156	1·54

TABLE II. COEFFICIENTS OF DIFFUSION IN MOIST GASES.

Gas	D for $+$ ions	D for $-$ions	Mean value of D	Ratio of D for $-$ to D for $+$ions
Air	·032	·035	·0335	1·09
Oxygen	·0288	·0358	·0323	1·24
Carbonic acid	·0245	·0255	·025	1·04
Hydrogen	·128	·142	·135	1·11

We see from these tables that the coefficient of diffusion for the negative ions is greater than that for the positive, the difference being much more marked in dry than in damp gases. The superior mobility of the negative ions was first observed by Zeleny[1], who measured by a method which we shall shortly describe the velocity of the ions when placed in an electric field, and found that the negative ions moved faster than the positive ones. The more rapid diffusion of the negative ions explains why in certain cases ionised gas, originally electrically neutral, acquires a charge of positive electricity. Thus, for example, if such a gas is blown through metal tubes, the gas emerging from the tubes will be positively electrified, as in the passage through the tubes it has lost more negative than positive ions. Zeleny (*loc. cit.*) has shown that this effect does not occur with carbonic acid gas in which the velocities of the two ions are very nearly equal. Some experiments made by Rutherford[2] seem to show that in addition to the effect produced by diffusion, there is a specific effect due to the metal, as he found that the excess of positive over negative ions was greater when the ionised gas passed through zinc tubes than

[1] Zeleny, *Phil. Mag.* v. 46, p. 120, 1898.
[2] Rutherford, *Phil. Mag.* v. 43, p. 241, 1897.

when it passed through copper. The difference in the rate of diffusion of the positive and negative ions causes a certain amount of electrical separation to take place when a gas is ionised; as the negative ions diffuse more rapidly than the positive ones, the region where ionisation takes place will have an excess of positive ions and be positively electrified, while in consequence of the diffusion of the negative ions the surrounding regions will have an excess of these ions and will therefore be negatively electrified.

The results given in Tables I and II show that the excess of the velocity of diffusion of the negative ions over that of the positive is much greater when the gas is dry than when it is moist; the effect of moisture on the velocity of diffusion is very remarkable, the results quoted in the table show that with the exception of ions in carbonic acid (where there is but little difference between the velocities of diffusion of positive or negative ions in either wet or dry gas) the effect of moisture is to produce a very considerable diminution in the rate of diffusion of the negative ions, while on the other hand it tends to increase the rate of diffusion of the positive ions, though the change produced in the positive ions is not in general as great as that produced in the negative.

The preceding experiments relate to ions produced by X-rays. Townsend[1] subsequently applied the same method to the determination of the coefficients of diffusion of ions produced by radioactive substances, by ultra-violet light and by discharges from electrified needle points; the results of these experiments are shown in the following table.

COEFFICIENTS OF DIFFUSION OF IONS PRODUCED IN AIR BY DIFFERENT METHODS.

Method	Dry air		Moist air	
	+ ions	− ions	+ ions	− ions
X-rays	·028	·043	·032	·035
Radioactive substances	·032	·043	·036	·041
Ultra-violet light		·043		·037
Point discharge	·0247	·037	·028	·039
	·0216	·032	·027	·037

[1] Townsend, *Phil. Trans.* A, 195, p. 259, 1900.

From these numbers we conclude that the ions produced by X-rays, by radioactive substances and by ultra-violet light are identical (but see below, p. 83).

Townsend[1] also investigated the coefficients of diffusion of ions produced by radioactive substances at a series of pressures ranging from 772 millimetres of mercury to 200 mm. and found that within this range the coefficient of diffusion was inversely proportional to the pressure; the Kinetic Theory of Gases shows that this would be true in a system where the diffusing systems do not change character with the pressure; as this result holds for ions we conclude that down to a pressure of at least 200 mm. the ions do not change. We shall see that at very low pressures the negative ions are very much smaller than at these high pressures.

24·1. Salles[2], and Franck and Westphal[3], have also made determinations of D by methods similar to that of Townsend. Salles finds that the material of which the fine tubes are made does not affect the rate of diffusion to them. He used tubes of German silver, brass and steel. Hensel[4] has tried in addition to metal tubes, tubes of non-conductors such as glass and wood, and finds no difference in the results.

The following table shows the values found by the various experimenters:

Gas	D for + ions	D for − ions	Author
Air, dry	·028	·043	Townsend
,, ,,	·032	·042	Salles
,, ,,	·029	·045	Franck and Westphal
,, moist	·032	·035	Townsend
Oxygen, dry	·025	·0396	Townsend
,, ,,	·030	·041	Salles
,, moist	·0288	·0358	Townsend
Carbon dioxide, dry	·023	·026	Townsend
,, ,, ,,	·025	·026	Salles
,, ,, moist	·0245	·0255	Townsend
Hydrogen, dry	·123	·190	Townsend
,, moist	·128	·142	Townsend
Nitrogen, dry	·029	·0414	Salles

[1] Townsend, *Phil. Trans.* A, 195, p. 259, 1900.
[2] Salles, *Ann. de Phys.* ix. t. 2, p. 273, 1914.
[3] Franck and Westphal, *Verh. d. Deut. Phys. Ges.* xi. pp. 146 and 276, 1909.
[4] Hensel, *Phys. Zeits.* xiii. p. 666, 1912.

25. It is of interest to compare the rates of diffusion of ions through a gas with those of the molecules of one gas through another. In the following table, taken from Winkelmann's *Handbuch der Physik*, I. pp. 645, 647, the coefficients of diffusion into each other for hydrogen, air, carbonic acid, and carbonic oxide, and for some vapours are given; it appears from the table that the gases diffuse very much more quickly than the ions, but that there are vapours whose coefficients of diffusion are of the same order as those of the ions.

Gas	D cm.^{2}/sec.	Gas	D cm.^{2}/sec.	Gas	D cm.^{2}/sec.
$CO-CO_2$	·13142	$CO-O_2$	·18717	ether $-CO_2$	·0552
air$-CO_2$	·13433	H_2-O_2	·66550	isobutylic $-H_2$	·1724
O_2-CO_2	·13569	H_2–air	·63405	amide	
H_2-CO_2	·53409	ether$-H_2$	·296	,, $-$air	·0426
air$-O_2$	·17778	ether–air	·0775	,, $-CO_2$	·0305

One cause of the slow diffusion of the ions is that the charged ion probably forms a nucleus round which the molecules of the gas condense, just as dust collects round a charged body, thus producing a complex system which diffuses slowly: this explanation is supported by the fact discovered by McClelland[1] that the coefficients of diffusion of the ions in the flame gases depend very much on the temperature of the flame and the distance of the ions from it; a comparatively small lowering of temperature producing a great diminution in the rate of diffusion of the ions, as if precipitation had occurred upon them. The view is also supported by the ability of the ions to act as nuclei for the precipitation of water vapour. It must be remembered also that an ion differs from an ordinary molecule in being charged with electricity and thus being surrounded by a strong electric field.

Rutherford[2] has also shown that the vapour of alcohol or ether, like that of water, produces a great diminution in the mobility of the negative ion.

[1] McClelland, *Camb. Phil. Soc. Proc.* x. p. 241, 1899.
[2] Rutherford, *Phil. Mag.* vi. 2, p. 210, 1901.

26. Relation between diffusion and mobility. The coefficient of diffusion of the ions through a gas is directly proportional to the speed with which the ions travel through the gas under the action of an electric field of given strength. The connection between this speed and the coefficient of diffusion can be established as follows. From the definition of the coefficient of diffusion D it follows that if n is the number of ions per cubic centimetre, the number of ions which in unit time cross unit area of a plane at right angles to x is equal to $D\dfrac{dn}{dx}$. We may thus regard the ions as moving parallel to the axis of x with the average velocity $\dfrac{1}{n}D\dfrac{dn}{dx}$. The ions being in the gaseous state will produce a partial pressure p which when the temperature is constant is proportional to the number of ions; we see therefore that the average velocity of the ions parallel to x is equal to $\dfrac{1}{p}D\dfrac{dp}{dx}$. Now dp/dx is the force acting parallel to the axis of x on unit volume of the gas; we may thus interpret the preceding expression as meaning that when the force acting parallel to the axis of x on the ions in unit volume is unity, the ions move parallel to the axis of x with a mean velocity of translation equal to D/p. Suppose now that the ions are placed in an electric field where the electric intensity parallel to the axis of x is equal to X, then the force on the ions in unit volume is equal to Xen; hence if u is the average velocity of translation of the ions parallel to the axis of x

$$u = Xen\frac{D}{p}.$$

Now n/p is the same for all gases at the same temperature, hence if N is the number of molecules of air in a c.c. at this temperature and at the atmospheric pressure Π, since $n/p = N/\Pi$, we have

$$u = XeD\frac{N}{\Pi},$$

or u_0 the velocity of the ions in a field of unit intensity is given by the equation

$$u_0 = D\frac{Ne}{\Pi}.$$

Thus u_0 is directly proportional to D, so that a knowledge of one of these quantities enables us at once to calculate the other, or if both are known we can find Ne.

27. Value of Ne. From Rutherford's experiments on the mean velocities of the ions in gases and the mean of the coefficients of diffusion given by Townsend we get

I.

Gas	$Ne \times 10^{-10}$
Air	1·35
Oxygen	1·25
Carbonic acid	1·30
Hydrogen	1·00

From Zeleny's values for the velocities of the ions and Townsend's for the coefficients of diffusion we get for $Ne \times 10^{-10}$

II.

Gas	Moist gas		Dry gas	
	Positive ions	Negative ions	Positive ions	Negative ions
Air	1·28	1·29	1·46	1·31
Oxygen	1·34	1·27	1·63	1·36
Carbonic acid	1·01	·87	·99	·93
Hydrogen	1·24	1·18	1·63	1·25

Since one electromagnetic unit or 3×10^{10} electrostatic units of electricity when passing through acidulated water liberates 1·23 c.c. of hydrogen at the temperature of 15° C. and pressure of 760 mm. of mercury, and since in 1·23 c.c. of gas there are $2 \cdot 46N$ atoms of hydrogen, we have, if E is the charge in electrostatic units on the atom of hydrogen in the electrolysis of solutions,

$$2 \cdot 46 NE = 3 \times 10^{10},$$

or $\qquad NE = 1 \cdot 22 \times 10^{10}.$

The mean of all the values of Ne in Tables I and II is $1 \cdot 24 \times 10^{10}$.

We conclude then (1) that the charges carried by the gaseous ions are the same whether the ions are produced in air, oxygen, hydrogen or carbonic acid, (2) that the charge on each is equal to the charge carried by the hydrogen atom in the electrolysis of solutions.

27·1. Townsend and his pupils have used the following method for comparing the coefficient of diffusion with the velocity acquired by an ion in a field of unit electric force[1]. A column of ions of

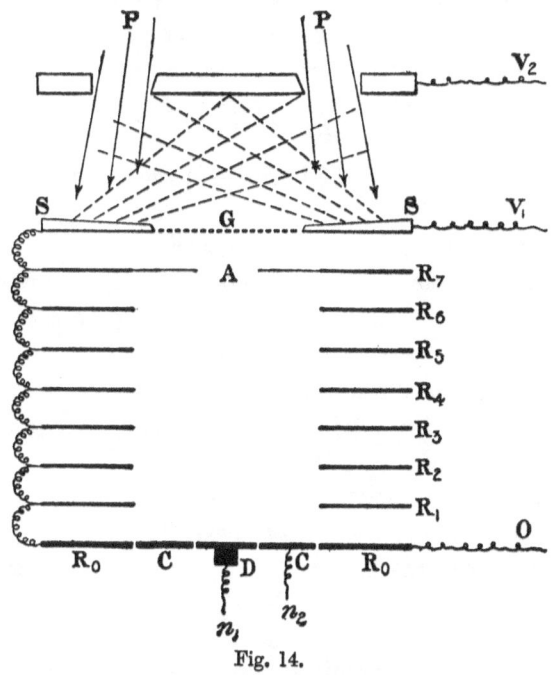

Fig. 14.

one sign passes through a circular aperture in a horizontal plate A and is driven downwards by a uniform vertical electric force Z towards a disc D, coaxial and parallel to the aperture A, and a ring C coaxial with D and insulated from it. D and C can be connected with an electrometer and the charges they receive measured. In consequence of diffusion the ions will spread out and this will increase the ratio of the charge received by C to that going to D. The longer the time they take to go from A to D, the greater will be the diffusion, but this time will depend upon the force Z; the greater this force the shorter the time. By

[1] Townsend, *Proc. Roy. Soc.* lxxx. p. 207; lxxxi. p. 464.

comparing the relative charges received by D and C under different electrical forces, the relation between the diffusion and the mobility may be deduced. To keep the electric force uniform a series of horizontal rings are arranged at equal distances apart and connected in series by high resistances.

Let the axis of z be the line joining the centre of the aperture A to the centre of the disc D, and the axes of x and y at right angles to z. If n is the number of ions per c.c. at the point x, y, z, then when things are in a steady state

$$D\left(\frac{d^2n}{dx^2} + \frac{d^2n}{dy^2} + \frac{d^2n}{dz^2}\right) = \frac{d}{dz}(vn) \quad \ldots\ldots\ldots\ldots(1),$$

where v is the velocity of the ion; if Z is the electric force,

$$v = kZ,$$

where k is the velocity of the ion under unit electric force. Since Z is constant v is constant and equation (1) may be written

$$\frac{d^2n}{dx^2} + \frac{d^2n}{dy^2} + \frac{d^2n}{dz^2} = \frac{kZ}{D}\frac{dn}{dz} = \frac{NeZ}{\Pi}\frac{dn}{dz}.$$

Or if r is the distance from the axis of z

$$\frac{d^2n}{dr^2} + \frac{1}{r}\frac{dn}{dz} + \frac{d^2n}{dz^2} = \frac{NeZ}{\Pi}\frac{dn}{dz}.$$

The equation may be solved by putting

$$n = \Sigma \mu J_0(\lambda r) \epsilon^{-\theta z},$$

where $J_0(x)$ represents Bessel's function of zero order and

$$\theta^4 + \theta^2 \frac{NeZ}{\Pi} = \lambda^2.$$

μ and λ are to be determined by the surface conditions when $z = 0$, i.e. that n is constant from $r = 0$ to $r = a$ and zero when r is greater than a. The values of n at the disc and the ring are got by putting $z = h$, where h is the distance between the aperture and the disc, and hence the ratio of the charges received by the ring and disc calculated; this ratio will be a function of (Ne/Π), and when it is known the value of Ne/Π can be determined. In the case of the negative ions under ordinary conditions of dryness Townsend found good agreement with the value of Ne for electrolysis, after a correction of about 10 per cent. had been made to allow for the mutual repulsion of the ions. Experiments were made with air at pressures of from 3 to 25 mm. and also with oxygen, hydrogen and CO_2. The ions were produced by X-rays incident on a metal surface. Haselfoot using α and β rays to produce the ionisation found the same result. For very dry gases at low pressures anomalous results were found. Under these conditions the negative ions have abnormally large mobilities (see below, § 37·3) and the assumption that their kinetic energy in the electric field is substantially the same as that of molecules at the same temperature ceases to be true. Thus the argument in § 26 breaks down. Townsend has considered this case in detail in *Electricity in Gases*, p. 172.

For the positive ions values of Ne considerably in excess of those found in electrolysis were obtained when the ionisation was produced by secondary X-rays, but not when it was due to radio-activity. This is attributed to the presence of a proportion of doubly charged ions. It is most marked at low pressures.

Franck and Westphal[1] found that they could separate the positive ions formed by X-rays by diffusion through fine wire gauze, so that those which passed through had a smaller diffusion coefficient than the original ions and, since the mobilities of all appeared to be the same, presumably contained a larger proportion of the doubly charged ions (see p. 79). After passage through three gauzes D was reduced from ·029 to ·0175. They estimated that 9 per cent. of all the ions were doubly charged.

When the ionisation was due to α, β or γ rays or point discharge, no evidence of double charges was found. Doubt has been thrown by Millikan on the existence of doubly charged ions in gases ionised by X-rays, as a consequence of his work on the determination of e (see p. 302).

Langevin[2] has devised a method of finding the value of k/D which has been used by Salles[3]. The principle of the method is as follows. From equation (2), p. 68, the number of ions of one sign between two plates, placed so near together that the loss of ions by recombination is negligible compared with that by diffusion, is given by

$$n_1 = \frac{q}{2D_1}(l^2 - x^2).$$

[1] Franck and Westphal, *Ver. der Deutsch. Phys. Ges.* xi. pp. 146, 276, 1909.
[2] Langevin, *Le Radium*, x. p. 113, 1913.
[3] Salles, *Ann. de Phys.* ix. t. 2, p. 273, 1914.

If an electrical field is applied between the plates too weak to affect appreciably the distribution of ions between the plates, the current through unit area carried by the ions of one sign, when $x = 0$, is $n_1 u e$, or
$$\frac{q e l^2 u}{2 D_1},$$
where u is the velocity of the ions due to the electric field.

The current carried by the ions of the opposite sign is
$$\frac{q e l^2 v}{2 D_2}.$$
The sum of these is equal to i, the current through unit area of the plate, so that
$$i = \frac{q e l^2}{2}\left(\frac{u}{D_1} + \frac{v}{D_2}\right).$$
If I is the saturation current through unit area, V the potential difference between the plates,
$$I = 2qel, \quad u = \frac{k_1 V}{2l}, \quad v = \frac{k_2 V}{2l},$$
hence
$$\frac{i}{I} = \frac{V}{8}\left(\frac{k_1}{D_1} + \frac{k_2}{D_2}\right);$$
hence the measurement of i/I will give
$$(k_1/D_1 + k_2/D_2) = 2Ne/\Pi.$$

Salles using this method found a value of Ne/Π slightly less than $1\cdot 25 \times 10^4$, indicating that there was no appreciable number of doubly charged atoms. The ionisation was caused by the γ-rays from radium.

CHAPTER III

MOBILITY OF IONS

28. If u is the average velocity parallel to x of ions acted on by an electric force X in this direction, u/X is called the mobility of the ions; it is as we shall see independent of X over a wide range of values of this quantity; it depends however on the sign of the ion and the pressure and nature of the gas through which it is moving.

The earliest systematic measurements of the mobilities of the ions were made in the Cavendish Laboratory in 1897 by Rutherford[1].

METHOD I.

29. One method used by Rutherford[1] is represented in Fig. 15. Two large metal plates A and B were placed parallel to

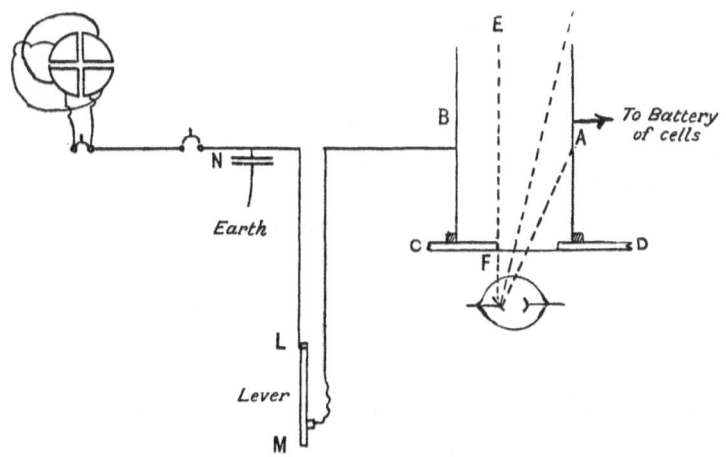

Fig. 15.

one another and 16 cm. apart on the insulating blocks C and D. The X-rays were arranged so as to pass through only one half of the gas included between the plates, thus no direct radiation

[1] Rutherford, *Phil. Mag.* v. 44, p. 422, 1897.

reached the air to the left of the line EF which is half-way between the plates. The plate A was connected with one terminal of a battery of a large number of small storage cells giving a potential difference of 220 volts, the other terminal of the battery being connected with the earth. The plate B was connected through a contact lever LM, mounted on an insulating block, to one pair of quadrants of an electrometer, the other pair being connected with the earth. A pendulum interrupter was arranged so as first to make the current in the primary of the induction coil used to produce the rays, then after a known interval to break the electrometer circuit by knocking away the lever LM, and then to break the battery circuit shortly afterwards. N is a condenser connected to the electrometer to increase its capacity. With this arrangement the ions have to travel over a distance of 8 cm. before they reach the plate B, and the object of the experiment was to find the time occupied by the rays in passing over this distance. It was found that there was only a very small deflection of the electrometer when the interval between putting on the rays and breaking the electrometer circuit was less than ·36 sec., but when the interval exceeded this value the deflection of the electrometer increased rapidly. Thus ·36 sec. was taken as the time required for the ions to pass over a distance of 8 cm. under a potential gradient of 220/16 volts per centimetre. This corresponds to a velocity 1·6 cm./sec. for the gradient of a volt per centimetre, and no difference was detected between the velocities of the positive and negative ions.

METHOD II.

Blast method. Electric force in direction of the blast.

30. The difference between the velocities of the positive and negative ions was discovered by Zeleny[1], who has made very valuable determinations of the velocities of the ions in an electric field. The method by which he discovered the difference of the velocities was by finding the electric force required to force an ion against a stream of gas moving with a known velocity parallel to the lines of electric force. Thus suppose A and B, Fig. 16,

[1] Zeleny, *Phil. Mag.* v. 46, p. 120, 1898.

represent two parallel plates made of wire-gauze and that between these plates we have a stratum of ionised gas, let the gas be moving through the plates from A to B with the velocity V, and let the potential gradient between the plates be n volts per centimetre, B being the positive plate. Then if the velocity of the positive ion under a potential gradient of 1 volt per centimetre be u, the velocity of the positive ion in the direction from B to A is $nu - V$ and this is proportional to the number of ions giving up their charges to A in unit time. Suppose now that we make B the negative plate, then if the potential gradient between the plates is n' volts per centimetre and the velocity of the negative ion under a potential gradient of 1 volt per centimetre is v, the velocity of the negative ion from B to A is $n'v - V$, and this is proportional to the number of negative ions giving up their charges to A in unit time. If we adjust the potential gradients so that the rate at which A receives a positive charge when B is positive is equal to the rate at which it receives a negative charge when B is negative, we have

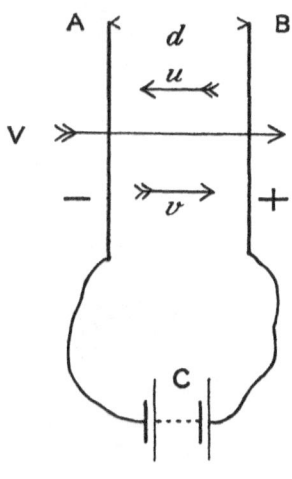

Fig. 16.

$$nu - V = n'v - V,$$

or
$$\frac{u}{v} = \frac{n'}{n}.$$

Thus from the measurement of the potential gradients we can determine the ratio $u : v$.

The apparatus used by Zeleny for carrying out this method is shown in Fig. 17. P and Q are brass plates 9 centimetres square. They are bored through their centres, and to the openings thus made the tubes R and S are attached, the space between the plates being covered in so as to form a closed box; K is a piece of wire-gauze completely filling the opening in the plate Q; T is an insulated piece of wire-gauze nearly but not quite filling the opening in the plate P and connected with one pair of quadrants of an electrometer E. A plug of glass-wool G filters out the dust from a stream of gas which enters the vessel by the tube D and leaves it by F; this plug has also the effect of making the velocity of flow of the gas uniform across the section of the tube. The X-rays to ionise the gas are produced by a bulb at O, the bulb and coil being in a lead-covered box fitted with an aluminium window through which the rays pass. Q is connected with one pole of a battery of cells, and P and the other pole of the battery

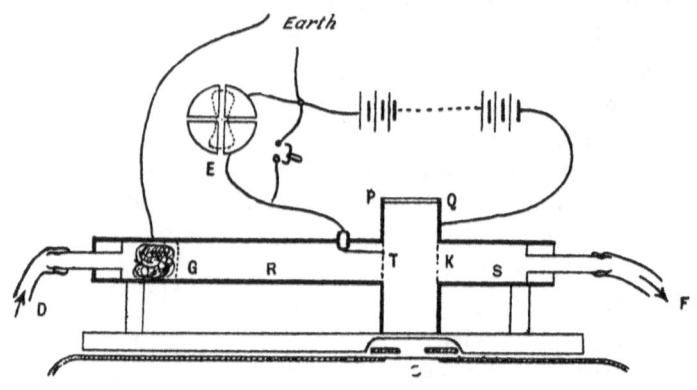

Fig. 17.

connected with earth. When the rays are entering PQ and the ions are travelling in opposite directions in the box, the charges they give to P, Q and K are conducted to earth, while those they give to T gradually change its potential at an approximately uniform rate, as long as this potential is small compared with that of Q. When the distribution of free charges in the gas has assumed a steady state all the changes in the potential of T are due to the charges given up by the ions striking against it.

The nature of the readings obtained with this apparatus are indicated by the curves shown in Fig. 18, where the ordinates represent the deflection of the electrometer in a given time and the abscissæ the potential difference in volts between the plates P and Q. Curve I is for the case when the negative ions, Curve II when the positive ions, are driven against the plate. It will be seen that after a point just above B the curves are for some distance straight lines, but that there is a curved portion to the left of B, indicating that some ions are delivered up to the gauze under smaller voltages than we should expect. This may possibly be explained by irregularities in the air blast, the deflections corresponding to the part of the curve about A arriving in the lulls of the blast. One way of treating the observations would be to produce the straight portion of the curves until they cut the horizontal axis; in the figure this would happen for Curve I at about 50 volts and for Curve II at about 60; we might then take 50 volts as the potential difference between the plates which would

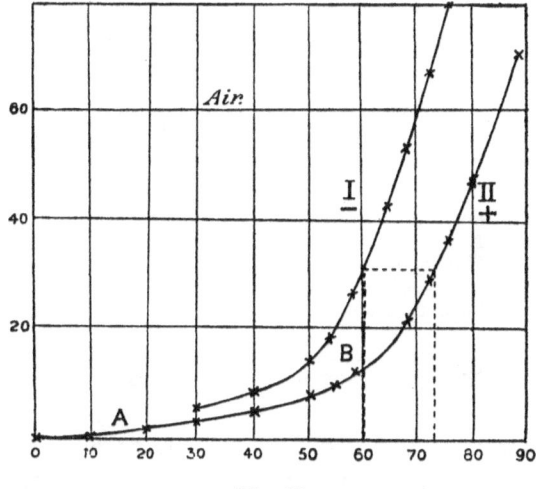

Fig. 18.

give to the negative ions a velocity equal to that of the blast, while 60 volts would be required to give the same velocity to the positive ions, so that under fields of equal strength the velocity of the negative ion would be to that of the positive as 6 to 5. The method actually adopted was different; the curves were regarded as merely a preliminary part of the experiment indicating the approximate values of the potential differences to employ in the final observations. Thus from the curves in Fig. 18 it is clear that to get the same deflection with the positive ions as is got with the negative ions for a potential difference of 60 volts, would require a potential difference of between 72 and 74 volts; a careful series of measurements with differences of potential between these values is taken and the true value of the potential difference found by interpolation. When this value, suppose for example 73·2, had been found, the ratio of the velocities of the negative to the positive ions was taken as 73·2 : 60.

The potential gradient between the plates was found to be not quite uniform owing to the accumulation of ions between the plates. The actual potential gradient was measured and a correction applied for the want of uniformity; this correction amounted to about 2 per cent. The results obtained by Zeleny are given in the following table.

RATIO OF VELOCITIES OF IONS.

Gas	Velocity of negative ion / Velocity of positive ion
Air	1·24
Oxygen	1·24
Nitrogen	1·23
Hydrogen	1·14
Coal gas	1·15
Carbon dioxide	1·00
Ammonia..........	1·045
Acetylene.........	0·985
Nitrogen monoxide .	1·105

Thus acetylene is the only gas in which the velocity of the negative ion is less than that of the positive, and here the difference is so small that it is within the limits of error of the experiment. The gases in this experiment were not specially dried; we have seen that moisture has a great effect in reducing the velocity of the negative ion.

Altberg[1] has applied this method with slight modifications to determine the mobilities of the ions separately. If A and B are the two gauze sheets, the electrometer is connected to a portion C in the middle of the gauze A; this portion is separated from the rest by a guard ring, so that the ions collected by the electrometer should move in as uniform a field as possible. The gas was ionised by a thin sheet of X-rays admitted through a slit, and the potential gradient needed to balance the speed of the air was found.

[1] Altberg, *Ann. der Phys.* xlvii. p. 849, 1912.

Method III.

31. In some later experiments Zeleny[1] has determined the absolute values of the velocity of both the positive and negative ions. The method he employed was a blast method, though in these experiments the blast was at right angles to the lines of electric force instead of along them. A method similar to the one described above was tried for a considerable time (it is evident that if we know the velocity of the blast and the points where the straight portions of the Curves I and II cut the horizontal axis we can deduce the velocity of both the positive and negative ions), but it had to be abandoned owing to the disturbance in the distribution of the velocity of the blast caused by the wire-gauze which in this method has to be used for the electrodes.

The theory of the method finally used is as follows. A stream of gas flows between two concentric metal cylinders which are

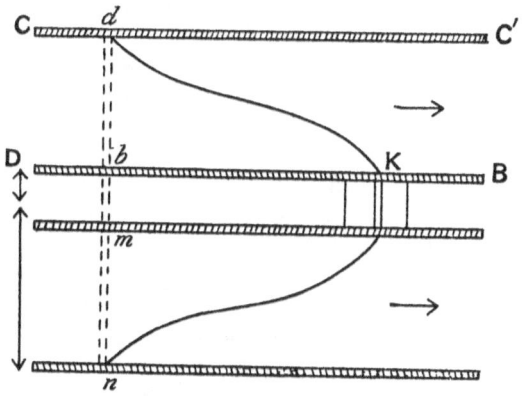

Fig. 19.

kept at different potentials, the gas at one place is traversed by a beam of X-rays at right angles to the axis of the cylinder; the ions thus produced are carried by the stream of gas parallel to the axis of the cylinder, while a velocity at right angles to this axis is imparted to them by the electric field. Let CC', Fig. 19, represent a section of the outer cylinder, DB that of the inner one, $dbmn$ the beam of X-rays ionising the gas. If CC' is at a

[1] Zeleny, *Phil. Trans.* A, 195, p. 193, 1900.

higher potential than DB, then a positive ion starting from d will move along a curved path between the cylinders, finally reaching the inner cylinder at a point K whose horizontal distance from d is one of the quantities measured in these experiments. This distance, X, can easily be expressed in terms of the velocity of the ion under unit electric force. For let b and a be respectively the radii of the outer and inner cylinders, A the potential difference between the cylinders, then the radial electric force R at a distance r from the common axis of the cylinders is given by the equation

$$R = \frac{A}{r \log_e (b/a)};$$

thus if v is the velocity of the ions under unit electric force, then on the assumption that the velocity is proportional to the electric force we have, if V is the radial velocity of the ion at a distance r from the axis of the cylinders,

$$V = \frac{Av}{r \log_e (b/a)}.$$

If u is the velocity of the gas parallel to the axis of the cylinders, which we shall take as the axis of x, the differential equation to the path of the ion is

$$\frac{dx}{dr} = \frac{u}{V}$$

$$= \frac{\log_e (b/a)\, ur}{Av};$$

hence X the horizontal distance from d at which the ion strikes the inner cylinder is given by the equation

$$X = \frac{\log_e (b/a)}{Av} \int_a^b ur\, dr.$$

Now $2\pi \int_a^b ur\, dr$ is the volume of gas which passes in unit time between the cylinders. We shall denote this quantity, which is easily measured, by Q; then we have

$$X = \frac{\log_e (b/a)\, Q}{2\pi Av},$$

or
$$v = \frac{\log_e (b/a)\, Q}{2\pi A X} \dots\dots\dots\dots\dots(1).$$

Thus if we know X we can easily determine v. The time T

taken by the ion to pass from one cylinder to the other is given by the equation

$$T = \int_a^b \frac{dr}{V} = \frac{\log_e (b/a)}{Av} \int_a^b r\, dr$$

$$= \frac{\log_e (b/a)}{2Av} (b^2 - a^2)$$

$$= \pi \frac{(b^2 - a^2) X}{Q} \quad \ldots\ldots\ldots\ldots (2).$$

These equations apply to ions starting from the inner surface of the outer cylinder. In practice the production of ions is not confined to the surface of the cylinder but extends throughout a layer db reaching from one cylinder to the other. The ions which start from a point in db, nearer to the surface of the inner cylinder than d, will evidently not be carried so far down the tube by the stream as an ion starting from d. Thus the preceding equations give us the position of the furthest point down the inner cylinder which is reached by the ions. In order to determine this point the inner cylinder is divided at K into two parts insulated from each other, the part D to the left being connected with the earth, while the part B to the right is connected with one pair of quadrants of an electrometer. If a constant stream of gas is sent between the cylinders, then when the potential of CC' is above a certain value, all the ions from the volume db which move inwards will reach DB to the left of K and will not affect the electrometer. By gradually diminishing the potential of CC' we reach a value such that the ions starting from the outer edge of d reach DB just to the left of K; when this stage is reached the electrometer begins to be deflected. If then in equation (1) we put for A the difference of potential corresponding to this stage and for X the horizontal distance of K from d, we shall be able to deduce the value of v.

Corrections. In consequence of the diffusion of the ions, all the ions starting from d will not follow exactly the line dK, and some of the ions will be found to the right of the line. The consequence of this is that the electrometer will begin to be deflected even when the potential difference A is theoretically sufficient to bring all the ions to the left of K; thus the observed potential difference when the deflections begin is slightly too large, and therefore the values of v determined by equation (1) are a little too small. Similar effects to those due to diffusion will be produced by the mutual repulsion of the ions. It is evident that the magnitude of these effects will depend upon the time it takes the ion to travel between the cylinders; if this time were zero, neither diffusion nor repulsion would have time to produce any effect; thus the longer the time taken by the ions to travel between the cylinders, the smaller would be the value of v as determined by this method. The time T, as we see from equation (2), depends upon the velocity of the air blast and the strength of the field; by altering these quantities it is possible to determine the values of

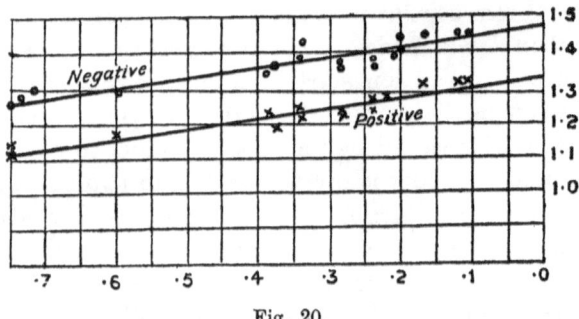

Fig. 20.

v for a considerable range of values of T; the values so found decrease, as was to be expected, slightly as T increases, the relation between v and T being found by experiment to be a linear one. Curves in which the ordinates were the ionic velocities and the abscissæ the time T were drawn, and the curve (which was found to be a straight line) prolonged until it cut the line $T = 0$; the corresponding value of v was taken as the ionic velocity. An example of such curves is given in Fig. 20, the o's and x's are the points determined by actual experiments. The points at which the lines intersect the line $T = 0$ give 1·48 cm./sec. for the velocity of the negative ion and 1·34 for the velocity of the positive, when the potential gradient is one volt per cm.

Smaller corrections have to be applied for the disturbance in the electric field produced by the presence of an excess of ions of one sign over those of the other in different parts of the field. It was proved by direct experiment that the effects due to surface ionisation were not appreciable.

The apparatus used to carry out this method is represented in section in Fig. 21. AA' was the outer cylinder; it had an internal diameter of 5·1 cm. and a total length of 142 cm. The parts to the right of V' and to the left of V were made of brass tubing; the part between VV' was aluminium tubing of the same diameter; this piece was inserted so as to permit the X-rays to pass through. The tubes were fastened together by air-tight joints and placed on insulating supports.

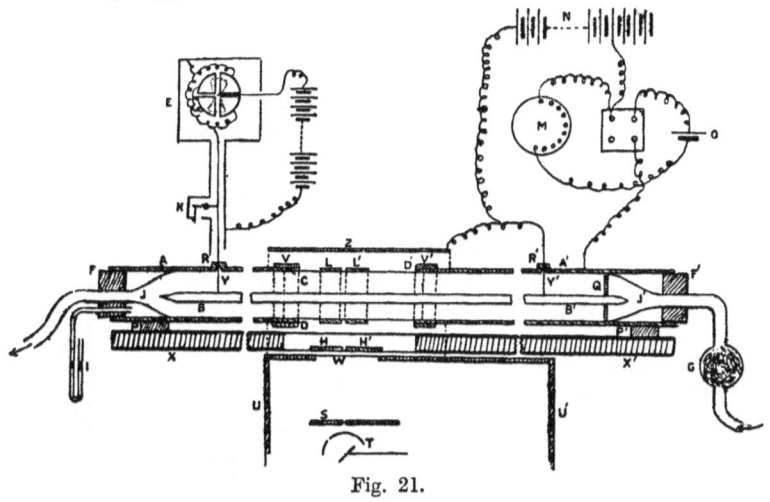

Fig. 21.

The inner cylinder BB' was an aluminium tube; in one set of experiments it was 1 cm. in diameter, in another it was 2·8 cm.; the ends of this tube were closed by conical pieces. The tube was divided at C and the two portions separated by ·5 mm. and insulated by ebonite plugs. The tube was supported by the ebonite rod Q and by the stiff brass wires Y and Y' which passed through ebonite plugs in the outer cylinder, and served to connect B' with the earth, and B with one pair of quadrants of the electrometer. The electrometer was a sensitive one, giving a deflection of 500 scale divisions for a potential difference of one volt. The narrow vertical beam of rays was adjusted and kept definite by the slits in the lead plates S, HH' and LL'. A constant and measurable supply of gas was sent through the tube by a gasometer. Experiments were made with gases carefully dried and with gases saturated with water vapour. Two series of experiments were made, one with an inner tube 1 cm. in diameter, the other with an inner tube 2·8 cm. in diameter; the results obtained in the one series agreed very well with those obtained in the other.

The values of the ionic velocities obtained by this method are given in the following table; they have been reduced to the uniform pressure of 760 mm. of mercury on the assumption (see p. 126) that the ionic velocity under a given potential gradient is inversely proportional to the pressure.

IONIC VELOCITIES.

Gas	Velocities in cm. per sec. under a potential gradient of one volt per cm.		Ratio of velocities of negative and positive ions	Temperature degrees centigrade
	Positive ions	Negative ions		
Air dry	1·36	1·87	1·375	13·5
Air moist	1·37	1·51	1·10	14
Oxygen dry	1·36	1·80	1·32	17
Oxygen moist	1·29	1·52	1·18	16
Carbonic acid dry .	·76	·81	1·07	17·5
Carbonic acid moist	·82	·75	·915	17
Hydrogen dry	6·70	7·95	1·19	20
Hydrogen moist ...	5·30	5·60	1·05	20

The intensity of ionisation was altered by causing the X-rays to pass through aluminium plates of different thicknesses, the ionic velocities were found to be independent of the intensity of the rays.

31·1. With this and other methods where currents of air are used it is necessary to guard against the velocity of the blast exceeding the critical velocity at which, as Osborne Reynolds showed, the steady flow of the air ceases and is replaced by irregular and turbulent movements.

Stanton and Pannell have shown that this change occurs when $D\bar{u}\rho/\eta$ is about 2500, where D is the diameter of the tube, $\bar{u}$ the mean velocity, ρ the density and η the viscosity of the gas. For a considerable range above this the flow is extremely erratic and appears to be quite unstable[1]. At much higher speeds it settles down again to a different type of steady flow, in which the velocity is nearly uniform across the pipe except close to the walls. This occurs when $D\bar{u}\rho/\eta$ is of the order 2×10^5.

31·2. Franck[2] has used this method to determine the mobility of the ions produced when a discharge passes between a tube and a very fine wire stretched along its axis. If the wire is very fine the electric field is too small to produce ions except in the immediate neighbourhood of the wire, so that all the ions start from close to the wire, and equation (1), p. 92, will apply. The fine wire which serves as one electrode takes the place of the tube BB' in Fig. 21, while the divided cylinder serves as the other electrode and the electrometer was connected with one of the divisions of the *outer* cylinder. As the ions are produced by the electric field itself, the parts of the apparatus connected with the X-rays are not needed; on the other hand, since there must be large differences of potential to produce the discharge additional precautions have to be taken so as to ensure adequate insulation.

This method gave very high values for the mobilities; for air at atmospheric pressure the mobility of the positive ion was 3·2 cm./sec. for 1 volt per cm. and that of the negative 10·26 cm./sec. under the same force.

Zeleny's blast method has been modified by Busse[3] so as to enable it to be used when only a small quantity of gas is available.

Erikson[4] has modified this method as follows. A rapid stream of air of the order of 2000 cm./sec. is drawn by the fan H (Fig. 22) between the plates A and B of a parallel plate condenser, kept at a difference of potential by the battery G. At C are a number of other plates kept at potentials varying in steps from that of A to that of B. These are to insure a uniform field. At D is

[1] See Bairstow, *Report of the Advisory Committee for Aeronautics*, 1913–14, p. 42.
[2] Franck, *Ann. der Phys.* xxi. p. 972, 1906.
[3] Busse, *Ann. der Phys.* lxxvi. p. 493, 1925.
[4] Erikson, *Phys. Rev.* xxiv. p. 502, 1924, and earlier papers.

TCE

a radioactive preparation which ionises the air drawn through E. The field drives the ions towards B and for a certain value of the field, ions of a given mobility will arrive at F and send a current

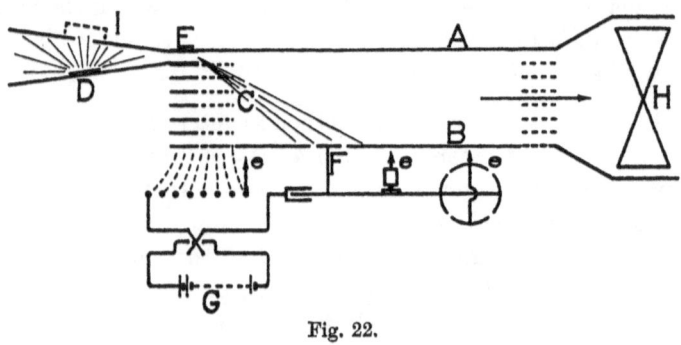

Fig. 22.

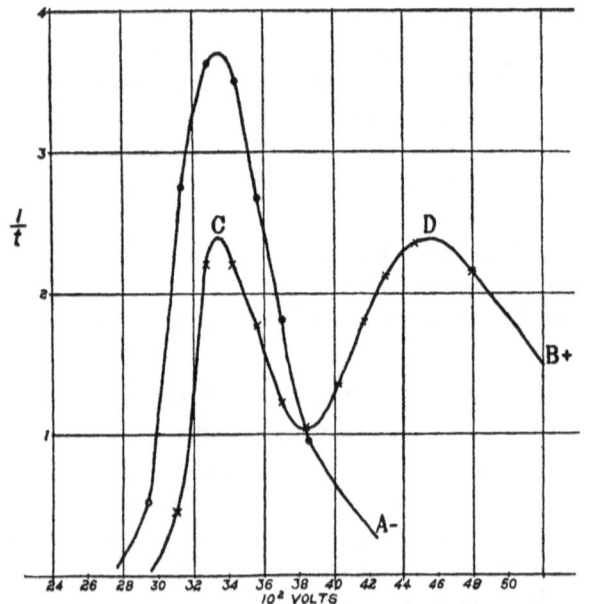

Curve A− refers to negative ions, curve CDB+ to positive ions.

Fig. 23.

to the electrometer. If this current is plotted against the potential between A and B, the existence of a group of ions of definite mobility will be shown by a hump on the curve. The distance between A and B was 3·5 cm. With this apparatus Erikson has found that the positive ions show *two* well defined maxima (Fig. 23), while the negative only show one. This is true both for air ions and for ions formed in CO_2 and drawn into air, the results being practically indistinguishable in the two cases. By altering the distance DE it can be shown that the two maxima for the positive ions are due to ions of different age, the ion starting with the larger mobility. Intermediate mobilities are not found. The mobility of the unaged positive ion is the same as that of the negative 1·87, while that of the aged ion is 1·36, giving a ratio of 1·36.

It is unfortunate that the speed of air blast used in these experiments gives a value of $D\bar{u}\rho/\eta$ of the order of 5×10^4, in the worst region of turbulence. There is thus a possibility of large eddies regularly formed producing a fictitious maximum, but the difference between positive and negative ions described above cannot be accounted for in this way, and must be real.

METHOD IV.

Method of determining the velocity by measuring the number of ions sent by a radial electric field to the sides of a tube of given length when traversed by a current of gas.

32. The principle of this method, which has been used by Rutherford[1] to measure the velocities of the ions produced by uranium radiation, is as follows. Suppose that ionised air is blown through a tube along the axis of which there is a wire charged positively; the electric field around the wire will drag the negative ions into the wire and thus rob the gas of a certain proportion of these ions; the number of these ions thus abstracted from the gas will depend upon the relation between the velocity of an ion in the electric field and the velocity of the air blast. If the ionic velocity were infinitely greater than the velocity of the blast, all the ions would be abstracted; while if the velocity of the blast were infinitely greater than the ionic velocity, they would all escape.

[1] Rutherford, *Phil. Mag.* v. 47, p. 109, 1899.

We see from equation (2), page 93, that t, the time taken by an ion to reach the wire, is given by the expression

$$t = \frac{r^2 - a^2}{2Au_2} \log_e \frac{b}{a} \dots\dots\dots\dots\dots(1),$$

where r is the distance from the axis of the tube of the point from which the ion starts, b the internal radius of the tube, a the external radius of the wire, A the difference of potential between the wire and the tube (the wire being at the higher potential), and u_2 the velocity of the negative ion under unit electric force. If in equation (1) we put t equal to the time taken by the air blast to pass from one end of the tube to the other, we see that all the ions whose distance from the axis of the tube is less than the value of r given by equation (1) will be dragged into the wire; hence if ρ is the ratio of the number of ions dragged from the gas to the whole number of ions, we have, assuming that the ions are uniformly distributed over the cross-section of the tube,

$$\rho = \frac{r^2 - a^2}{b^2 - a^2} = \frac{2Au_2 t}{(b^2 - a^2) \log_e (b/a)},$$

or

$$u_2 = \frac{\rho (b^2 - a^2) \log_e (b/a)}{2At} \dots\dots\dots\dots\dots(2).$$

The arrangement used by Rutherford is represented in Fig. 24.

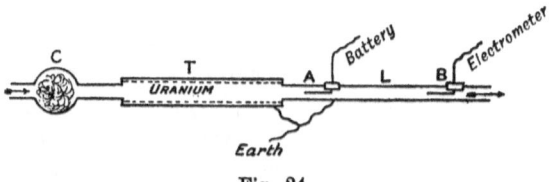

Fig. 24.

A paper tube coated with uranium oxide was fitted into a metal tube T 4 cm. in diameter. A blast of air from a gasometer, after passing through a plug of cotton-wool C to remove the dust, passed through a long metal tube AB connected with the earth; into this tube cylindrical electrodes A and B were fastened by insulating supports so as to be coaxial with the tube. The electrode A was charged up by a battery, and the electrode B was connected with one pair of quadrants of an electrometer. If B were charged initially to a potential of the same sign as A

(suppose positive) large enough to saturate the gas, then the rate of leak of the electrometer when the air blast was passing would measure the number of negative ions which escaped being dragged into the electrode A; by comparing the rate of leak when the electrode A is not charged, with the rate when it is charged to a known potential, we can determine the value of ρ in equation (2). Rutherford did not use this arrangement to measure directly the velocity of the ions produced by the uranium radiation, but proved by means of it that the velocities of these ions were the same as those of the ions produced by X-rays. For this purpose, after measurements of ρ had been made with the uranium cylinder in place, this cylinder was removed and replaced by an aluminium one exposed to X-rays, the strength of these rays being adjusted so that the amounts of ionisation in the two cases were approximately equal; measurements of ρ were then made with the X-rays on and were found to be identical with those obtained when the ionisation was produced by uranium radiation, thus proving that the ionic velocities are the same in the two cases.

The method, with various modifications, has been used by Gerdien[1], by Mache[2] and by Becker[3]. Kohlrausch[4] has pointed out the importance, when the velocity of the stream depends on the distance from the axis, of the condition mentioned above that the ions should be uniformly distributed over the cross-section of the tube. This is liable not to be true if the ionisation is produced by radiation.

33. A method which is the same in principle as this was first used by McClelland to measure the velocities of the ions produced by flames[5], and by arcs and incandescent wires[6]: the results of these experiments showed that the velocity of the ions diminishes very greatly when they get into the cooler parts of the flame, suggesting that there is a rapid condensation round the

---

[1] H. Gerdien, *Phys. Zs.* iv. p. 632, 1903.
[2] H. Mache, *Phys. Zs.* iv. p. 717, 1903.
[3] A. Becker, *Ann. der Phys.* xxxi. p. 98, 1910; xxxvi. p. 209, 1911.
[4] K. W. F. Kohlrausch, *Wiener Ber.* cxxiii. p. 1929, 1914.
[5] McClelland, *Phil. Mag.* v. 46, p. 29, 1898.
[6] McClelland, *Proc. Camb. Phil. Soc.* x. p. 241, 1899.

ions of some of the products of combustion in the flame. The diminution of velocity is clearly shown in the following table given by McClelland.

Distance of point where velocity was measured from the flame	Temperature at this point	Velocity of ion under a force of one volt per centimetre
5·5 cm.	230° C.	·23 cm./sec.
10 cm.	160° C.	·21 cm./sec.
14·5 cm.	105° C.	·04 cm./sec.

These velocities are all of them small compared with the velocities of the ions produced by X-rays or by radioactive substances. In the case of the ions from flames, as in other cases, the negative ions move faster than the positive. McClelland applied the same method to the determination of the velocities of the ions produced by arcs or incandescent wires; he found in these cases the same variability in the velocity as he had previously observed in the ions from flames; in the case of the arcs and wires, however, he found that the hotter the flame or wire the smaller the velocity of the ion. We shall return to the consideration of these phenomena when we discuss the electrical properties of flames and arcs.

METHOD V.

Determination of the ionic velocities by means of an alternating electric field.

34. This method, which in its original form can only be applied when the ionisation is confined to a thin layer of gas, and when moreover all the ions are of one sign, is a very convenient and accurate one. It was used by Rutherford[1] to determine the velocity of the negative ions which are produced close to a metallic plate when that plate is illuminated by ultra-violet light. The principle of the method is as follows. AB (Fig. 25) is a horizontal plate made of well-polished zinc, which can be moved vertically up and down by means of a screw; it is carefully insulated, and is connected with one pair of quadrants of an electro-

[1] Rutherford, *Proc. Camb. Phil. Soc.* ix. p. 401, 1898.

meter, the other pair being connected with the earth. CD is a base plate with a hole EF cut in it; this hole is covered in with fine wire-gauze, ultra-violet light from the source S passes through the quartz plates Q_1, Q_2, and this gauze, and falls on the plate AB. CD is connected with an alternating current dynamo or any other means of producing an alternating difference of potential proportional to a simple harmonic function of the time; the other

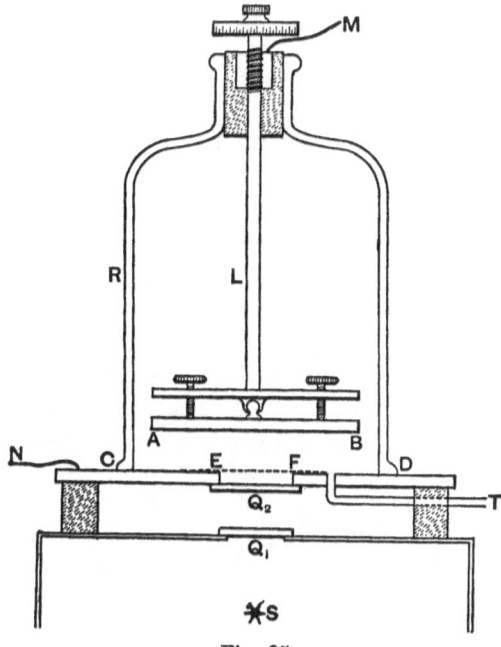

Fig. 25.

pole of this instrument is put to earth. Suppose now that at any instant the potential of CD is higher than that of AB; the negative ions at AB will be attracted towards CD, and will continue to move towards it as long as the potential of CD is higher than that of AB. If however the potential difference between CD and AB changes sign before the negative ions reach CD these ions will be driven back to AB, and this plate will not lose any negative charge. AB will thus not begin to lose negative electricity until the distance between the plates AB and CD is less than the distance passed over by the negative ion during the time the potential of CD is greater than that of AB. The distance between the plates is altered until AB just begins to lose a negative charge, then if we know this distance and the frequency and maximum value of the potential difference we can deduce the ionic velocity of the negative ion. For let the potential difference between CD and AB at the time t be equal to $a \sin pt$; then if d is equal to the distance between these plates, the electric force is equal to $(a/d) \sin pt$, and if u is the velocity of the ion under unit electric force, the velocity of the negative ion in this field will be

$$u\,(a/d) \sin pt;$$

hence if x is the distance of the ion from the plate AB at the time t we have

$$\frac{dx}{dt} = \frac{ua}{d} \sin pt,$$

or

$$x = \frac{ua}{dp}(1 - \cos pt),$$

if $x = 0$ when $t = 0$.

Thus the greatest distance the ion can get from the plate AB is equal to $2ua/pd$. If the distance between the plates is gradually reduced, the plate AB will begin to lose a negative charge when

$$d = \frac{2ua}{pd}, \text{ or } u = \frac{pd^2}{2a} \quad \ldots\ldots\ldots\ldots\ldots(1).$$

Hence if we measure p, a and d we can determine u.

In this way Rutherford found the following values for the velocities, under a potential gradient of 1 volt per cm. of the negative ions produced by the incidence of ultra-violet light on a zinc plate, in dry gases at atmospheric pressure.

Gas	Ionic velocity
Air............	1·4 cm./sec.
Hydrogen.....	3·9 cm./sec.
Carbonic acid..	·78 cm./sec.

These values differ but little from those obtained when the ionisation was produced by X-rays.

Rutherford found that the velocity of the ions was independent of the metal of which the plate AB was made; and he proved by this method that the velocities of the ions under a constant potential gradient vary inversely as the pressure, at any rate down to pressures of 34 mm. of mercury, which was the lowest pressure at which he worked.

34·1. The electromotive force between the plates need not necessarily be represented by a simple harmonic term; we may use the electromotive force from a battery of storage cells and reverse it periodically by a rotating commutator or by a tuning fork interrupter. If d is the greatest distance between the plates when the electrometer receives a charge,

$$d = k \int_0^T X dt,$$

if 0 to T covers the whole of the time during which the force acts in one direction If the force is reversed by a commutator this should be designed so that there is no considerable interval during which the ions are free from electric force of one sign or the other, for in this interval ions might diffuse towards the plate connected with the electrometer and thus when the force is next applied have a shorter distance than d to travel before they reach the plate. Lattey[1] used an arrangement of this kind for measurements at low pressures. Bowman[2] has devised for this purpose a method of producing by means of triode valves, an electromotive force which is nearly constant in magnitude and periodically is reversed in sign.

METHOD V (a).

34·2. This method has been modified by the following device introduced by Franck and Pohl[3]. The upper plate is made of wire-gauze and the ionisation is confined to a layer of gas AB, CD (Fig. 26) above this plate. This layer can be ionised by X-rays or by rays from radioactive substances (polonium is very suitable for this purpose). The number of ions produced by either of these means is proportional to the pressure of the gas, and at low pressure there may be

[1] R. T. Lattey, *Proc. Roy. Soc.* A, lxxxiv. p. 173, 1910.
[2] L. L. Bowman, *Phys. Rev.* (2), xxiv. p. 31, 1924.
[3] Franck and Pohl, *Ver. der Deut. Phys. Ges.* ix. p. 69, 1907.

too few ions to produce any effect on the electrometer. In such cases it is advisable to place in the region above the gauze a strip of platinum heated by an electric current, and coated with calcium or barium oxide if negative, or with aluminium phosphate if positive ions are desired. A *small* constant difference of potential is maintained between AB and CD so as to cause positive or negative ions to trickle through the gauze. An alternating electromotive force is applied between AB and the plate connected with the electrometer as before. With this or the following method the mobilities of the ions may be determined in gases of which only a small quantity is available. It is also suitable for determining the variation of mobility with pressure.

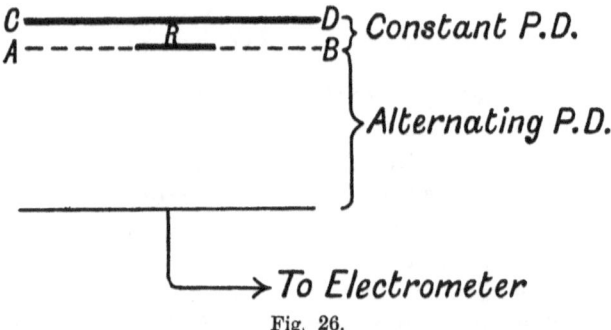

Fig. 26.

Loeb[1] has shown however that the method is open to objection owing to the lines of force due to the potential difference between AB and CD spreading out through AB and disturbing the electric field through which the ions have to pass. The values found for the mobility depended on the potential difference between AB and CD.

METHOD VI.

Langevin's method of measuring the velocities of the ions.

35. Langevin[2] has devised a method of measuring the velocities of the ions which has the advantage of not requiring the use of a uniform source of X-rays. The theory of the method is as follows. Suppose that we ionise, say by X-rays, the gas between two parallel plates A and B, then stop the rays and

[1] Loeb, *Phys. Rev.* xxi. p. 720, 1923; see also C. Zimmerscheid, *ibid.* p. 721.
[2] Langevin, *Ann. de Chimie et de Physique*, t. xxviii. p. 289, 1903.

apply a uniform electric field to the region between the plates. If the force on the positive ion is from A to B the plate B will receive a charge of positive electricity. After the field has been on in one direction for a time T, reverse it; B will now begin to receive negative electricity and, if the force is not reversed again, will continue to do so until the ions are exhausted. If n is the number of ions of one sign per unit volume between the plates when the rays are cut off, X the electric force, k_1, k_2 the velocities of the positive and negative ions respectively under unit electric force, then during the time T the plate B will receive per unit area nk_1XT positive ions, supposing the field to be so strong that the loss of ions by recombination may be neglected, and also that k_1XT is less than l, the distance between the plates. The number of negative ions which in this time have gone to A is nk_2XT provided k_2XT is less than l, so that the number of negative ions left between the plates is $n(l - k_2XT)$ and these when the field is reversed are driven to B. Thus if Q is the positive charge received by each unit area of the plate B,

$$Q = nk_1XT - n(l - k_2XT) = n(k_1 + k_2)XT - nl,$$

provided both k_1XT and k_2XT are less than l.

This will hold until T is equal to the smaller of the two quantities $\dfrac{l}{k_1X}$, $\dfrac{l}{k_2X}$; if k_2 is greater than k_1 the second limit will be reached first; when this limit is passed there are no negative ions left to be driven against B when the field is reversed, thus

$$Q = nk_1XT;$$

this will hold until $T = l/k_1X$; for this and greater values of T,

$$Q = nl.$$

Thus when $k_2 > k_1$

$Q = e\{n(k_1 + k_2)XT - nl\}$; $T < l/k_2X$;

$Q = enk_1XT$; $T > l/k_2X$, $< l/k_1X$;

$Q = enl$; $T > l/k_2X$, $> l/k_1X$;

while if $k_1 > k_2$ we see that

$Q = e\{n(k_1 + k_2)XT - nl\}$; $T < l/k_1X$;

$Q = e\{nl - n(l - k_2XT)\} = enk_2XT$; $T > l/k_1X$, $< l/k_2X$;

$Q = enl$; $T > l/k_2X$.

Thus if we represent the relation between Q and T graphically, the curves will be portions of straight lines as in Fig. 27 intersecting at the points corresponding to $T = l/k_1 X$, $T = l/k_2 X$; if we find these points of intersection we can deduce the values of k_1, k_2.

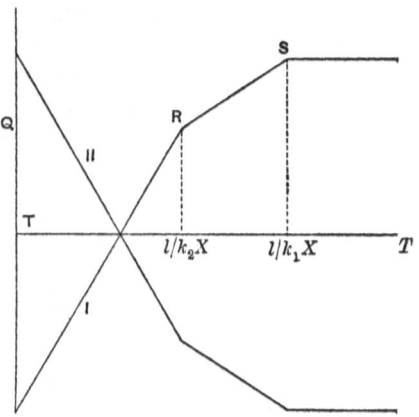

Fig. 27. In curve I the electric force is initially from A to B, in curve II from B to A.

In consequence of inequalities in the ionisation between the plates, and from the recombination and diffusion of the ions, the

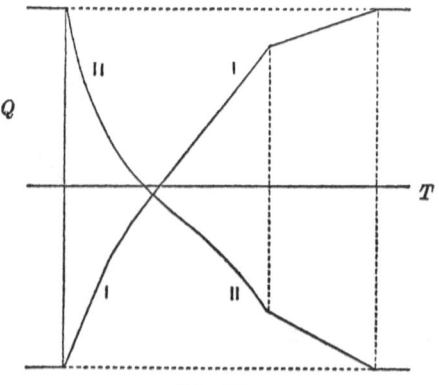

Fig. 28.

curve obtained in practice does not consist of three straight lines, but of three curves which intersect at well defined nicks; an example of such a curve is shown in Fig. 28, which is taken from

Langevin's paper. If $k_1 = k_2$ two of the nicks coincide. If we had a mixture of different gases with several kinds of ions, there would be a nick corresponding to each positive ion and also one corresponding to each negative ion.

35·05. In the preceding investigation we have supposed that the ionisation was uniform between the plates; let us now consider the more general case when the ionisation is variable. Suppose that originally A is charged positively.

During the time T, B will receive all the positive ions which were originally in the layer of gas between B and L, where $BL = k_1 XT$; during this time the negative ions in the region between A and M, where $AM = k_2 XT$, are withdrawn from the gas, and after reversal the negative ions between B and M are driven to B. The total negative charge received by B will therefore be the charge on the negative ions which were originally between L and M. After L has crossed M the charge received will be positive and equal to the charge on the positive ions between L and M. Take the case when the velocity of the negative ion is greater than that of the positive, so that M reaches B before L reaches A. The rate of increase of Q with T just before M reaches B will be $n_B k_2 + n_L k_1$, where n_B and n_L are the densities of the ions at the plate B and at L; just after M has reached B the rate of increase is $n_L k_1$, thus the change in angle at the nick R will be $n_B k_2$ and thus proportional to the density of the ionisation at the plate B. Again, just before L reaches A the gradient of Q will be $n_A k_1$, while just after it will be zero. Thus the sharpness of the nick S will be proportional to the density of the ions at A. The sharpness of the nicks depends essentially upon the initial densities of ionisation at the plates, though the *position* of the nicks R and S does not depend upon the distribution of the ionisation. If the changes in the initial distribution of the ionisation are very abrupt, there may be nicks in the Q and T curve due to this distribution in addition to those we have been considering. Thus if the ionisation were confined to a very thin layer at a distance d_1 from B and d_2 from A, the curve would have the form shown in Fig. 29 with nicks at times corresponding to $d_1/k_1 X$ and $d_2/k_2 X$, while those corresponding to R and S would disappear.

If instead of reversing the field in this case we applied an alternating potential represented by $X_0 \cos pt$, and plotted the values of Q against X_0, we should get a curve of the type of that in Fig. 29 with nicks at O, C, D corresponding to $X_0 = d_1 p/k_2$, $d_1 p/k_1$, $d_2 p/k_2$, supposing these to be in ascending order of magnitude; in this way we might determine k_1 and k_2. At $t = 0$ the force on the positive ions is supposed to act towards B.

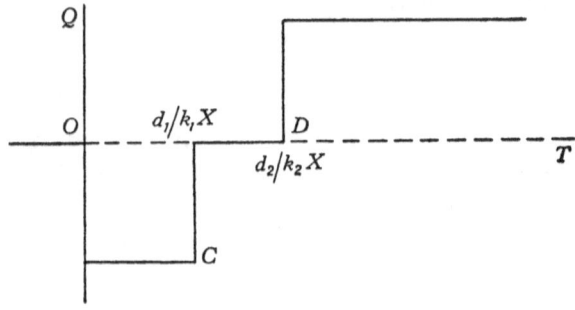

Fig. 29.

Phillips in his experiments on the effect of temperature on the mobility[1] increased the sharpness of the nicks by localising the ionisation to the neighbourhood of one of the plates. Thus suppose the ionisation were confined to an exceedingly thin layer close to the plate B. The curves (I) and (II) of Fig. 27 will now resemble (I) and (II) of Fig. 30. In (I) the nick at R is exceedingly sharp and enables us to determine k_2, while in (II) S is sharp and gives the value of k_1.

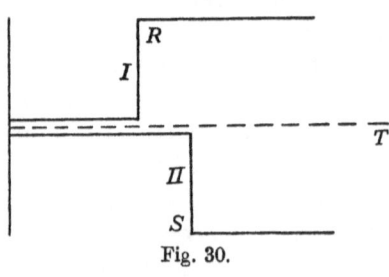

Fig. 30.

He also made the method a null method to some extent, by connecting one quadrant of his electrometer to B, the other to one electrode of a condenser ionised by the same flash of X-rays, and arranged to balance the condenser when the field was reversed if the interval before reversal was very large.

[1] Phillips, *Proc. Roy. Soc.* lxxviii. p. 167, 1906.

Method VII.

Tyndall and Grindley's method of measuring the velocities of the ions.

35·1. Tyndall and Grindley[1] have made a large number of measurements of mobility, using the following method.

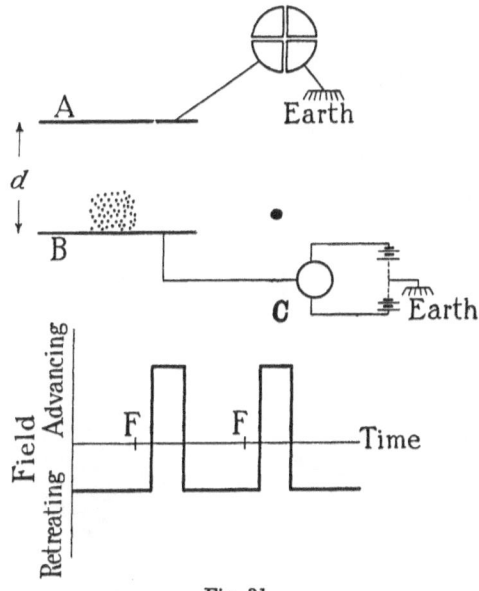

Fig. 31.

A and B are two parallel metal plates. A is connected with one pair of quadrants of an electrometer, B through a commutator with a battery of storage cells; thus there is a uniform electric field between A and B which is sometimes in one direction, sometimes in the opposite. A flash of ionisation is produced near B by a strip of polonium placed on a rotating wheel on the same shaft as the commutator. When the apparatus was used to measure the mobility of negative ions, the flash was arranged to occur shortly before the end of a period when the electric force tended to make the negative ions move towards B, and the positive ones towards A. On the reversal of the field, if this occurs soon enough

[1] Tyndall and Grindley, *Proc. Roy. Soc.* A, cx. p. 341, 1926.

after the flash, the positive ions are dragged back to B, and those negative ones which have been left in the field are driven towards A, which they will all reach if the field remains in this direction long enough. If the field reverses again before all the negative ions have reached A, those which are left will be dragged back to B, and the commutator is arranged so that the time the field drags negative ions back is greater than that during which it pushes

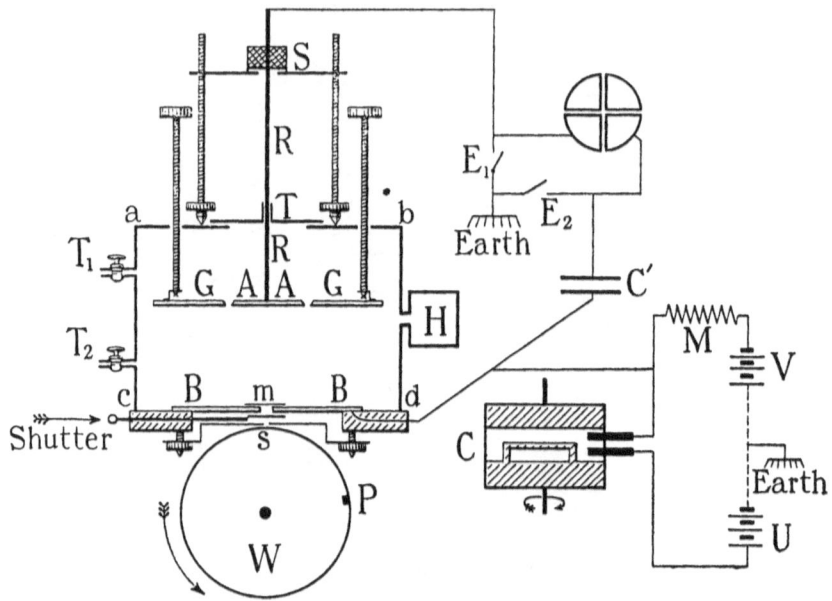

Fig. 32. Details of Tyndall and Grindley's apparatus.

them forward, so that the space between A and B is cleared of negative ions before the next flash of ionisation occurs. The interval between the flash of ionisation and the reversal of the field, the time the field pushes the negative ions towards A, and that during which it drags them back again, are all proportional to the time occupied by one revolution of the commutator. If this time is very large all the negative ions produced by the flash will be drawn into B before the field is reversed, so that no negative ions will reach A. When the time of rotation is diminished, some of the negative ions will escape being dragged into B and will reach A before the field is reversed; as this time diminishes the number of negative ions will at first increase, but will begin to fall off when the time the field acts in the direction B to A is not sufficient to allow all the negative to reach A before the field is reversed. When this stage is reached the number reaching A will fall off very rapidly, as the time of rotation of the commutator diminishes. As the number of ions reaching A vanishes when the time of rotation is zero and when it is infinite there must be some speed when the number is a maximum: this will depend upon the mobility, so that it may be used to determine this quantity.

A first approximation to the relation between the mobility and the time of rotation may be obtained as follows. Let us suppose that the ionisation is confined to a thin layer near B, whose thickness is h. Let T be the time of rotation of the commutator, pT the interval between the flash and the reversal of the field, X the force between the plates, k_1 the mobility of the negative ions; then before the field is reversed the negative ions can travel towards B through a distance pTk_1X; hence if pTk_1X is greater than h all the negative ions will be driven into B and none will reach A. If, however, pTk_1X is less than h, a number of negative ions proportional to $h - pTk_1X$ will be left over and these will all be driven into A, provided l the distance between the plates is less than qTk_1X, where qT is the time the force acts in the direction from B to A; if qTk_1X is less than l, no negative ions will reach the plate. Hence N, the number of negative ions reaching the plate for one flash of ionisation, will have the following values:

$$N = 0, \text{ when } T > \frac{h}{pk_1X},$$

$$N = 0, \text{ when } T < \frac{l}{qk_1X},$$

N proportional to $h - pTk_1X$ for values of T between these values.

Hence the graph representing the relation between N and T will have a sharp maximum when

$$T = \frac{l}{qk_1X}.$$

If there are negative ions of different kinds, the graph will be got by superposing several graphs of this type and will have several maxima. This method would not detect electrons among the negative ions, for these move so quickly that they would all be dragged into B before the field is reversed.

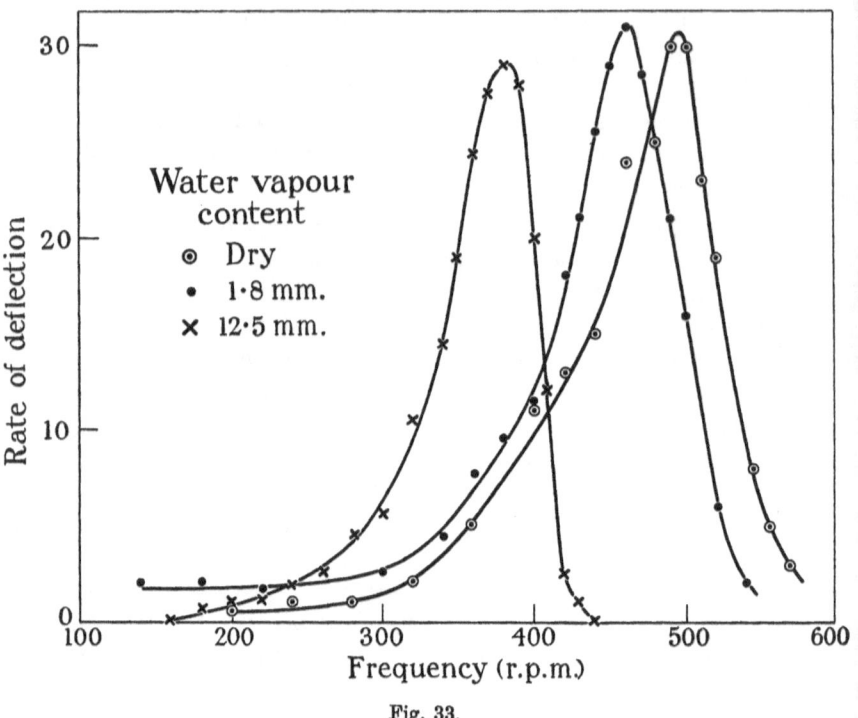

Fig. 33.

Tyndall and Grindley plotted the charge received by A against the number of revolutions per second, which is inversely proportional to T. Some of these graphs are represented in Fig. 33; the three graphs are for negative ions in air of different humidities. The mobility of both positive and negative ions was found to diminish with increasing humidity. They also confirmed Erikson's result that the positive ions decrease in mobility on ageing, and find that the rate of ageing is more rapid the drier the air. With this method an accuracy of the order of 1 % can be reached.

Method VIII.

35·2. When the ionisation is localised in a very thin layer of intense ionisation close to one of the electrodes, or when the electrodes themselves emit ions, the mobility can be determined by measuring the current produced by a known potential difference between the electrodes; the mobility k is (see p. 208) given by the expression

$$k = \frac{32\pi}{9V^2} il^3,$$

where V is the potential difference between the plates, l their distance apart and i the current per unit area supposed small.

This method has been used by Rutherford, Child, Garrett, Kovarik and others; it has not however given satisfactory results, except at high pressures.

Method IX.

Revolving slit method.

35·3. Laporte[1] has used a method based on the following principles. A flash of ionisation from a strip of polonium passes through a radial slit in a circular disc A which rotates about a horizontal axis; another parallel disc B mounted on the same axis has also a radial slit cut in it making a finite angle with the radial slit in A. A uniform electric field is maintained between the discs A and B. The ions going through the slit in A would not be able to get through the one in B and reach a plate behind it, connected with an electrometer, unless the time taken by the ions to move from one plate to another is just equal to the time taken by the discs A and B to rotate through the angle between the slits in A and B. Thus if the slits were exceedingly narrow and there was no diffusion of the ions, the electrometer would only receive a charge when the speed of rotation had a particular value depending on the time the ions take to travel from one plate to another, and therefore upon the electric force between the plates. When the speed of rotation and the electric force is known the mobility can be deduced at once. In this case the graph representing the relation

[1] Laporte, *Thesis for Doctorate*, Paris, 1927. See also *Comptes Rendus*, clxxii. p. 1028, 1921; clxxxii. pp. 620, 781, 1926; clxxxiii. pp. 119, 287; *Ann. de Phys.* viii. p. 466, 1927.

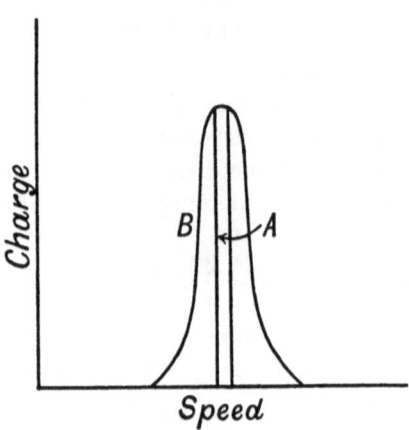

Fig. 34.

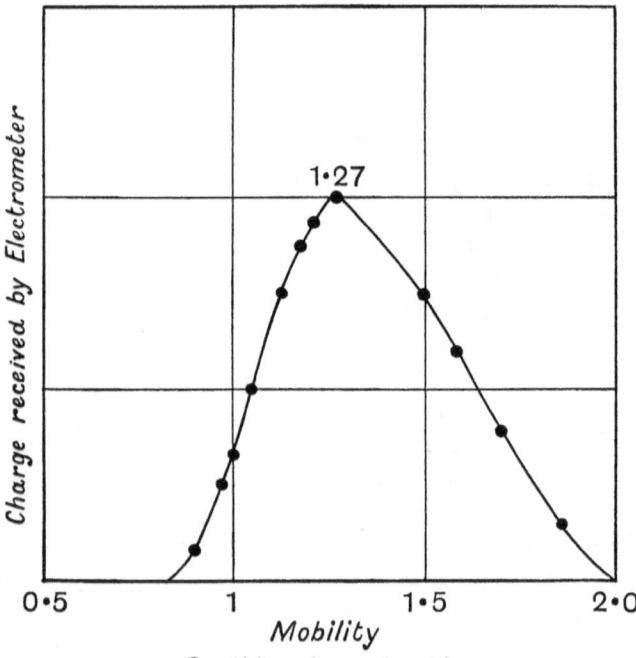

Fig. 35.

between the charge received by the electrometer and the angular velocity of rotation would resemble the curve A in Fig. 34. The finite width of the slits, the diffusion of the ions, and eddy currents in the gas, prevent the change in the charge received by the electrometer being as abrupt as this, and we might expect the graph would be represented by a curve like that marked B. The curves obtained by Laporte have this shape, as is seen from Fig. 35, which is taken from his paper.

Laporte is of the opinion from the tests he has applied that the causes to which we have alluded are not sufficient to explain the flatness of the maximum shown by his curves, and he considers that his experiments prove that in the gases he investigated (air, CO_2, argon) ions exist which may have any mobilities between certain rather wide limits, and that there are not merely one or two kinds of ions but a large number of kinds gradually merging into each other. The limits of mobility for the positive ions in air range from ·8 to 2·1. There are theoretical reasons, as we shall see, for thinking that for positive ions the range of values cannot be as great as this.

Lafay[1] used a similar arrangement, except that he detected the ions by collecting them on a disc covered with resin, where their position was shown up by dust figures obtained by powdering the disc with a mixture of sulphur and red lead. The disc rotated on the axis behind the plate B of the condenser, the hole in which was enlarged so as always to allow the passage of the ions. He found only one speed of ion of each sign in air, and also in a mixture of hydrogen and carbon dioxide.

METHOD X.
Magnetic deflection of the ions.

35·4. Townsend and Tizard[2] have used a method that is specially suitable for measurements at low pressures. It consists in deflecting the ions, moving under a uniform electric field, by a magnetic field at right angles to the former. If W is the velocity of the ions in the direction of the field Z, and θ the angle which

[1] Lafay, *Comptes Rendus*, clxxiii. p. 75, 1921.
[2] Townsend and Tizard, *Proc. Roy. Soc.* lxxxviii. p. 336, 1913.

the resultant direction of motion makes with Z, then since the side force due to the magnetic field is $eW.H$, and that due to Z is Ze, we have $\tan\theta = HW/Z$, assuming the resultant velocity to be in the direction of the force. The apparatus used is shown in Fig. 36. The ions are released from the plate A by photoelectric action, pass through a small slit in the plate B and then come under the influence of the main electric and magnetic fields. In order to make the electric field as uniform as possible, the rings R are maintained at potentials varying step by step from that of B to that of R_0 which was earthed. In the plane of R_0 were three insulated segments c_1, c_2 and c_3 as shown, each of which could

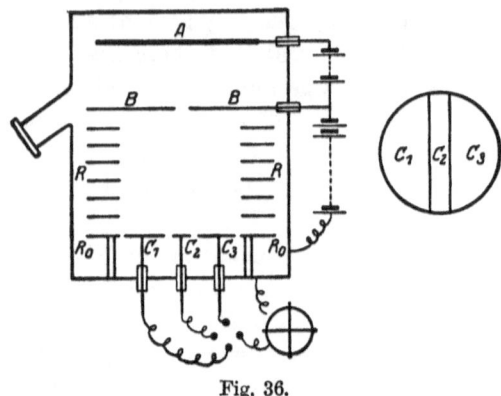

Fig. 36.

be connected in turn with the electrometer. In the absence of the magnetic field the current received by c_1 would be equal to that received by c_3, both being usually small. In the presence of the magnetic field the ions are deflected to one side, say c_1, and the field can be adjusted so that the current received by c_1 is equal to that received by c_2 and c_3 connected together. When this is the case the centre of the ionic stream has moved from the centre of c_2 to the middle of the gap between c_1 and c_2, and the value of θ can readily be found. The method was only used for pressures below 20 mm. At these low pressures W is no longer simply proportional to Z for the fields used, so that the 'mobility' ceases to have a definite meaning.

Method XI.

Chattock's method of measuring the velocities of ions produced by the discharge of electricity from a sharp point.

36. The preceding methods would be very inconvenient in the case when the electric field is as strong, and the velocities of the ions therefore as great, as they are when electricity is discharging from a pointed conductor. For this case, in which the ions at some little distance from the point are all of one sign, Chattock[1] has devised a very ingenious method by means of which he has been able to measure the velocities of these ions. The principle of the method is as follows. Let P represent a vertical

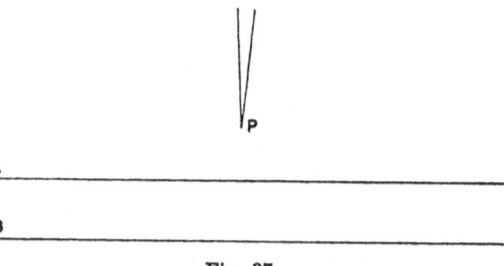

Fig. 37.

needle discharging electricity from its point into the surrounding air; consider the force acting on the ions included between two horizontal planes A and B, Fig. 37. If Z is the vertical component of the electric intensity, ρ the density of the electrification, the resultant force F on the ions included between A and B is vertical and equal to

$$\iiint Z\rho \, dx\, dy\, dz.$$

If the velocity of the ion under unit electric force is u, then w the vertical velocity of the ion is equal to uZ. If all the ions are of one sign so that u is the same for all the ions, we have, since $Z = w/u$,

$$F = \frac{1}{u}\iiint w\rho \, dx\, dy\, dz.$$

[1] Chattock, *Phil. Mag.* v. 48, p. 401, 1899; Chattock, Walker and Dixon *Phil. Mag.* vi. 1, p. 79, 1901.

Since the ions are all of one sign $\iint \rho w\, dx\, dy$ is the quantity of electricity streaming across a horizontal plane in unit time; this is the same for all horizontal planes, and is equal to i, where i is the current of electricity flowing from the needle-point; hence we have
$$F = \frac{i}{u}\int dz = \frac{i}{u}(z_B - z_A),$$
where $z_B - z_A$ is the vertical distance between the planes A and B. This force F must be balanced by the difference of the gaseous pressures over A and B, hence if p_B and p_A denote respectively the total pressures over the planes A and B we have
$$F = p_B - p_A,$$
and hence
$$u = \frac{i(z_B - z_A)}{p_B - p_A} \quad \ldots\ldots\ldots\ldots\ldots\ldots(1).$$

Thus, by the measurement of these pressures and of the current flowing from the point (the latter measurement is easily made by inserting a galvanometer in series with the needle-point), we can deduce the value of u.

The apparatus used by Chattock to carry out this method is represented in Fig. 38. The discharging needle is supported in a

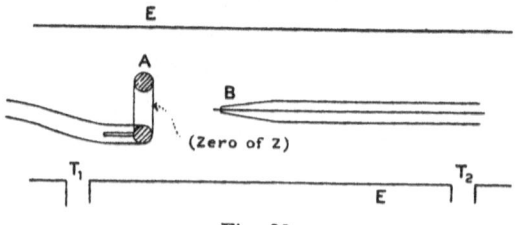

Fig. 38.

narrow sliding glass tube drawn out at the end B; it discharges to a ring A made of smooth metal; the needle and ring are enclosed in a wide glass tube E, the ends of which are connected by tubes T_1 and T_2 with the ends of a U-tube pressure gauge containing water; the ring A can be moved along the tube by means of a screw. In this apparatus, since there is no current to the left of the ring or to the right of the point, if we put $z_B - z_A$ equal to z, the distance of the point from the ring, and if ω is the difference of pressure in dynes per sq. cm. measured by the pressure gauge, A the area of cross-section of the tube, then

$$p_B - p_A = \omega A + p',$$

where p' is the part of the pressure which is borne by the ring. We have, by equation (1),

$$\omega = \frac{i}{Au} z - \frac{p'}{A}.$$

It was assumed that when the point was a considerable distance from the ring p' became independent of z; on this supposition we have, if $\Delta\omega$, Δz are corresponding changes in ω and z,

$$\frac{\Delta\omega}{\Delta z} = \frac{i}{Au},$$

and it was from this relation that u was calculated. Chattock found for the velocities of the negative and positive ions in air under a potential gradient of a volt per cm. the values 1·8 cm./sec. and 1·38 cm./sec., which agree well with those found for the ions produced by X-rays, and we conclude that the ions in the two cases are the same. In the second paper Chattock extends the method to hydrogen, oxygen, and carbonic acid as well as air, and again finds close agreement between the velocities of the ions produced by the point discharge and those produced by radio-active substances. He points out that while the determinations of the ionic velocities of the positive ions showed in all gases great consistency, considerable variations which could not be attributed to errors of experiment were found in the values of the velocities of the negative ions. This was especially the case in hydrogen, where the values of the ionic velocity of the negative ion varied from 6·8 to 8·5; in the other gases the variation is not so marked. Chattock ascribes this variation to the gases occluded by the discharging point; when this point is negative some of these occluded gases are given off and help to carry the discharge, and as the velocity of the hydrogen ions is very large compared with that of other ions, it is urged that a small admixture of other more slowly moving ions might produce a considerable lowering of the average velocity. When the point is positive the occluded gas is supposed either not to be given off, or if given off, not to take any part in carrying the discharge. The values obtained for the velocities of the ions produced by the point discharge were as follows; $\overset{+}{u}$, $\overset{-}{u}$ denote the velocities of the positive and negative ions respectively under an electric force of a volt per centimetre.

Gas	$\overset{+}{u}$	$\overset{-}{u}$	$\overset{-}{u}/\overset{+}{u}$
Hydrogen...	5·4	7·43	1·38
Carbonic acid	0·83	0·925	1·11
Air.........	1·32	1·80	1·36
Oxygen.....	1·30	1·85	1·42

36·1. Ratner[1] has used a method depending on the same principles to measure the mobility of ions from salts heated on platinum. The ions are driven by the electric field through a parallel plate condenser with gauze plates. The reaction of the air is transmitted through the gauze to a special pressure gauge. Only relative values can be found, the positive ion at atmospheric pressure being taken as standard.

Franck[2] determined the velocity of the ions from a point discharge by a modification of Method V. The point was placed in a closed vessel above a hole in a plate A, the electrometer was connected with a parallel plate B; an alternating potential difference was applied between A and B and adjusted so that the ions just reached the plate B, the mobility of the ions was then calculated by equation (1), p. 104. For air Franck found

$$\overset{+}{u} = 1·34, \quad \overset{-}{u} = 1·79.$$

Experimental Results.

37·1. The mobilities of the ions in different gases, as far as they have been determined, are given in the following table, for most of which we are indebted to Professor Fulcher. It will be seen that there is a considerable divergence between the values obtained by different observers, suggesting that the mobility may be affected to the extent of several per cent. by circumstances such as the purity

[1] Ratner, *Phil. Mag.* v. 32, p. 441, 1916.
[2] Franck, *Ann. der Phys.* xxi. p. 972, 1906.

MOBILITY OF IONS

Dry Gases.

Gas	M	Observer		$\bar{u}$	$\overset{+}{u}$	Method	Ioniser
Air	29	Zeleny	1900	1·87	1·36	III	X-rays
		Langevin	1903	1·70	1·42	VI	,,
		Phillips	1906	1·79	1·39	VI	,,
		Blanc	1908	2·00	1·27	V	,,
		Wellisch	1909	1·78	1·54	VI	,,
		Lattey	1910	—	1·47	V	,, *
		Wellisch	1915	1·93	1·23		,,
		Franck and Pohl	1907	1·79	1·37	V	a-rays
		Dempster	1912	2·20	1·36	V	,, †
		Kovarik	1912	1·89	1·35	VIII	,, †
		Rothgieser	1913	1·93	1·33	V	,,
		Tyndall and Grindley	1926	2·15	—	VII	,,
		Erikson	1924	1·87	{1·87 / 1·36}	III	,,
		Chattock	1899	1·80	1·38	XI	Point discharge
		Chattock, Walker and Dixon	1901	1·80	1·32	XI	,, ,,
		Franck	1907	1·79	1·34	V	,, ,,
		Kovarik	1910	2·04	—	V	Ultra-violet light
		Hughes	1910	?	—	—	,, ,,
		Todd	1911	—	1·36	V	Heated aluminium phosphate*
		Loeb	1923	2·18			Ultra-violet light
		Probable mean		2·1	1·36	—	
H	2	Zeleny	1900	7·95	6·70	III	X-rays
		Chattock, Walker and Dixon	1901	7·43	5·40	XI	Point discharge
		Franck and Pohl	1907	7·68	6·02	V	a-rays
		Blanc	1908	(10·00)	5·33	V	X-rays
		Franck	1909	—	6·21	IV	Recoil atoms
		Chattock and Tyndall	1910	7·6	5·8	XI	Point discharge
		Todd	1911	—	5·30	V	Heated phosphate*
		Lattey and Tizard	1912	8·12	5·10	V	X-rays*
		Ratner	1912	—	5·65	V	Recoil radiation atoms RaB
		Kovarik	1912	8·19	6·20	VI	a-rays†
		Rothgieser	1913	8·26	5·91	V	,, ,,

$M =$ molecular weight.   * Low pressures.   † High pressures.

Gas	M	Observer		$\bar{u}$	$\overset{+}{u}$	Method	Ioniser
CO_2	44	Zeleny	1900	·81	·76	III	X-rays
		Chattock, Walker and Dixon..	1901	·92	·83	XI	Point discharge
		Langevin ..	1903	·90	·86	VI	X-rays
		Blanc ..	1908	1·03	·83	V	,,
		Wellisch ..	1909	·85	·81	VI	,,
		Todd ..	1911	—	·99	V	Heated phosphate
		Kovarik ..	1910	1·01	—	V	Ultra-violet light
		Lattey and Tizard	1912	1·05	·84	V	X-rays
		Rothgieser ..	1913	·99	·76	V	a-rays
		Mean value		·98	·84	—	
O_2	32	Zeleny	1900	1·80	1·36	III	X-rays
		Chattock, etc.	1901	1·85	1·30	XI	Point discharge
		Franck ..	1910	1·79	1·29	V	a-rays
		Mean value		1·80	1·31		
He	4	Franck and Pohl	1907	6·31 to 500	5·09 5·09	V V	a-rays ,, (pure gas)
N_2	28	Franck ..	1910	1·84 to 145	1·27 1·27	V V	a-rays ,, (pure gas)
N_2O	44	Wellisch ..		0·90	0·82	VI	X-rays
CO	28	Wellisch ..	1909	1·14	1·10	VI	,,
SO_2	64	Wellisch ..	1909	·41	0·44	VI	,,
		Todd ..	1911	—	0·57	V	Heated phosphate
		Wellisch ..	1917	·41	·41	V	a-rays
		Yen	1918	·41	·41	V	,,
CH_4	16	Todd ..	1911	—	1·84	V	Heated phosphate
Ar	40	Franck ..	1910	1·70 to 206	1·37 1·37	V V	a-rays ,, (pure gas)

of the gas, the age of the ion, the method of ionisation and other conditions which would naturally be different in different experiments and which, if their influence were not suspected, might not have been specified.

The measurements in the above table were made at the ordinary room temperature. In case measurements were not made at 760 mm. they have been reduced to 760 by multiplying by $p/760$. The gases were dried by the ordinary methods but contained, no doubt, traces of moisture. For the effect of the complete removal of moisture and other impurities, see p. 133.

It is seen that the mobility of ions formed by X-rays, α-rays, point discharge and heated aluminium phosphate is the same. The results obtained with the various methods are concordant, though rather surprising variations occur between the results of different observers.

The following table of results for various vapours is taken

Gas	$\bar{u}$	$\overset{+}{u}$	Author
Ammonia	0·80	0·74	Wellisch
Ammonia	0·66	0·56	Loeb
Ethane	1·30	—	Wahlin
Methylene	0·91	—	Wahlin
Pentane............	0·35	0·36	Wellisch 1909
Pentane............	0·45$_1$	0·38$_2$	Yen
Methyl alcohol 66° ...	0·38	0·37	Przibram
Ethyl alcohol	0·27	0·34	Wellisch 1909
Ethyl alcohol	0·41	0·39	Wellisch 1917
Ethyl alcohol	0·37$_3$	0·36$_3$	Yen
Ethyl alcohol 79°.....	0·35	0·34	Przibram
Propyl alcohol 97° ...	0·22	0·22	Przibram
Isobutyl alcohol 105° .	0·21	0·21	Przibram
Isoamyl alcohol 130°..	0·23	0·19	Przibram
Ethyl formate	0·31	0·30	Wellisch 1909
Methyl acetate	0·36	0·33	Wellisch 1909
Methyl acetate 58° ...	0·24	0·19	Przibram
Ethyl acetate........	0·28	0·31	Wellisch 1909
Ethyl acetate........	0·24$_7$	0·22$_6$	Yen
Ethyl acetate 77°	0·19	0·16	Przibram
Propyl acetate 100° ..	0·17	0·15	Przibram
Aldehyde............	0·30	0·31	Wellisch 1909
Aldehyde............	0·33$_3$	0·30$_7$	Yen
Ethyl ether..........	0·31	0·29	Wellisch 1909
Ethyl ether..........	0·35	0·27	Wellisch 1917
Ethyl ether..........	0·22	0·19	Loeb
Ethyl chloride	0·31	0·33	Loeb
Ethyl chloride	0·31$_7$	0·30$_4$	Yen
Methyl bromide	0·28	0·29	Wellisch 1909
Methyl iodide........	0·22	0·21	Wellisch 1909
Methyl iodide........	0·23	0·24	Wellisch 1917
Methyl iodide........	0·22$_6$	0·21$_6$	Yen
Ethyl iodide	0·16	0·17	Wellisch 1909
Ethyl iodide	0·18$_1$	0·18$_1$	Yen
Carbon tetrachloride..	0·31	0·30	Wellisch 1909
Water 100°	0·95	1·1	Przibram
Sulphuretted hydrogen	0·56	0·62	Loeb and Cravath[1]
Sulphuretted hydrogen	0·55	0·61	Loeb and Cravath[1]

[1] Loeb and Cravath, *Phys. Rev.* xxvii. p. 811, 1926.

from K. Przibram's article in the *Handbuch der Physik*, 1926, XXII. p. 324. His own measurements[1] refer to vapours at their boiling point, the others are at room temperature reduced to atmospheric pressure, the vapours being mostly unsaturated.

VARIATION OF MOBILITY WITH PRESSURE.

Positive Ions.

37·2. All observers agree that down to a pressure of at most 1 mm. of mercury the mobilities of the positive ions are inversely proportional to the pressure, provided the electric forces acting on the ions are not so large as to make the velocities of the ion comparable with those due to thermal agitation. In the cases of air, hydrogen and helium this has also been tested and found to be approximately true for pressures much exceeding atmospheric. Kovarik[2] made experiments on the first two gases with pressures up to 70 atmospheres and Dempster[3] tested air up to 102 atmo-

Observer		Gas	Range of pressure	Max. $\overset{+}{up}$	Min. $\overset{+}{up}$
Langevin ..	1903	Air	1435–75 mm.	1·46	1·40
Wellisch ..	1909	Air	760–100 ,,	1·56	1·49
		CO	360–35 ,,	1·16	1·03
		CO_2	750–60 ,,	·82	·78
		N_2O	760–50 ,,	·86	·80
Lattey	1910	Air	29–14 ,,	1·52	1·44
Todd	1911	Air	9–1·1 ,,	1·38	1·30
		Air	12–4 ,,	1·43	1·30
		SO_2	3–0·5 ,,	·59	·53
		CO_2	9–3 ,,	1·04	·95
Lattey and Tizard	1912	H_2	72–10 ,,	5·21	4·89
		CO_2	13–4 ,,	·86	·82
Kovarik ..	1912	Air	75–13 atm.	1·39	1·30
		H_2	70–20 ,,	6·38	6·01
Dempster ..	1912	Air	102–5 ,,	1·42	1·28
McLennan and Keys	1915	Air	181–67 ,,	1·71	1·32

$\overset{+}{u}$ is the velocity for an electric force of one volt per cm. and p the pressure in atmospheres.

[1] K. Przibram, *Wiener Ber.* cxviii. p. 331, 1909.
[2] Kovarik, *Phys. Rev.* xxx. p. 415, 1910; *Proc. Roy. Soc.* lxxxvi. p. 154.
[3] Dempster, *Phys. Rev.* xxxiv. p. 53, 1912.

spheres, while Todd[1] verified the law for air down to a pressure of 1 mm. McLennan and Keys[2] tested air up to 181 atmospheres and found that the mobility was somewhat greater at high pressures than if it were inversely proportional to the pressure. McLennan and Evans[3] worked with helium up to 81 atmospheres, and found variations from the inverse pressure law. The results obtained by different observers are given in the above table, for which we are indebted to Professor Fulcher.

These results show that $\overset{+}{p}u$ is approximately constant

for air from 102 atm. to 1·1 mm.
,, H_2 ,, 70 ,, 10 ,,
,, CO ,, 1 ,, 35 ,,
,, CO_2 ,, 1 ,, 4 ,,
,, SO_2 ,, 3 mm. to ·5 ,,
,, N_2O ,, 1 atm. to 50 ,,

At very low pressures Todd found a very considerable increase in the product, $\overset{+}{p}u$. The positive ions he used were generated by heating a platinum strip covered with aluminium phosphate. He used Method V and found that at these low pressures the values found for the mobilities depended upon the rate of reversal of the electric forces; with high rates of reversal the mobilities came out higher than with low ones. He found also that the increase in $\overset{+}{p}u$ at low pressures depended upon the size of the apparatus, it was much more marked when the distance between the plates was 5 than when it was 20 centimetres. He suggests that these results indicate that when the positive ion is first formed it is a simple molecule or atom and that the growth of this into a cluster takes an appreciable time, thus the mobility of the ion when young would be greater than when it is old. In the experiments when the rate of alternation of the electric force was very rapid or when the distance between the plates was small, the ion had not time to form a cluster and so possessed more than its normal mobility. On the theory of the mobility of ions given in §§ 39·1–

[1] Todd, *Proc. Camb. Phil. Soc.* xvi. pp. 21, 653; *Phil. Mag.* xxii. p. 791; xxv. p. 163, 1916.
[2] McLennan and Keys, *Phil. Mag.* xxx. p. 484, 1915.
[3] McLennan and Evans, *Proc. Roy. Soc. Canada*, xiv. p. 19, 1921.

39·4, the effect of clustering is confined to that due to the increase in mass, and as the mobility of a single molecule would only be $\sqrt{2}$ times that of the most complicated cluster, the value of $\overset{+}{pu}$ could not increase by more than about 50 per cent.

Becker[1], who used the blast method, found that in air the mobilities of ions whose average age was ·3 sec. were about 20 per cent. greater than those whose age was ·9 sec.

Erikson[2] has shown that the mobility of freshly formed positive ions in air is about $\sqrt{2}$ times that of positive ions which have lived for a considerable fraction of a second. He found however no change in the mobility of the negative ions with age; his experiments were made at atmospheric pressure.

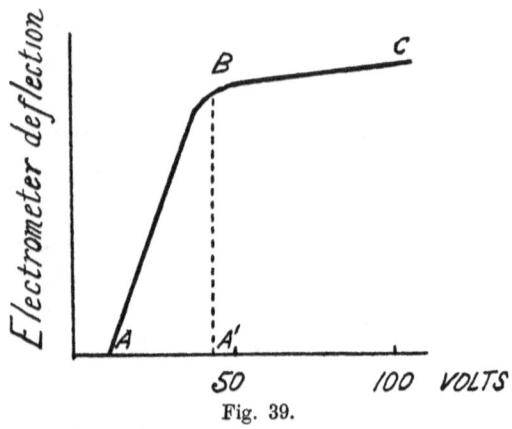

Fig. 39.

Ratner, using an electric wind method, found that $\overset{+}{pu}$ was constant in air down to 5 mm. Loeb[3], who used fields of the order of 10,000 volts per cm., found no sign of change between 304 cm. and atmospheric pressures. The values were the same as those found with small fields. Wellisch found $\overset{+}{pu}$ a constant down to ·05 mm. with a field of 1·15 volt/cm. He used air, CO_2 and hydrogen.

Altberg considers that he has been able to detect even at atmospheric pressures ions whose mobility is greater than twice the normal, and ascribes the higher velocities to freshly formed ions. He used the blast method (see p. 90) and the relation he

[1] Becker, *Ann. der Phys.* xxxvi. p. 208, 1912

[2] Erikson, *Phys. Rev.* xx. p. 117, 1922; xxiv. pp. 502, 622, 1924; xxvi. pp. 465, 629, 1925.

[3] Loeb, *Proc. Nat. Acad.* ii. p. 345, 1916.

obtained between the deflection of his electrometer (the ordinates) and the electric force (the abscissæ) is represented by the curve ABC in Fig. 39; if the mobilities of the ions were all the same and the experiment working ideally the curve should be of the type $A'BC$. As however any irregularity in the velocity of the air blast would produce a curve like ABC even if the mobilities were all equal, we cannot regard the existence of positive ions with such great mobilities at ordinary pressures as established.

The interpretation of the results of experiments on the mobilities of ions at very low pressures is somewhat ambiguous, because the velocities which the rays acquire under the electric field are so large that they are no longer negligible in comparison with the velocities due to thermal agitation, and when this is the case the ordinary theory ceases to apply.

Variation of the Mobility of Negative Ions with Pressure.

37·3. Langevin's experiments, made as long ago as 1902, established the fact that while there is a considerable range of pressure over which the mobility varies inversely as the pressure, yet at lower pressures than about 2 cm. in the case of air, the mobility when the pressure is diminished increases much faster than is indicated by this law, so that pu the product of the pressure and the mobility increases as the pressure diminishes. How large this increase is may be seen from the following table taken from Langevin's paper[1]; p represents the pressure in centimetres of mercury and k the velocity of the ions through air under the electric force of 300 volts per centimetre.

Negative ions			Positive ions		
p	k	$pk/76$	p	k	$pk/76$
7·5	6560	647	7·5	4430	437
20·0	2204	580	20	1634	430
41·5	994	530	41·5	782	427
76·0	510	510	76·0	480	420
142·0	270	505	142·0	225	425

[1] Langevin, *Ann. de Chim. et de Phys.* xxviii. p. 289, 1903.

Kovarik[1] has carried the determination of the mobility of the negative ions in air and CO_2 down to a pressure of 1 cm. and has found that at these low pressures the value of pu increases with great rapidity; his results are represented in Fig. 40.

Lattey[2], who worked with very carefully dried air, found that the increase of pu was very much more marked than in air which contained an appreciable amount of moisture. Lattey and Tizard[3] found a rapid increase of pu at low pressures in dry hydrogen and CO_2. They used Method V a.

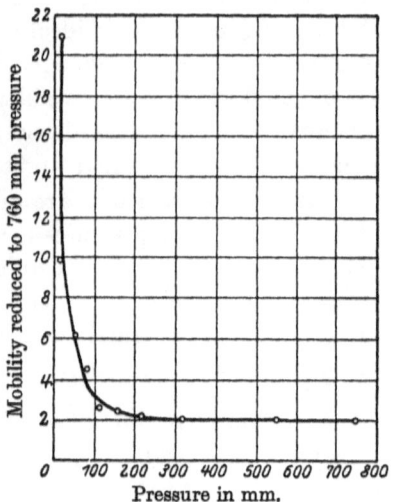

Fig. 40.

The abnormal increase in the velocity of the negative ion may very easily be explained by supposing that at low pressures the negative ion is an electron for part of its life. Owing to the small mass of the electron its mobility will be very large compared with that of the normal negative ion, and if the ion is a free electron for even a small part of the time its average mobility will be large.

Difference between the Mobilities of Positive and Negative Ions.

37·4. A very interesting property of the ions in dry air and other gases was discovered by Zeleny in 1900; it is that in these gases the mobility of the negative ion is greater than that of the positive. The numerous determinations of mobilities which have been made since that time have furnished other examples of it, in fact it seems to be a property possessed by all elementary gases. The following table gives the ratio r of the mobility of the negative to that of the positive ion at the same temperature and pressure for all the elementary gases whose mobilities are known.

Gas	H_2	He	N_2	O_2	Argon
r	1·29	1·24	1·46	1·38	1·24

[1] Kovarik, *Phys. Rev.* xxx. p. 415, 1910.
[2] Lattey, *Proc. Roy. Soc.* lxxxiv. p. 173, 1910.
[3] Lattey and Tizard, *ibid.* lxxxvi. p. 349, 1912.

An inspection of the tables of mobilities given on pages 124, 125 will show that for compound gases r is very much nearer unity than for elementary ones: thus r for CO_2 is 1·17, for CO $r = 1$, for N_2O $r = 1\cdot1$, for NH_3 $r = 1\cdot1$, while for ethyl alcohol it is slightly less than one; there is no compound whose mobilities have yet been measured for which r is as large as its least value for an elementary gas.

Erikson finds that the mobilities of the freshly formed positive and negative ions in air are equal, and the same as that of the normal negative ion. After about 1/50 sec. the former falls to its normal amount. Under certain conditions positive ions of both mobilities may be simultaneously present.

Effect of Moisture and Vapours on the Mobilities.

37·5. Zeleny[1] found that the presence of moisture produces a large diminution in the value of r: this diminution is occasioned by the diminution in the mobility of the negative ion by the moisture, the mobility of the positive ion in air being unaffected. Thus he found that the mobility of the positive ions in moist air was 1·37 as compared with 1·36 in dry air, while for the negative ions the mobility in moist air was only 1·51 as compared with 1·87 for dry air. The value of r was reduced by the moisture from 1·38 to 1·1. In hydrogen he found that moisture diminished the mobility of the positive as well as of the negative ion, though not to so great an extent. Lattey[2] and Lattey and Tizard[3] found that at a pressure of 18 mm. the presence of large quantities of moisture diminished the mobility of the positive ions by about 50 per cent. At low pressures, where the mobility no longer varies inversely as the pressure, it was found that moisture produced an enormous effect on the mobility of the negative ion, a trace of moisture decreasing the mobility to one-tenth of its value in very carefully dried gas. They found too that pu was constant to much lower pressures in moist gases than in dry.

Mayer[4] found that the mobility of both positive and negative ions in oxygen was diminished by moisture, the latter the most.

[1] Zeleny, *Phil. Trans.* cxcv. p. 193, 1900.
[2] Lattey, *Proc. Roy. Soc.* lxxxiv. p. 173, 1910.
[3] Lattey and Tizard, *ibid.* lxxxvi. p. 349, 1912.
[4] Mayer, *Phys. Zeits.* xxviii. p. 637, 1927.

Rutherford[1] found that the vapour of ethyl alcohol produced a more marked effect even than water on the value of r, while methyl iodide also produced a very marked effect.

Przibram[2] made a valuable and extensive series of experiments on the effects of vapours of liquids of various chemical types on the mobility of ions, and came to the conclusion that the vapours of alcohols, fatty acids and chloroform decrease $\bar{u}$ more than $\overset{+}{u}$, that on the other hand the vapours of the fatty esters decrease $\overset{+}{u}$ more than $\bar{u}$, while hydrocarbons, carbon tetrachloride and ethyl iodide affect the mobility merely by increasing the density of the medium. Those substances seem to produce the greatest effects which are most associated when in the fluid condition and whose vapours have abnormally high specific inductive capacities, *i.e.* which have a finite electrostatic moment.

Bělǎr[3] has confirmed some of these results and finds that formaldehyde and propionic aldehyde have a considerable effect on the negative ions, and ammonia on the positive.

Wellisch[4] also tried the effects of adding organic vapours to air; he found that a vapour pressure of 10 mm. of ethyl alcohol reduced $\overset{+}{u}$ to ·91, $\bar{u}$ to 1·10; acetone also affected both mobilities to a considerable extent, ethyl bromide reduced $\overset{+}{u}$ but not $\bar{u}$, while methyl iodide had no effect; this is opposed to Rutherford's observations that methyl iodide had a very considerable effect.

Tyndall and Phillips[5] have made a detailed study of the effect of vapours of the normal aliphatic alcohols on the mobility of negative ions in air: the diminution in the mobility increases with the position of the alcohol in the series, the effects of the higher alcohols being very marked. Similar effects were observed for the mobilities of the positive ions, though these were not so well marked as those for the negative.

[1] Rutherford, *Phil. Mag.* ii. p. 210, 1901.
[2] Przibram, *Wien. Ber.* cxviii. p. 331, 1909.
[3] Maria Bělǎr, *ibid.* cxxx. p. 373, 1921 (quoted by Przibram, *Handbuch der Physik*, xxii. p. 331).
[4] Wellisch, *Phil. Trans.* ccix. A, p. 249, 1909.
[5] Tyndall and Phillips, *Proc. Roy. Soc.* cxi. p. 577.

Mayer[1] has found that traces of chlorine diminish the mobility of ions of either sign in hydrogen and oxygen, but Erikson[2] finds no effect on air ions.

Effect of traces of Electronegative Gases on $\bar{u}$.

37·6. Even more remarkable than the effect produced by moisture is that produced by exceedingly minute traces of the electronegative gases. In 1900 Warburg[3] found that the removal of the last traces of oxygen from nitrogen increased the current from a negatively electrified point in that gas fifty times. Chattock and Tyndall[4] found that a trace of oxygen greatly diminished the mobility of the negative ions produced by the discharge from a fine point in hydrogen; the value of u increased from 7·6 to 206 as the percentage of oxygen diminished from one to zero.

Franck, who ionised the gas by α-rays from polonium, found[5] that while the mobility of negative ions in argon which contained 1 per cent. of oxygen was only 1·7, it increased to 206 when great care was taken to remove every trace of that gas; he subsequently obtained a similar result with nitrogen, the mobility of the negative ion in that gas increasing to 145 when the oxygen was removed. In pure helium Franck and Gehlhoff found that $\bar{u}$ reached 500 cm./sec. On the other hand hydrogen did not show any increased mobility on purification, nor did oxygen. Chlorine has the same effect as oxygen.

Haines[6] found large mobilities, up to 509, in carefully purified nitrogen, and also in hydrogen, contrary to the last mentioned authors. Very small traces of electronegative gases destroyed the effect. Yen[7] also finds abnormally high mobilities in nitrogen and hydrogen. Loeb[8] has made quantitative measurements of these high mobilities, and finds that they depend on the strength

1 Mayer, *Phys. Zeits.* xxviii. p. 637, 1927.
2 Erikson, *Phys. Rev.* xxx. p. 339, 1927.
3 Warburg, *Ann. der Phys.* xi. p. 295, 1900.
4 Chattock and Tyndall, *Phil. Mag.* xix. p. 449, 1910.
5 Franck, *Verh. Deutsch. Phys. Gesell.* xii. p. 291, and p. 613.
6 W. B. Haines, *Phil. Mag.* vi. 30, p. 503, 1915; 31, p. 339, 1916.
7 Yen, *Phys. Rev.* p. 337, 1918.
8 Loeb, *Phys. Rev.* xix. p. 244, 1922; xx. p. 398, 1922; xxiii. p. 157, 1924.

of the field. For vanishingly small fields he finds at atmospheric pressure:

Nitrogen	30,500	cm./sec. per volt/cm.
Hydrogen	7,800	,, ,,
Helium	22,000	,, ,,

He gives empirical formulae for the mobility as a function of X/p, where X is the field strength. Wahlin[1] finds 18,000 for nitrogen under the above conditions and the variation with the field follows a theoretical relation of Compton's[2]. Owing to the great difficulty of getting the gas perfectly pure the above results are probably not very accurate, but they undoubtedly give the order of magnitude; see also mobility of electrons in flames, Chap. X. Wahlin[3] finds that CO must be added to the list if very carefully purified. The highest mobility for a weak field was 11,800. It follows that the older measurements of $\bar{u}$ for these gases refer to the mobility of negatively charged molecules, or larger aggregates, of some impurity.

It would seem that the electrons attach themselves with great ease to the electronegative gases such as chlorine and oxygen, while they only unite with atoms of argon, helium and nitrogen with great difficulty, if at all. A further illustration of this is given by experiments made with positive rays or Kanalstrahlen; among these are found molecules of oxygen and chlorine carrying a negative charge, but as far as our experience goes a negative charge is never found even on atoms of helium, nitrogen or argon.

Dependence of Mobility on Field Strength.

37·7. Besides the above-mentioned variation of the mobility of free electrons in certain gases, with the strength of the field, several other workers have found a similar effect. Thus Franck, using the ions derived from the high tension discharge from a wire (see p. 97), found abnormally large values for the mobility of both positive and negative ions. Chattock however and Ratner, who both used large fields, found normal values for the positive ions, and it seems possible that Franck's result was due to ionisation

---

[1] Wahlin, *Phys. Rev.* xxiii. p. 169, 1924.
[2] K. T. Compton, *ibid.* xxii. p. 333, 1923.
[3] Wahlin, *ibid.* xxi. p. 517, 1923.

throughout the volume of the gas by easily absorbable radiation from the region of intense ionisation near the wire.

In the case of negative ions at pressures below atmospheric there is no doubt that the apparent mobility depends on the field. Thus Lattey (Method V a), Townsend and Tizard (Method X), and Ratner (Method XI) all find effects of this kind in air. Lattey, working at pressures (p) round 20 mm. and forces (X) of 1·6 to 2·7 volts per cm., finds an *increase* of the mobility with X/p. Townsend and Tizard, working at from ·25 to 18·5 mm. and X from 2 to 50 volts per cm., find a *decrease* with increasing X/p. Ratner[1], working at atmospheric pressures and below, found a maximum mobility for a certain value of X (620 volts/cm. at 100 mm.) but some of the results could not be expressed as a function of X/p alone. In all these cases the mobilities are abnormally large for the pressure, being from 10 to 1000 times the normal if reduced to atmospheric pressure and forces of 1 volt/cm. on the assumption that the velocities varied as X/p.

While there is no doubt that we have to do in this case with ions which are electrons for part of the time, the exact interpretation is a matter of some difficulty. Assuming that the ions start as electrons, as is certainly the case when the ionisation is due to X-rays or the photoelectric effect, the variation of this mobility with X and p may be due to any one of the following causes: (1) The electron may after a time attach itself to a molecule (perhaps of an impurity) and remain so attached for the rest of its life. (2) The ion may alternate more than once during the time it is under observation, between being a free electron and being a charged molecule or group of molecules. (3) The electron may have its velocity reduced by inelastic collisions whose nature depends on the energy of collision; this view is taken by Townsend. (4) If the velocity of the electron due to the field approaches that due to thermal agitation it will, on any theory, cease to be proportional to the force. Of course some of these causes may be acting simultaneously, and their relative importance may vary in different conditions.

It seems practically certain that either (1) or (2) must take place for some conditions at least. The difference between the

[1] Ratner, *Phil. Mag.* vi. 32, p. 441, 1916.

velocities of the positive and negative ions in most gases at high pressures is far too small to let us suppose that the latter is a free electron for any considerable part of the time. On the other hand it is certain that it is one, at low pressures under large forces. Under intermediate conditions, when the 'abnormal' mobility is beginning, it must be an electron for part, but not all, of its life.

Wellisch[1], using Method V a and ionising the gas by polonium, made the very interesting discovery that when the pressure of the gas is low enough to make the mobility of the negative ion abnormal, an appreciable fraction of the carriers of negative electricity are electrons from start to finish. The variations of the mobility of the negative ion with pressure, so far as it is abnormal, may be regarded as due to variations in the fraction of the whole number of carriers which cross as electrons. Fig. 41 is a reproduction of a curve given by Wellisch; it represents the relation between the quantity of negative electricity received by the plate (the ordinate) and the voltage between the plates (the abscissa), the rate of reversal of the voltage being constant. The point A represents the stage at which the electrons get across, and B that at which ions whose mobility is normal (i.e. has the value determined by experiments at high pressures) can get across.

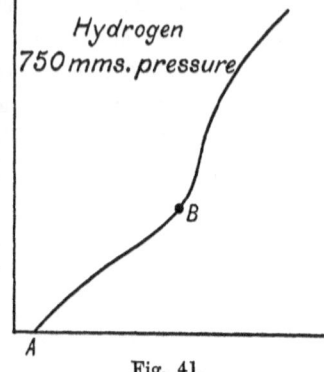

Fig. 41.

Wellisch interpreted the curves as showing that the carriers of electricity retained during the whole of their journey the state in which they started, i.e. if they started as electrons they remained so, if they started as ions they remained ions. This would mean that an electron never becomes associated with a molecule, which is contrary to evidence from other sources and is, as we shall see, not a necessary consequence of Wellisch's curves.

37·71. We shall see, in fact, that on either of assumptions (1) or (2) most of the usual methods of measuring mobility break down. Take first assumption (1). The case we shall consider is when the charged particles move between two parallel plates under a uniform

[1] Wellisch, *American Journal of Science*, May 1915.

electric force at right angles to the plates; these are the conditions under which the mobility of the negative ion is usually determined. We shall suppose that in their passage between the plates some of the electrons become attached to molecules and form negative ions. The first problem we shall investigate is that of finding the expectation of an electron starting from one plate reaching the other without uniting with a molecule to form a negative ion. We shall suppose that when an electron strikes a molecule the chance of its uniting with the molecule to form a negative ion is $1/n$. We shall denote the mobility of the electron by K, then if X is the value of the electric force, KX is the velocity of drift of the electron. We shall suppose that this is small compared with V the velocity of the electron due to thermal agitation.

To calculate the chance of the electron passing over a distance x parallel to the electric force without becoming attached to a molecule, we proceed as follows: the time taken to pass over this distance is x/KX and in this time the length of the path described by the electron is Vx/KX if V is large compared with KX; we shall denote this distance by ax, where $a = V/KX$.

By the Kinetic Theory of Gases the expectation of the electron traversing this distance without a collision is $\epsilon^{-ax/\lambda}$, where λ is the mean free path of the electron; this is also the expectation that it should travel over this distance without either making a collision or becoming a negative ion.

The chance that it makes one collision and no more in passing over a distance x may be found as follows: the chance that it makes a collision between ξ and $\xi + d\xi$ is $a d\xi/\lambda$, the chance that it reaches ξ without a collision is $\epsilon^{-a\xi/\lambda}$, and the chance that it does not make a collision during the rest of the journey is $\epsilon^{-a(x-\xi)/\lambda}$, hence the chance that a collision should occur between ξ and $\xi + d\xi$ and that this should be the only one is

$$\frac{a d\xi}{\lambda} \epsilon^{-a\xi/\lambda} \epsilon^{-a(x-\xi)/\lambda} = \epsilon^{-\frac{ax}{\lambda}} \frac{a}{\lambda} d\xi;$$

the chance that it should make one and only one collision in a distance x is therefore

$$\int_0^x \epsilon^{-\frac{ax}{\lambda}} \frac{a}{\lambda} d\xi = \epsilon^{-\frac{ax}{\lambda}} \frac{ax}{\lambda}.$$

The chance that this collision does not result in the union of the electron and the molecule is $1 - \dfrac{1}{n}$, the expectation of the electron getting across with one collision is

$$\epsilon^{-\frac{ax}{\lambda}} \frac{n-1}{n} \frac{ax}{\lambda}.$$

Consider now the electron which makes two collisions and no more. The chance that the electron makes just two collisions in its journey is

$$\int_0^x \epsilon^{-a\xi/\lambda} \frac{a\xi}{\lambda} \epsilon^{-a(x-\xi)/\lambda} \frac{a d\xi}{\lambda},$$

for $a d\xi/\lambda$ is the chance that it makes a collision between ξ and $\xi + d\xi$, $\epsilon^{-a\xi/\lambda} a\xi/\lambda$ the chance that it makes one collision and no more before reaching ξ, and $\epsilon^{-a(x-\xi)/\lambda}$ the chance that it makes the rest of the journey without a collision. The value of the integral is

$$\epsilon^{-\frac{ax}{\lambda}} \frac{1}{2} \left(\frac{ax}{\lambda}\right)^2;$$

the chance that neither collision should result in the combination of the ion and molecule is $\{(n-1)/n\}^2$.

Hence the expectation of a two collision electron crossing in a free state is

$$\epsilon^{-ax/\lambda} \frac{1}{2} \left(\frac{n-1}{n} \frac{ax}{\lambda}\right)^2.$$

Similarly the expectation of a three collision electron crossing in a free state is

$$\epsilon^{-ax/\lambda} \frac{1}{2.3} \left(\frac{n-1}{n} \frac{ax}{\lambda}\right)^3,$$

and so on. Hence since all the electrons must make an integral number of collisions between nought and infinity, the expectation of an electron crossing in a free state is

$$\epsilon^{-ax/\lambda} \left\{ 1 + \frac{n-1}{n} \frac{ax}{\lambda} + \frac{1}{2} \left(\frac{n-1}{n} \frac{ax}{\lambda}\right)^2 + \frac{1}{2.3} \left(\frac{n-1}{n} \frac{ax}{\lambda}\right)^3 + \ldots \right\}$$
$$= \epsilon^{-ax/\lambda} \epsilon^{(n-1) ax/n\lambda} = \epsilon^{-ax/n\lambda}.$$

Hence if N electrons start from the plate $x = 0$, the number which reach the plane $x = d$ in the free state is

$$N\epsilon^{-ad/n\lambda}.$$

In many of the experiments on the mobilities of the negative carriers the quantity measured is the charge of electricity received within the time T by a plate $x = d$, by ions starting from the plane $x = 0$ at the beginning of the time. Some ions may get across even though they only travel part of the way as electrons, completing the journey as negative ions. In this case x the smallest distance they can travel as free electrons is given by the equation

$$\frac{x}{KX} + \frac{d-x}{k_2 X} = T \quad \text{or} \quad x = \frac{d - k_2 XT}{1 - \frac{k_2}{K}} \dots \dots (1),$$

where k_2 is the mobility of the negative ion; all the electrons traversing the distance x given by this equation will arrive in time, no others will do so, hence if N electrons start at $t = 0$ from $x = 0$ the number which reach the plate $x = d$ within the time T will be equal to
$$N\epsilon^{-ax/n\lambda},$$
where x has the value defined by equation (1). Since $a = V/KX$, $ax/n\lambda$ may be written $\frac{\beta}{X}(d - k_2 XT)$, where $\beta = V/(K - k_2) n\lambda$, and the number reaching the plate will be

$$N\epsilon^{-\frac{\beta}{X}(d-k_2XT)}.$$

The smallest value of X which will drive any electrons to the plate is given by $X = d/KT$; when X has this value the number of electrons reaching the plate is

$$N\epsilon^{-\frac{VT}{n\lambda}};$$

when X is equal to or greater than $d/k_2 T$ all the electricity which leaves $x = 0$ will reach $x = d$, and the charge received by the plate will be Ne, where e is the charge on an ion. The graph of the charge received by the plate and the electric force X is represented by $ABCD$ in Fig. 42, the curves I and II representing the limiting form, at very low and very high pressures respectively. There is a

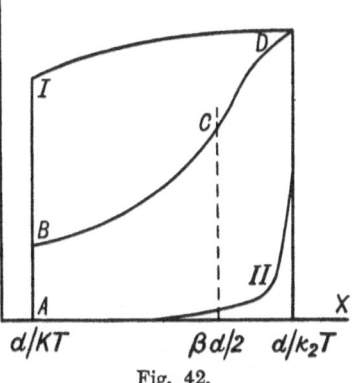

Fig. 42.

point of inflection where $X = \beta d/2$, if this falls between d/KT and d/k_2T. If Q_1 and Q_2 are the charges received by the plate for values X_1, X_2 of X, we see from the preceding equation that

$$\log (Q_1/Q_2) = \beta d \left(\frac{1}{X_2} - \frac{1}{X_1}\right),$$

an equation which would enable us to determine $n\lambda$ if K and k_2 are known. In this case since some ions get across for all values of X greater than d/KT, there is no definite value for the mobility, there are carriers possessing any mobility between K and k_2.

The value of X which brings half the charge across is given by

$$\frac{\beta}{X}(d - k_2 XT) = \cdot 693;$$

regarding d/XT as a kind of mean mobility, and substituting from the equation, we find

$$d/XT = \frac{k_2}{1 - \cdot 693 X/\beta d}.$$

Thus this mean mobility is equal to k_2 until $\cdot 693X/\beta d$ becomes appreciable; it rapidly increases however when this quantity becomes comparable with unity. Substituting the value for β we find that

$$\cdot 693 X/\beta d = \cdot 693 X (K - k_2) n\lambda/Vd.$$

Since both λ and $(K - k_2)$ are inversely proportional to p, the pressure of the gas, this is of the form X/Vdp^2, so that the ratio of this mean mobility to its normal value k_2 is a function of X/Vp^2d; it thus depends on the value of the electric force and the distance between the plates as well as the pressure, thus for example if the distance and the pressure remain constant the mean mobility will increase with the electric force.

In some of the methods for determining the mobility we have instead of a flash of ionisation at the time $t = 0$ a continuous supply of ions extending over the whole time T. If $k dt$ is the number of electrons emitted from the plate $x = 0$ between t and $t + dt$, the charge of electricity they will give to the plate will be

$$ek\,dt\,\epsilon^{-\frac{\beta}{X}\{d - Xk_2(T-t)\}};$$

the whole charge will be the integral of this. We have however to consider the limits of integration. The lowest value of $T - t$

will always be d/KX. The upper limit will depend on the value of $d - Xk_2T$; if this is positive the upper limit of $T - t$ will be T and, performing the integration, the quantity of electricity received by the plate will be

$$\frac{ek}{\beta k_2}\left\{\epsilon^{-\frac{\beta(d-Xk_2T)}{X}} - \epsilon^{-\frac{Vd}{n\lambda}\frac{1}{KX}}\right\} \quad \ldots\ldots\ldots\ldots\ldots(2).$$

When $d - Xk_2T$ is negative, then until $T - t = d/k_2X$, all the electrons leaving the plate $x = 0$ will give up their charge; the amount given in this time is $ek\left(T - \dfrac{d}{k_2X}\right)$, the amount for the rest of the time is

$$\frac{ek}{\beta k_2}\left(1 - \epsilon^{-\frac{Vd}{n\lambda}\frac{1}{KX}}\right).$$

Thus in this case the quantity Q received by the plate will be

$$ek\left(T - \frac{d}{k_2X}\right) + \frac{ek}{\beta k_2}\left(1 - \epsilon^{-\frac{Vd}{n\lambda}\frac{1}{KX}}\right).$$

There is no discontinuity in either Q or dQ/dX when X passes through the value d/k_2T, but d^2Q/dX^2 is discontinuous at this point, the change being equal to

$$\frac{ek}{k_2}\frac{\beta d^2}{X^4},$$

where X has the critical value d/k_2T.

If some of the negative carriers started from the plane $x = 0$ as negative ions, *i.e.* if the agent employed to ionise the gas produces at the instant of ionisation negative ions as well as electrons, there will be a discontinuity in dQ/dX as well as in d^2Q/dX^2 when $X = d/k_2T$.

We see from these curves that when the mobility of the negative ion becomes abnormal, *i.e.* as soon as the pressure is low enough to make the curve depart appreciably from the type (II), the mobility not merely becomes abnormal, it becomes indefinite. If we use a method which depends on determining the electric force when the charge first becomes appreciable, we shall get values which depend on the sensitiveness of the electrometer; if we use methods which depend on finding the value of X when the charge reaches its maximum value, we shall get quite a different set of values which would also depend very appreciably on the sensitiveness of the electrometer.

Similarly if the mobility of a carrier oscillates between very different values, as supposed in assumption (2), p. 135, no reliance can be placed on results found by methods such as V or Va when the mobility is measured by the greatest distance which can be traversed in an alternating electric field. For, to take the case of a negative ion oscillating between an electron and a cluster: suppose that as an electron it is being driven towards the plate connected with the electrometer by the electric field, but that before it reaches the plate the field is reversed; if it remains an electron it will under the reversed field lose the ground it previously gained: if however when the field is reversed it joins a cluster, its mobility will be much reduced and it will only retreat a short distance. If when the electric force is reversed again it becomes an electron it will advance much further than it retreated and thus may be able to gain the plate connected with the electrometer by a succession of stages, whereas the calculation of the mobility by that method is founded on the assumption that the ion reaches the plate in one stage. If the reversals of the electric field are so infrequent that the ion has time to change from the electron to the cluster many times while the electric force is acting in one direction, this effect will not be serious; but if the rate of reversal of the electric force is so rapid that the time the force acts in one direction is comparable with the time an electron is free or in a cluster, the effect may be very large and the values of the mobilities quite erroneous.

37·72. From experiments made on similar lines to that suggested by the above investigation Loeb[1] and Wahlin[2] find the following values for n, the number of collisions the electron makes before being captured.

Loeb used Method V, with electrons derived by the action of ultra-violet light on one plate of the condenser, and interpreted the variation of the charge received by the electrometer with

[1] Loeb, *Phys. Rev.* xvii. p. 89, 1921; *Phil. Mag.* xliii. p. 229, 1922.
[2] Wahlin, *Phys. Rev.* xix. p. 173, 1922.

potential difference, pressure, and frequency of commutation, on the theory given above.

Gas	n
N_2	Infinite
H_2	Infinite
CO	1·0 to 6·0 × 10⁷*
NH_3	0·7 to 1·6 × 10⁷
CO_2	2·3 to 4·3 × 10⁶
N_2O	0·8 to 6·4 × 10⁵
Air	0·7 to 6·4 × 10⁴
O_2	1·4 to 5·7 × 10³
Cl_2	less than 2·1 × 10³

* Since shown to be infinite.

He found that the number of collisions made by an electron before union took place was independent of the pressure of the gas. The experiments agreed well with the above theory. He found no sign of the permanently free electrons considered by Wellisch.

No great weight can be attached to the numerical values given in the above table, as they depend on knowing the mobility of the free electrons and their free path. Both these are uncertain and the latter in particular is probably a complicated function of the speed of the electron. It is also possible that n is a function of the field strength. Thus Bailey[1], using a development of Townsend's method for finding the free path, finds that in dry air $\frac{1}{n}$ depends on the field strength Z (volts per cm.) and pressure p (in mm.) as follows:

Z/p	·5	1	2
$\frac{1}{n}$	3·3 × 10⁻⁶	2 × 10⁻⁶	·7 × 10⁻⁶

These are larger values of n than found by Loeb, but Z/p was smaller in the latter case, where the experiments were made at atmospheric pressure. Nielsen[2], in some experiments on the attachment of electrons to mercury atoms to form negative ions,

[1] Bailey, *Phil. Mag.* l. p. 825, 1925.
[2] Nielsen, *Phys. Rev.* xxvii. p. 716, 1926.

finds also that the chance of attachment in general diminishes with increasing energy of the electron, but becomes abnormally large at certain critical values.

37·73. A somewhat similar method might be used to investigate the time a molecule retained an electron after capturing it, a very important physical constant about which we have no information.

If instead of starting a swarm of electrons from the plate A we start one of negative ions, then if the negative ions did not dissociate in their journey from one plate to the other, they would all get across in the same time, and the electrometer would not receive any charge until $t = \dfrac{d}{k_2 X}$, and then the charge would be constant whatever the value of t. If however on the journey a negative ion broke up into an electron and a molecule, since the mobility of the electron is much greater than that of the ion, the negative charge would reach the electrometer sooner than if no dissociation had taken place and charges will arrive before $t = d/k_2 X$.

If the negative charge is to get across in a time less than $d/k_2 X$ it must decompose into an electron and a molecule on its journey; if x is the greatest distance it can go before decomposing, then

$$\frac{x}{k_2 X} + \frac{d-x}{KX} = t.$$

If T_1 is the life of the negative ion the chance that it travels over a distance x without splitting up is

$$\epsilon^{-\frac{x}{k_2 X T_1}};$$

thus the chance that it will split up in this distance is

$$1 - \epsilon^{-\frac{x}{k_2 X T_1}}.$$

Since
$$x\left(\frac{1}{k_2} - \frac{1}{K}\right) = Xt - \frac{d}{K},$$
this is equal to
$$1 - \epsilon^{\frac{d}{T_1 X (K - k_2)}} \epsilon^{-\frac{K}{K - k_2}\frac{t}{T_1}}.$$

Thus if N is the number of ions which start, the number which will reach the electrometer in time t is

$$N\left(1 - \epsilon^{\frac{d}{T_1 X (K - k_2)}} \epsilon^{-\frac{K}{K - k_2}\frac{t}{T_1}}\right),$$

an equation which gives the means of finding T_1. No experiments of this kind seem to have been made; it would be a matter of great interest to know how T_1 was affected by pressure, temperature and the nature of the gas. The decomposition of the negative ion into the electron and molecule is a monomolecular reaction of an interesting type, because the electron, one of the products of dissociation, has a mass very small compared with the mass of the other product, the molecule. Owing to this, very little energy will be communicated to the electron by collisions with other molecules. Thus we should expect collisions with molecules to have much less effect on the rate of this reaction than in ordinary chemical reactions when the masses concerned are not greatly different from each other.

In the absence of experiments the only light that can be thrown on this question is from thermodynamical considerations. The energy required to detach an electron from a neutral molecule is always several volts. If we assume that the energy to detach one from a negatively charged molecule is at least of the order of a volt, then if there is a possibility of the complex ion forming at all, since the energy of thermal agitation at room temperature is only about $\frac{1}{30}$ volt, in the equilibrium state almost all the electrons will be attached to molecules, and the process of dissociation will be even slower than that of the formation of the complex. Thus once the ions have become attached to a molecule they will remain so, and the first of our assumptions will hold rather than the second, but this point must be regarded as still quite unsettled. The question to be considered is whether this idea of ions starting as free electrons and then becoming loaded up by combination with a molecule of the gas, is sufficient to account for the results obtained at low pressure. We have seen that on this view the average speed of an ion would not be a function of X/p alone, and it is important to see if it is so in practice. Ratner found that it was not. Lattey's experiments are more doubtful. He found that in the case of a carefully dried gas at pressures low enough to make pu abnormally high, the velocity of the negative ion was not proportional to the electric force but increased much more rapidly than the force acting upon it. The magnitude of the effect may be seen in the following table taken from his paper[1]. Here p is the pressure in mm., X the electric force in volts per centimetre and v the velocity in cm. per sec. of the negative ion in carefully dried air.

p	X	X/p	v
14·3	·60	·0421	118
	·888	·0621	352
	1·165	·0815	1244
	1·415	·0990	2670
18·2	1·185	·0651	389
	1·42	·0809	1075
24·5	1·385	·0566	218
	1·54	·0630	339
	1·76	·0717	451
	2·03	·0829	1340
	2·33	·0952	2550

Thus we see that at the pressure 14·3 the velocity is increased tenfold when the electric force is doubled, while at the pressure 24·5 the same increase in velocity is produced by an increase of about 75 per cent. in the electric force. Theory indicates that the proportionality between velocity and electric force would cease when the increase in the velocity produced by the electric field became comparable with the average velocity due to the thermal agitation, and in this case there would be a further tendency to high mobilities, as the greater energy in the ions would make their collisions with the molecules of the gas more violent and might increase the tendency of the clusters to break up into simpler systems; but in the case just cited the increase in energy is very slight, not more than would be produced by a rise in temperature of a few degrees centigrade.

In the region of pressures where pu is constant the velocity of the ion remains constant when X/p is constant, and in this case is proportional to X/p; this proportionality ceases when the pressure is so low that pu is above its normal value, but Lattey considers that in this region when X and p are altered so that their ratio remains constant the velocity u of the ion is unaffected. The evidence for this is set out in the following table.

[1] Lattey, *Proc. Roy. Soc.* lxxxiv. p. 179, 1910.

p mm.	$X/p=0.04$	0.05	0.06	0.07	0.08	0.09	0.1
	u	u	u	u	u	u	u
14.3	107	175.5	310	580	1126	(2200)	(4210)
18.4	103	163	279	514.5	1006	(2050)	(4120)
24.5	(119.5)	172.5	286	509	936.5	1799	(3480)
28.8	116	180	298.5	519.5	926	(1652)	—

The numbers in brackets were obtained by extrapolation. There are variations of nearly 30 per cent. in the values of the velocity for the same value of X/p and the range of pressure is small.

Townsend and Tizard, using Method X, found that the velocities were a function of X/p only; but as Townsend has pointed out[1], the mean velocity found by this method, when the mass is not constant, is different from that found by a direct method. The range of forces was different from that used by Lattey and a direct comparison does not seem possible. Townsend and Bailey[2] found that the velocities in oxygen did not depend only on X/p, though those in nitrogen and hydrogen did, but in these last two cases there is no attachment of electrons to molecules, according to Loeb. The same remark applies in the case of argon and helium, which these authors investigated later. Loeb[3] has found that the value of X/p at which abnormal mobilities began in his experiments was quite different from that found by Lattey and attributes this to his pressures being different. On the whole it may be said that there is no definite evidence against the view we have given, and Loeb and Wahlin's work is strongly in its favour.

Energy of Electrons in an Electric Field.

37·8. This is perhaps the most convenient place to mention the work of Townsend and his pupils on the energy acquired by an electron in an electric field. By measurements of diffusion, using the apparatus described on p. 81, they have been able to show that in the case of a gas showing the 'abnormal' mobility, *i.e.*

[1] Townsend, *Electricity in Gases*, p. 186.
[2] Townsend and Bailey, *Phil. Mag.* xlii. p. 873, 1921.
[3] Loeb, *Proc. Nat. Acad.* ii. p. 345, 1916.

carefully dried and at a low pressure, the mean kinetic energy of the ions, electrons, may become many times that of thermal agitation, even under moderate fields where there is no appreciable ionisation by collision. This energy does not show itself as a velocity in the direction of the field, but is transformed by collisions with the gas molecules into a general spreading sideways of the beam of ions. It first showed itself as an abnormally small value of Ne when calculated in the usual way. This could be explained on the assumption that the velocities of the ions were larger than those to be expected from the energy of thermal agitation, and Townsend calculated the factor k by which this energy must be multiplied to explain the results[1]. For a force of 2 volts per cm. and a pressure of 11 mm. in dry air, k was 2·90, to give a typical example. The value of k increases greatly as the force increases and the pressure diminishes. Townsend and Bailey[2] have measured this quantity, and also the velocity in the direction of the field, in oxygen, nitrogen, hydrogen, argon and helium. In all but the first it was a function only of X/p, as is to be expected if the electrons retain their state throughout. Using a theory of Pidduck's[3] they calculate the average loss of energy in a collision between an electron and a molecule. Except in the case of helium the result is greater than could be explained by elastic collisions. In view of the large values of k found, of the order 100, it seems probable that the chief cause is the occasional loss of a large amount of energy in the exceptional collisions for which the electron has acquired sufficient energy to bring the molecule to an excited state, but the authors do not take this view.

Ions of more than one Mobility.

37·9. Several authors have claimed to have discovered that positive ions in the same gas can have mobilities of different and definite amounts, and that other negative ions exist besides the free electron and the normal ion. Thus Haines finds in hydrogen a negative ion of mobility 18·9 and various positive ions. Nolan[4]

[1] See *Electricity in Gases*, p. 172.
[2] Townsend and Bailey, *Phil. Mag.* xlii, xliii, xliv, xlvi.
[3] Pidduck, *Proc. Roy. Soc.* A. lxxxviii. p. 296, 1913.
[4] Nolan, *Phys. Rev.* xxiv. p. 16, 1924; *Proc. Roy. Irish Acad.* xxxv. p. 38, xxxvi. p. 74. Nolan and Harris, *Proc. Roy. Irish Acad.* xxxvi. p. 31.

finds seven types of ion in dry air from a mobility of 1·87 to 2·24 and others outside this range, the mobilities of positive and negative being the same. He finds similar results with moist air, CO_2 and hydrogen, in which among others he finds some mobilities agreeing with those of Haines. The most persistently appearing ions in air are those of mobilities 2·04, 1·79, 1·52 and 1·37. Erikson has found mobilities of 4·35 and 1·55 for the recoil atoms of actinium. Wahlin has found indications of two mobilities of positive ions in helium and of a number in air, some of which agree with Nolan's values.

In most of these cases the mobilities were measured either by Method V or Method III, and the various mobilities are identified by 'nicks' in the experimental curves. In many cases these nicks are not well marked, and the method is one which lays itself open to error. While the results are interesting as indicating a possibility they are not conclusive as they stand, particularly as Yen[1] and Wellisch[2] have failed to find the ions claimed by Haines, and Blackwood[3] has repeated some of Nolan's experiments and considers that his results are due to irregularities in working of the air blast and other experimental errors. Tyndall and Grindley also find no sign of more than one kind of negative ion.

This criticism does not apply to Erikson's work on the ageing of ions, for which the evidence is much more definite.

Effect of Temperature on the Mobility of the Ions.

37·91. The question of the influence of temperature on the mobility is important from the light it throws on the nature of the collisions between the ions and the molecules through which they pass. Phillips[4] made measurements of the mobility of the positive and negative ions in air between the temperatures 94° and 411° absolute. The gas was kept at constant pressure, the ions were produced by X-rays and the mobilities measured by Langevin's method. His results are given in the following table, the values

[1] Yen, *Phys. Rev.* xi. p. 337, 1918.
[2] Wellisch, *Phil. Mag.* xxxiii. p. 33, 1917
[3] Blackwood, *Phys. Rev.* xx. p. 499, 1922.
[4] Phillips, *Proc. Roy. Soc.* lxxviii. p. 167, 1906.

for constant density being obtained by multiplying those at constant pressure by $273/T$.

T	Constant pressure		Constant density	
	$+$ u	$-$ u	$+$ u	$-$ u
94	·235	·235	·68	·68
209	·945	1·23	1·233	1·61
285	1·39	1·78	1·330	1·71
333	1·60	2·00	1·312	1·64
348	1·67	2·12	1·310	1·67
373	1·81	2·21	1·324	1·62
383	1·85	2·30	1·318	1·64
399	1·95	2·40	1·334	1·64
411	2·00	2·49	1·328	1·65

These results are represented in Fig. 43. The results indicate that above the temperature of about 250 abs. the mobility at constant density is independent of the temperature.

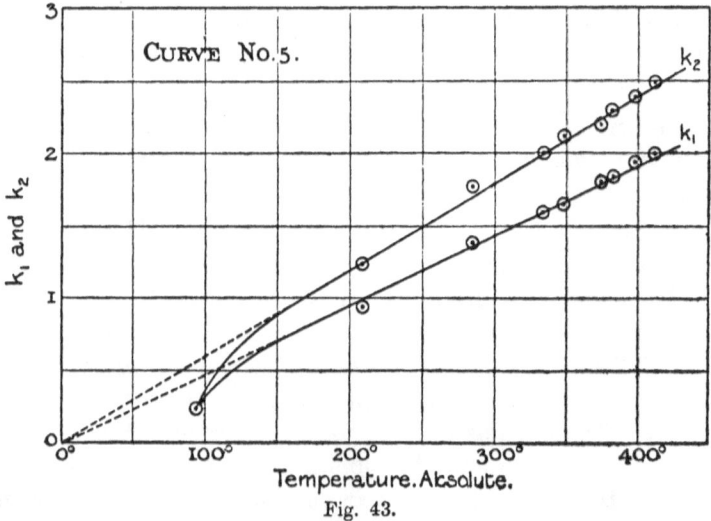

Fig. 43.

Kovarik[1] measured through a wide range of temperatures the mobility of the negative ions liberated from a metal by the

[1] Kovarik; *Phys. Rev.* xxx. p. 415, 1910.

action of ultra-violet light; the measurements were made at constant pressure, but if reduced to constant density by multiplying by $273/T$ they also indicate that the mobility at constant density is through a considerable range independent of the temperature. Kovarik's measurements are represented in Fig. 44.

Sutherland has suggested[1] that the mobility is proportional to $\dfrac{T^{\frac{1}{2}}}{a+T}$ and has shown that with a suitable value of a, Phillips' results may be represented with some accuracy by a formula of this type. We are led to this law if we suppose that the

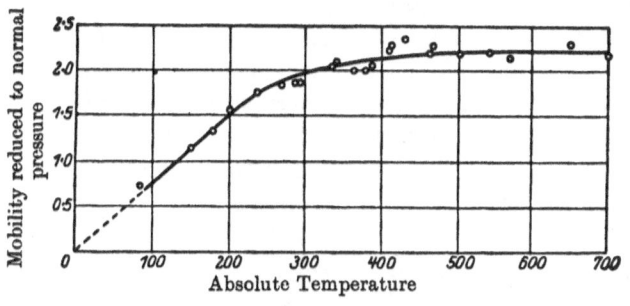

Fig. 44.

dimensions of the ion do not vary with the temperature, but take into account the effect of electrical attraction in increasing the number of collisions made by the ion and thus diminishing its free path. See below, p. 177.

Erikson[2], who has made determinations of the variations of mobility with temperature at constant density, finds that his results can be expressed by a formula of this type and that the mobility is a maximum at about room temperature for the positive ion, and at about 40° C. for the negative. Przibram has pointed out that there is a tendency for the mobilities in a gas or vapour to be abnormally small near the saturation point. Instances of this can be seen in the table, p. 125, *e.g.* methyl acetate, ethyl acetate and perhaps ethyl alcohol.

[1] Sutherland, *Phil. Mag.* xviii. p. 341, 1909.
[2] Erikson, *Phys. Rev.* iii. p. 151, 1914; vi. p. 345, 1915.

Mobility in Mixed Gases.

37·92. A very remarkable effect is that, according to most investigators, in a mixture of gases ionised by X-rays where there would be ions of different kinds, the ions have all the same mobility. Thus the mobility seems to depend on the nature of the gas through which the ions move, but not on the nature of the ion. Blanc found that the mobility of an ion made in CO_2 and driven through air was the same as that of an ion made in the air itself.

Grindley and Tyndall[1] have recently made a number of measurements in which they measured the velocity in air of ions formed in a number of other gases: H_2, NH_3, CO_2, $(C_2H_5)_2O$ and $CHCl_3$. In all cases the mobilities were within 1 per cent. of those of ordinary air ions. The accuracy of measurement reached in these experiments makes them very strong evidence. It should be mentioned, however, that Rothgieser has found that ions formed in CO_2 have only about half the mobility in hydrogen of normal ions, and conversely hydrogen ions moved abnormally fast in CO_2.

Erikson[2] finds that argon, CO_2 and hydrogen ions in air have the same mobility as air ions, and the positives age in the same way: the negative hydrogen ion is however rather more mobile than the corresponding air ion. Acetylene is exceptional in that it shows no ageing effect on the positive ion, which is always slightly less mobile than the negative, the mobility of the latter being normal. This holds whether acetylene ions are mixed with air, or a mixture of air and acetylene is ionised. If acetylene is mixed with air previously ionised, the slower ('aged') air ion disappears in about ·3 sec., leaving a positive ion of about the mobility of the negative, as when the ionisation is started in acetylene. This is strong evidence for an interchange of charge from air to acetylene.

In a later paper[3] he has found that 0·01 % of acetylene produces a marked effect, which depends on the length of time the acetylene has been mixed with the air.

[1] Grindley and Tyndall, *Phil. Mag.* xlviii. p. 711, 1924.
[2] Erikson, *Phys. Rev.* xxvi. p. 465, 1925; xxviii. 372, 1926.
[3] *Ibid.* xxx. p. 339, 1927.

The mobilities of the 'recoil atoms' from radioactive substances give further important evidence. These are heavy atoms, atomic weight over 200. Rutherford[1] found that the recoil products from radium and thorium emanation had a mobility in air of 1·3, approximately that of an ordinary positive ion. Schmidt[2] found a similar result. Franck[3] and, by a different method, Franck and Meitner[4], found for atoms of Thorium C" mobilities of 1·56 in air, 1·54 in nitrogen, and 6·21 in hydrogen. Erikson[5] alone finds *two* mobilities, 4·35 and 1·55, for the atoms from radium, thorium and actinium emanations, the former being an abnormally large value for a positive ion. It might perhaps be explained by supposing that some of the ions carried a multiple charge. In almost all these experiments the arrival of the radioactive ion was detected by the radioactivity produced on the electrode. Thus the identity of the ion was established. These results are very important, as they show that the mobility of an ion of mass at least 200 is not sensibly different from that of an ordinary ion. Also they dispose of what would otherwise be a very attractive way of explaining this curious result, namely to suppose that the charge is readily transferred from molecule to molecule, so that whatever an ion may start as, it is soon a molecule of the gas through which it is moving.

An extremely important series of experiments by Wellisch[6] has shown that when ions are formed in mixtures of sulphur dioxide and oxygen, or ethyl ether and air, only one kind of ion, of either sign, is formed in each case, all the positive ions which can be detected have the same mobility and the same holds for the negative ions. The existence of more than one type of ion would, when Langevin's method is used, give rise to a kink for each type on the curve representing the relation between the charge received by the electrometer and the time T (see p. 108). Wellisch, using this method and paying special attention to the number of such kinks which could be detected, was never able to find more

[1] Rutherford, *Phil. Mag.* v. p. 95, 1903.
[2] Schmidt, *Phys. Zeits.* ix. p. 184, 1908.
[3] Franck, *Verh. der Deut. Phys. Gesell.* xi. p. 397, 1909.
[4] Franck and Meitner, *Verh. der Deut. Phys. Gesell.* xiii. p. 671, 1911.
[5] Erikson, *Phys. Rev.* xxiv. p. 622, 1924; also xxvi. p. 629, 1925.
[6] Wellisch, *Proc. Roy. Soc.* lxxxii. p. 500, 1909.

than one, showing that the mobility of the ion is determined quite definitely by the gas through which it has to move.

He further found that the mobility of an ion formed from a heavy gas in a light one was practically the same as that of an ion formed from the light gas. This was done as follows. He found that hydrogen at atmospheric pressure was too little ionised under the conditions of experiment to give a measurable effect, while methyl iodide at 6 mm. pressure gave quite a strong one. He ionised a mixture of the two and concluded that the ions were, at least initially, ions of methyl iodide. The mobilities found however for both positive and negative ions were near those found for hydrogen ions in hydrogen. A similar result was found with ions formed in carbon tetrachloride and mercury methyl and then passed into hydrogen.

It should be mentioned that Wahlin[1] in a recent paper has come to a different conclusion. In mixtures of two of the three, hydrogen, ethyl chloride and nitrogen, he finds in each case *two* kinds of ions of mobilities in the ratio of the square roots of the molecular weights of the gases used. He used Method Va, which has been shown not to be entirely reliable, and the different kinds of ions showed up as kinks in the voltage-current curve. They do not seem very well marked in the cases illustrated in the paper, but he says that the curves could be repeated at will. These were newly formed ions, of the order of $\frac{1}{100}$ sec. life.

Erikson[2] has found that the mobilities of negative, initial positive, and 'aged' positive ions in air are decreased in the same proportion by the addition of CO_2 and increased in the same proportion by that of hydrogen.

Discussion of Experimental Results.

38·1. Though the mobility of gaseous ions has been the subject of many investigations during the last thirty years, there is still very much that is obscure or uncertain about this matter. The results obtained by different experimenters show wide variations and in some cases are contradictory. This is largely due to the

[1] Wahlin, *Phys. Rev.* xxv. p. 630, 1925.
[2] Erikson, *Phys. Rev.* xxx. p. 339, 1927.

fact that the mobilities of the ions are profoundly influenced by what might be regarded as trivial differences in the conditions under which the experiments are made. Thus, for example, the presence of a mere trace of an impurity in the gas may diminish the mobility of the negative ion to a fraction of its value in the pure gas, *e.g.* Franck found that the mobility of the negative ion in nitrogen carefully freed from any trace of oxygen was 150, while a mere trace of oxygen reduced it to 1·84. Again, it has been known from the first that the presence of water vapour diminished the mobility of the negative ion; most observers thought that it had little effect on the mobility of the positive. Blanc[1] and Laporte[2] however found that for oxygen and nitrogen the mobility of the positive ion was greater in damp than in dry gas. Busse on the contrary finds that it is much less, and he is supported by Tyndall and Grindley and by Mayer. Recent investigations by Erikson[3] have shown that at any rate in some gases the mobility of the ions when they are first formed is greater than when they have existed for some time. Thus according to him freshly formed ions in air have the same mobility whether they are positively or negatively charged, but whereas the mobility of the negative ion does not alter with age, that of the positive diminishes and reaches a value about $1/\sqrt{2}$ that of the freshly formed ion.

The mobility of the ions is thus a function of a great many variables, some of which have only recently been suspected. This and the fact that few of the methods employed can legitimately claim to have any very great precision, makes it impossible to apply any vital test to the various theories which have been proposed.

The chief interest in the values of the mobilities is the light they may throw on the nature of the ions; are these electrons, atoms, molecules or even more complex aggregates? Is the ion a stable system or does it pass through a series of phases, the observed mobility being merely a mean and not necessarily one corresponding to any particular phase, the mobilities observed depending upon the duration of the phases as well as upon their nature?

[1] Blanc, *Journ. de Phys.* 1908, p. 830.
[2] Laporte, *Ann. de Phys.* viii. p. 466, 1927.
[3] Erikson, *Phys. Rev.* xx. p. 117, 1922.

The chief results which have emerged from investigations on the mobility of the ions are:

1. The mobility of the ions, especially in the lighter gases, is much less than it would be if the ion consisted of a single molecule, having the same free path as an uncharged molecule.

2. The mobility of the ions depends only on the gas through which they move, so that though in a mixture of gas there may be ions of different kinds all the ions of the same sign have the same mobility. An exception to this is the discovery by Erikson that aged positive ions have a smaller mobility than fresh ones (see, however, p. 154).

3. In the lighter gases the mobility of the negative ions is greater than that of the positive.

4. The mobility of the negative ions is diminished by traces of the vapours of water and various alcohols.

5. For positive ions the product of the mobility and pressure is constant over a very wide range of pressure; for negative ions, on the other hand, the product, when the pressure gets low, rapidly increases as the pressure diminishes.

6. The results of experiments with radioactive ions show that a heavy ion has the normal mobility and retains its identity during the motion through the electric field.

It is helpful to consider from the point of view of the Kinetic Theory of Gases the various conditions which might be expected to affect the mobility of the ions. The mobility of an ion through a gas is proportional to the coefficient of diffusion of the ion through the gas. This coefficient[1] is equal to

$$\epsilon \lambda \sqrt{\frac{1}{\pi h}\left(\frac{1}{m_1} + \frac{1}{m_2}\right)} \quad \ldots \ldots \ldots \ldots \ldots (1),$$

where ϵ is a numerical constant, λ the free path of the ion through the gas, m_1, m_2 the masses of the ion and molecule respectively, and $h = RT/2$, where T is the absolute temperature and R the gas constant. Let us consider what would be the effect of the electric charge on the free path supposing for the moment that the mass of the ion is the same as that of a molecule. The attraction between an ion and a molecule would in consequence of the charge

[1] Jeans, *Dynamical Theory of Gases*, p. 332.

be greater than that between uncharged molecules; thus the ion and the molecule will be dragged together and the free path of an ion will be shorter than that of a molecule. Thus λ for the ion will be less than that for the molecule, so that the coefficient of diffusion will be less.

If λ' is the free path for the charged ion supposed monomolecular, λ that of the uncharged molecule, w the work required to separate an ion from a molecule with which it is in contact, we can show that

$$\lambda' = \frac{\lambda}{1 + \alpha\,(w/RT)^{\frac{1}{2}}},$$

if the force between the ion and the molecule, like that between a charged point and a metal sphere, varies inversely as the fifth power of the distance: in this formula α is a numerical constant.

If the force between the ion and the molecule varies inversely as the cube of the distance, as it will if the molecule be polar, then

$$\lambda' = \frac{\lambda}{1 + \alpha' w/RT}.$$

We see from these expressions that if we are to explain by the charge alone the slow diffusion of the ion, then w must be large compared with RT, or the work required to separate an ion from a molecule must be large compared with the kinetic energy of a molecule due to thermal agitation. It follows however from the principles of thermodynamics that when this is the case the ions will unite with the molecules and form aggregates, and thus any considerable diminution in the free path is inconsistent with monomolecular ions.

Let us consider some other consequences of supposing that the diminution in the mobility is due to the charge on a monomolecular ion. When w/RT is large, on the inverse fifth power law,

$$\lambda' = \frac{\lambda}{\alpha\,(w/RT)^{\frac{1}{2}}}.$$

If s_1 and s_2 are the radii of the ion and molecule respectively, since the force varies inversely as the fifth power of the distance and as the volume of the molecule, w is proportional to $(s_1 + s_2)^{-4} s_2^{3}$, while λ is proportional to $(s_1 + s_2)^{-2}$, $\lambda/w^{\frac{1}{2}}$ is proportional to $s_2^{-\frac{3}{2}}$ and does not depend on the size of the ion but only upon that of the molecule. Thus since λ' does not depend upon the nature of the ion, the variation of the mobility with the nature of the ion will by equation (1) be proportional to

$$\sqrt{\frac{1}{m_1} + \frac{1}{m_2}}.$$

Thus when the mass of the ion is greater than that of the molecule, the mobilities, when the mass of the ion is 1, 2, 3 ... ∞ times that of the molecule, are in the ratio of 1·41, 1·22, 1·15 ... 1. Though these differences are not very large they are much greater than are consistent with one mobility for one type of ion, while the mobility of a light ion through a heavier gas will be much greater than that of an ion formed by a molecule of that gas; thus if the mass of the light ion is 1/4 of that of the molecule its mobility will be 1·6 times that of the molecule, while the mobility of a hydrogen ion through methyl iodide will be six times that of a methyl iodide ion.

The case we have considered when λ' is independent of the nature of the ion, is the one which gives the greatest chance for the mobility of the ion to be independent of its nature; we have seen, however, that even in this case there would be much greater variations in the mobility than are consistent with the experiments. We are forced to the conclusion that the ions if monomolecular cannot remain unaltered during their path through the gas, but go through various phases, in which they have different mobilities, and that the mobility measured by the usual methods is an average and not its value in any particular phase.

38·2. Thus to take the case of a negative ion in a mixture of two gases A and B. It begins by being an electron, then it unites with a molecule of A to form a negative ion, then perhaps it unites with one or more molecules of A or B to form an aggregate, which may be itself transitory and dissociate; thus the process is repeated many times during the course of the measurement of the mobility, that measurement giving the average mobility and not the mobility in any special phase.

Again to take the case of the positive ion in a mixture, suppose this commences as a molecule of A; after a time it joins up with other molecules and forms an aggregate, this aggregate breaks up and another one is formed, and so on, the mobility measured being again an average. It does not necessarily follow that the positive charge always remains associated with the same molecule; when in an aggregate the positively charged molecule might abstract an electron from another molecule and become neutralised and break away from the aggregate as a neutral molecule, its positive charge having been transferred to the molecule from which it abstracted the electron. The probability of this would depend on the electrochemical properties of the charged molecule, a positively electrified molecule of a strongly electronegative element like iodine would be more likely to abstract electrons from the other molecules in the aggregate than the positively electrified molecule of an electropositive element. It appears from Erikson's work that an air ion can transfer a charge to acetylene, but radioactive recoil ions certainly do not lose their charge in air.

The effect of age on the mobility of positive ions might be explained in some such way as the following. Let us suppose that in the gas there is a small quantity of some impurity which forms with the ion an aggregate of exceptional stability. To take an extreme case let us suppose that the aggregate is permanent until it is neutralised by combination with an ion of the opposite sign. As these aggregates are so much more permanent than others, the ions of one sign would ultimately all be found in these aggregates, and as these do not break up the mobility measured would be the mobility of this particular aggregate. This view is in accordance with the fact that the time taken to 'age' the ion is a considerable fraction of a second, and is therefore enormously greater than the interval between collisions of the ion with the molecules of the gas at a pressure approaching atmospheric.

We may hope to get direct evidence of the existence of different phases in the life of an ion by experiments of the type of those discussed on p. 144; these methods have already been applied to give a measure of the life of an electron in a free state, they might with little modification be applied to find the life of any particular phase of an ion.

The influence of different phases is shown most clearly by the carriers of negative electricity, for which one of the phases is an electron, having a mobility many hundred times that of the other phase, the negative ion. Thus even though the life of an electron is but a small fraction of the life of an ion, the existence of the electronic phase may make the average mobility considerably greater than that of the negative ion. If t_1, t_2 are respectively the lives of an electron and a negative ion, K_1, K_2 the mobilities of the electron and negative ion respectively, then the observed mobility of the negative carrier, K, will be given by the equation

$$K = \frac{K_1 t_1 + K_2 t_2}{t_1 + t_2}.$$

The electron passes from the free state when it unites with a molecule to form a negative ion. Loeb has shown that the number of collisions an electron makes before uniting with a molecule to form a negative ion is independent of the pressure of the gas. The time between each collision will vary inversely as the pressure of the gas, so that t_1 will also vary inversely as the pressure. The time t_2 is the time the electron remains combined with the molecule. The separation of the electron from the molecule is represented by the monomolecular reaction

negative ion $\rightleftarrows$ electron + molecule,

and $1/t_2$ is the rate of this reaction.

The mechanism of monomolecular reactions is a subject of profound interest; the reaction which has been most closely studied is the decomposition of N_2O_5. The rate of this does not vary with the pressure, and the dissociation seems to be spontaneous, in the sense that it does not depend on such external agents as collisions with other molecules or the effect of radiation. Assuming that t_2 is independent of the pressure p, we have $t_1/t_2 = \beta/p$,

$$K = \frac{K_1 \dfrac{\beta}{p} + K_2}{\dfrac{\beta}{p} + 1}.$$

K_1 and K_2 vary inversely as p; let $K_1 p$, $K_2 p$ be denoted by A and B respectively, then

$$Kp = \frac{A \dfrac{\beta}{p} + B}{\dfrac{\beta}{p} + 1}.$$

The graph representing the relation between Kp and corresponding to this equation is shown in Fig. 45. Fig. 40 shows the graph obtained by Kovarik by his experiments, p. 130. It will be seen that the two graphs are of the same general character, indicating a large increase in the value of Kp at low pressures. We may take this as an indication that the life of the negative ion, if not sensibly infinite, does not vary rapidly with the pressure. If the splitting off of the electron were due to the impact of other

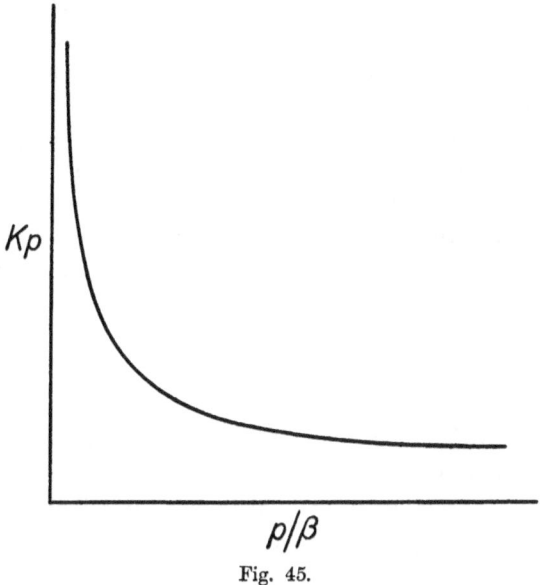

Fig. 45.

molecules we might expect that the life of the negative ion would be inversely proportional to the number of impacts per unit time, and so inversely proportional to the pressure. If both t_1 and t_2 varied inversely as the pressure, we see that Kp would be constant, and this is contradicted by the experiments.

The mobility of the negative carrier would on this view be very sensitive to the duration of the complex phase, and if any impurity was present which formed complex phases of great stability the mobility of the negative ion would diminish. If this substance were present in very small quantities, it would take some time before the electron entered into a complex containing this impurity, so that the mobility of the negative ion would be greater when it was fresh than it would be later. Lattey, as we have seen, found that the mobility was greater under strong electric forces, when the observations would be made while the ions are fresh, than when the electric forces were weak, when the ions would be older before they passed out of observation.

If we measure the mobility by timing the ion across a given space the abnormality comes in when the time of transit drops below a certain value; one way of diminishing the time is by diminishing the pressure, and thus we should expect to reach the abnormal stage by diminishing the pressure. This is a well-known effect. Another method would be by increasing the electric field. Thus by continual increase in the electric field we ought to reach a stage where the mobility is abnormal, *i.e.* where the apparent velocity increases more rapidly than the electric force (see p. 135). Another method would be by diminishing the distance passed over by the ion in the experiment; if we measured the mobility over very short distances we should expect the abnormal increase in the mobility to come in at higher pressures than when the distance was large. Anything that diminished the life of the negative ion as a free electron would tend to retard the period at which abnormal mobilities began to appear; this is in accordance with the observation of Lattey[1] that pu increased rapidly in dry gases at a pressure where it remained constant when moisture was present. Water vapour and the electronegative gases have exceptionally strong affinity for the electron and so curtail its life in the free state. Hitherto we have been considering the case of the negative ion. Somewhat similar considerations will apply to the positive one which may alternate between a single molecule and a cluster; inasmuch however as the difference between the mobilities of a single molecule and a cluster is at most about 50 per cent., while the mobility of the free electron is some thousands of times that of the cluster, the effects on the positive ion will be insignificant compared with those on the negative one. Since the time taken by an electron to pass over a constant distance d is directly proportional to the

---

[1] Lattey, *Proc. Roy. Soc.* lxxxiv. p. 179, 1910.

pressure p and inversely proportional to the electric force X, it will be unaltered if X/p is unchanged; thus the stage at which the mobilities become abnormal will be determined by X/p having a certain value or, if the distance d varies, by X/pd having a definite value.

38·3. The mobility of the negative carrier when in the electronic phase will depend on the velocity the electron possesses before the electric force is applied. If this velocity is not that corresponding to the energy due to thermal agitation, *i.e.* at 0° C. about 10^7 cm./sec., then the ordinary expressions for its mobility will not apply. It seems however unlikely that the electron in its short life should get into thermal equilibrium with the gas, for owing to the enormous difference between the masses of the electrons and atoms the transference of energy from one to the other is an exceedingly slow process. Direct evidence of this is afforded by an interesting experiment made by Franck and Hertz[1]. They started the electrons from a hot wire placed in one part of a vessel containing helium, with a velocity corresponding to a fall of potential of several volts at low pressure, and then collected them in another part of the vessel at a considerable distance from the place whence they started, and found that an appreciable fraction of them retained velocities approaching those of their projection, though in the interval they must have made a very considerable number of collisions. The velocities so retained were large multiples of those corresponding to the thermal agitation in the gas. The difficulty in transferring energy between the electrons and the molecules makes the initial velocity of the electrons an important matter in connection with the question of mobility, for there seems considerable probability that during the life of the electron this will not be greatly altered by the collisions. The connection however between the velocity of the ion and the electric force and the pressure is very different according as the velocity of the ion, apart from that due to the electric force, is small or great compared with that generated in the electric field between two collisions. In the former case it is proportional to $\sqrt{X\lambda}$ or $\sqrt{X/p}$ and in the latter to X/p, where X is the electric force, λ the free path and p the

[1] Franck and Hertz, *Verh. Deutsch. Phys. Gesell.* xv. p. 373, 1913.

pressure. In the first case the velocity w of a negative ion would be of the form

$$w = a\sqrt{X/p} + b\frac{X}{p} \quad \text{or} \quad pw = a\sqrt{Xp} + bX,$$

in the latter
$$w = a'\frac{X}{p} + b'\frac{X}{p},$$

where a, b, a', b' are constants. It is only in the latter case that the 'mobility' of an ion has a simple meaning, for in the other the velocity is not directly proportional to the electric force. In the latter case too pw is independent of p, while in the former it increases, though it may be slightly, with p. Now as far as our knowledge goes at present, there is no evidence that the velocity of the negative ion at not very low pressures is not simply proportional to the electric force, even in those gases where there is considerable difference between the velocities of the negative and positive ions. The evidence too seems to show that in these gases there is a considerable range of pressure over which pw is constant. Both these results point to the conclusion that the free electron has a very considerable amount of energy independent of that given to it by the electric field. We could not have assumed this result without collateral evidence. When the negative ion is alternating between the electron and the cluster the energy independent of the electric field which it possesses will be its energy when the cluster breaks up. According to the theory of the equipartition of energy, this energy should be that possessed by a monatomic molecule at the temperature of the gas, which would satisfy the condition of being, under normal conditions, large compared with that given to the electron when describing a free path under the electric field. The law of the equipartition of energy is however a somewhat treacherous guide; it certainly does not hold, for example, for the numerous electrons which form a part of the structure of the atom. In the case of an electron in the cluster however the forces acting on it are much less intense than on those inside the atom, and the looser the connection of the electron with other systems the more likely is it, given sufficient time, to acquire the energy corresponding to equipartition: it seems quite possible that in the cluster we have a system in which there is at least an approximation to equipartition.

Theory of the Mobility of Ions.

39·1. When an ion with a charge e moves under an electric force X and attains a constant velocity v, the momentum Xe communicated to the ion in unit time by the force must equal the momentum in the opposite direction given to it by the molecules through which it is moving.

Maxwell[1] has shown that when a gas A containing N molecules per unit volume moves through a gas B containing N' molecules per unit volume, the momentum P communicated to the N molecules of A per unit time by the collisions with the B molecules is equal to

$$\frac{4\pi M_2 M_1}{M_1 + M_2} \int \ldots (\xi_2 - \xi_1) \sin^2\theta f_1(\xi_1, \eta_1, \zeta_1) f_2(\xi_2, \eta_2, \zeta_2)$$
$$Vb\, db\, d\xi_1 d\eta_1 d\zeta_1 d\xi_2 d\eta_2 d\zeta_2,$$

where M_1 and M_2 are the masses of an A and a B molecule respectively; ξ_1, η_1, ζ_1; ξ_2, η_2, ζ_2 are the components of the velocities of the A and B molecules respectively before a collision;

$$f(\xi, \eta, \zeta)\, d\xi,\, d\eta,\, d\zeta$$

the number of molecules which have velocities whose components are between $\xi + d\xi$, $\eta + d\eta$, $\zeta + d\zeta$, V is the relative velocity of two particles before the collision, so that

$$V^2 = (\xi_2 - \xi_1)^2 + (\eta_2 - \eta_1)^2 + (\zeta_2 - \zeta_1)^2,$$

2θ is the angle through which the relative velocity of the two molecules is turned by the collision, b is the perpendicular let fall from one molecule B on the line through A parallel to the direction of relative motion before the collision. If the collision is like that between two elastic spheres, the angle between the direction of relative motion and the line of centres is $\sin^{-1}\dfrac{b}{\sigma}$, where σ is the

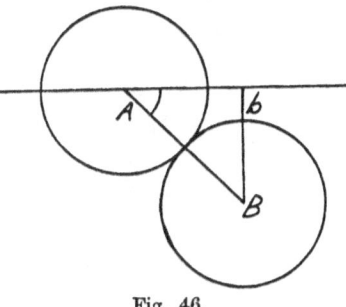

Fig. 46.

[1] *Collected Papers*, vol. II. p. 26.

sum of the radii of the spheres; this angle has the same value before and after collision, so that

$$\frac{\pi}{2} - \theta = \sin^{-1}\frac{b}{\sigma}.$$

If instead of collisions between perfectly elastic spheres we suppose that the molecules repel each other with a force varying inversely as the $(n+1)$th power of the distance, we can obtain the value of θ as follows. Since the centre of gravity of the two particles will move uniformly in a straight line, if we superpose on each of the particles a velocity equal and opposite to the initial velocity of the centre of gravity the relative motion of the two will not be affected, while the centre of gravity G will remain at rest. The initial velocity of A relative to the centre of gravity is ωV, where $\omega = M_2/(M_1 + M_2)$; the perpendicular from G on the line of this velocity is ωb; and if ρ is the distance of A from G at any time, r the distance AB at the same time, then $\rho = \omega r$.

The repulsive force acting on A is K/r^{n+1} or K'/ρ^{n+1}, where $K' = K\omega^{n+1}$. Thus A may be regarded as moving under a central repulsive force from G equal to K'/ρ^{n+1}. If ϕ is the angle which GA makes with the line joining G to the apse of A's orbit, we have

$$\rho^2 \frac{d\phi}{dt} = \omega^2 V b = h;$$

by the energy equation we have

$$\tfrac{1}{2} M_1 \left\{ \left(\frac{d\rho}{dt}\right)^2 + \rho^2 \left(\frac{d\phi}{dt}\right)^2 \right\} = \tfrac{1}{2} M_1 \omega^2 V^2 - \frac{1}{n}\frac{K'}{\rho^n},$$

or

$$\tfrac{1}{2} M_1 \left(\frac{d\rho}{dt}\right)^2 = \tfrac{1}{2} M_1 \omega^2 V^2 - \tfrac{1}{2} M_1 \frac{h^2}{\rho^2} - \frac{1}{n}\frac{K'}{\rho^n},$$

which is equivalent to

$$\left(\frac{dr}{dt}\right)^2 = V^2 - V^2 \frac{b^2}{r^2} - \frac{2}{n} \frac{K}{M r^n} \quad \ldots\ldots\ldots(2),$$

where $\qquad M = M_1 M_2/(M_1 + M_2).$

At the apse $dr/dt = 0$, so that if r' is the least value of AB, r'

must be a value of r which makes the right-hand side of equation (2) vanish,

$$\phi = \int \frac{h}{\rho^2} dt = \int \frac{h}{\rho^2} \frac{dr}{\frac{dr}{dt}} = \int \frac{Vb}{r^2} \frac{dr}{\frac{dr}{dt}}$$

$$= \int \frac{b\,du}{\left\{1 - b^2 u^2 - \frac{2Ku^n}{nMV^2}\right\}^{\frac{1}{2}}}, \text{ if } u = \frac{1}{r}$$

$$= \int \frac{dx}{\left\{1 - x^2 - 2\left(\frac{x}{a}\right)^n\right\}^{\frac{1}{2}}},$$

if $\qquad a^n = nMV^2 b^n / K.$

The value of x corresponding to the apse is a root of the equation

$$1 - x^2 - 2\left(\frac{x}{a}\right)^n = 0.$$

Let us call this value x', then $\frac{\pi}{2} - \theta$, the angle the initial relative velocity makes with the apse line, is given by the equation

$$\frac{\pi}{2} - \theta = \int_0^{x'} \frac{dx}{\left\{1 - x^2 - 2\left(\frac{x}{a}\right)^n\right\}^{\frac{1}{2}}},$$

thus θ for a given value of n depends only on the value of a.

We have $\qquad b\,db = \left(\frac{K}{nMV^2}\right)^{\frac{2}{n}} a\,da,$

so that $\qquad \int_0^\infty \sin^2\theta\, b\,db = \left(\frac{K}{nMV^2}\right)^{\frac{2}{n}} \int_0^\infty \sin^2\theta\, a\,da.$

The integral, which we shall denote by C, is a number depending only on the value of n

$$V\int_0^\infty \sin^2\theta\, b\,db = \left(\frac{K}{nM}\right)^{\frac{2}{n}} V^{1-\frac{4}{n}} C \quad\ldots\ldots\ldots(3).$$

Thus P the momentum communicated in unit time to the A

molecules in unit volume by the B molecules is given by the equation

$$P = 4\pi \left(\frac{K}{n}\right)^{\frac{2}{n}} M^{1-\frac{2}{n}} C \int (\xi_2 - \xi_1) f_1(\xi_1, \eta_1, \zeta_1) f_2(\xi_2, \eta_2, \zeta_2)$$
$$V^{1-\frac{4}{n}} d\xi_1 d\eta_1 d\zeta_1 d\xi_2 d\eta_2 d\zeta_2 \ldots (4);$$

when $n = 4$, V disappears from this expression, which becomes

$$2\pi K^{\frac{1}{2}} M^{\frac{1}{2}} CNN'u,$$

where u is the velocity of the gas A through B. In this case the result is the same whatever may be the distribution of velocities among the molecules of the gas; it does not for example require them to be distributed in accordance with Maxwell's Law.

If the gas A consists of ions with a charge e moving under an electric force X, the momentum given to the ions per second is XNe; equating this to the momentum lost by the ions in the same time we have, when $n = 4$,

$$XNe = 2\pi K^{\frac{1}{2}} M^{\frac{1}{2}} CNN'u,$$
or $$u = Xe/2\pi K^{\frac{1}{2}} M^{\frac{1}{2}} CN'$$

If k is the mobility of the ion $u = kX$, so that

$$k = \frac{e}{2\pi K^{\frac{1}{2}} M^{\frac{1}{2}} CN'}.$$

In this case the mobility is always independent of the electric force and inversely proportional to the pressure.

Another case which can be treated by equation (4) without evaluating the functions f_1 and f_2, is when the velocity of the A' ions due to the electric field is very great compared with the velocity due to thermal agitation of either the A or the B molecules. In this case V may be regarded as constant and if the direction of motion of the A ions is along the axis of x, $\xi_2 - \xi_1 = V$; on these suppositions

$$P = 4\pi \left(\frac{K}{n}\right)^{\frac{2}{n}} M^{1-\frac{2}{n}} CV^{2-\frac{4}{n}} \cdot NN',$$

and since $P = XeN$

$$V = \left(\frac{aXe}{N'}\right)^{\frac{n}{2n-4}},$$

where
$$a = \frac{1}{4\pi CM}\left(\frac{K}{nM}\right)^{-\frac{2}{n}}.$$

Thus unless $n = 4$, V will not be proportional to X, and there will be no definite mobility in the ordinary acceptation of the term.

In the case $n = \infty$, which is that of collision between elastic spheres, V is proportional to $X^{\frac{1}{2}}$. When $n = 2$, i.e. when the force between the ions and the molecules varies inversely as the cube of the distance, P is independent of V, so that there is no steady state unless $X = 2\pi KCN/e$. When $n = 1$, i.e. when the force varies inversely as the square of the distance, V varies as X^{-2}.

The case of collisions analogous to those between elastic spheres is an important one and gives a simple expression for the relation between V and X. In this case
$$\sin^2\theta = 1 - \frac{b^2}{\sigma^2},$$
where σ is the sum of the radii of the spheres; when b is greater than σ, θ vanishes, hence
$$\int_0^\infty \sin^2\theta\, b\, db = \frac{\sigma^2}{4},$$
and
$$P = \pi\sigma^2 NN'V^2 M = XeN,$$
hence
$$V = \left\{\frac{Xe}{MN'\pi\sigma^2}\right\}^{\frac{1}{2}}.$$

$\frac{1}{N'\pi\sigma^2}$ is proportional to λ the mean free path of a molecule of A through B, hence
$$V \propto \left(\frac{Xe\lambda}{M}\right)^{\frac{1}{2}}.$$

Langevin's Evaluation of P.

39·2. When the velocities of the molecules are distributed according to Maxwell's law, Langevin[1] has shown that P reduces to the form
$$P = \tfrac{32}{3}NN'\left\{\frac{\pi M}{h}\right\}^{\frac{1}{2}} u \int_0^\infty \phi(v)\, \epsilon^{-z^2} z^5\, dz,$$

[1] Langevin, *Ann. de Chim. et de Phys.* viii. t. 5, p. 245, 1905.

where u is the average velocity along x of the A ions relative to the B molecules,

$$\phi(v) = \int_0^\infty \sin^2\theta\, b\, db,$$

$$z = V\{hM\}^{\frac{1}{2}},$$

and $h = 3/4aT$, where T is the absolute temperature of the gas and aT the mean kinetic energy of a molecule of any gas at this temperature.

When the collisions are like those between perfectly elastic spheres, $\phi(v) = \sigma^2/4$ and

$$P = \tfrac{8}{3} NN' \left\{\frac{\pi M}{h}\right\}^{\frac{1}{2}} \sigma^2 u \int_0^\infty \epsilon^{-z^2} z^5 dz$$

$$= \tfrac{8}{3} NN' \left\{\frac{\pi M}{h}\right\}^{\frac{1}{2}} \sigma^2 u \quad\ldots\ldots\ldots\ldots\ldots\ldots(5),$$

since $\int_0^\infty \epsilon^{-z^2} z^5 dz = 1.$

Equating this value to NXe we see that k the mobility of the ion is given by the equation

$$k = \frac{3}{8} \frac{e}{\sigma^2 N'} \left\{\frac{h}{\pi M}\right\}^{\frac{1}{2}} \quad\ldots\ldots\ldots\ldots\ldots\ldots(6).$$

When the ions and molecules act on each other with forces represented by K/r^{n+1},

$$\phi(v) = \left(\frac{K}{nM}\right)^{\frac{2}{n}} V^{-\frac{4}{n}} C$$

(see equation (3), p. 167) and

$$P = \tfrac{32}{3} NN' u \left\{\frac{\pi M}{h}\right\}^{\frac{1}{2}} \left\{\frac{Kh}{n}\right\}^{\frac{2}{n}} C \int_0^\infty \epsilon^{-z^2} z^{5-\frac{4}{n}} dz,$$

and k the mobility of the ion is given by the equation

$$k = \frac{3e}{32 N'} \left\{\frac{h}{\pi M}\right\}^{\frac{1}{2}} \left(\frac{n}{Kh}\right)^{\frac{2}{n}} \frac{1}{C} \cdot \frac{1}{\int_0^\infty \epsilon^{-z^2} z^{5-\frac{4}{n}} dz} \quad\ldots\ldots(7).$$

A case worthy of notice is when the deflections of the ions are supposed to be due mainly to elastic collisions between the ions and the molecules, while the number of collisions is supposed to be increased by the forces between the two. We can take this into account in the following way. If the deflections are due to collisions

$$\sin\left(\frac{\pi}{2} - \theta\right) = \frac{p}{\sigma},$$

where p is the perpendicular from the centre of the molecule on the path of the ion when it strikes the molecule. If V is the relative velocity of the ion and molecule at impact

$$vp = Vb,$$

and

$$v^2 = V^2 + \frac{2R}{M},$$

where R is the work required to move the ion from the surface of the molecule to an infinite distance away;

$$\sin^2 \theta = 1 - \frac{p^2}{\sigma^2} = 1 - \frac{b^2}{\sigma^2(1 + 2R/MV^2)}.$$

Thus θ has finite values from $b = 0$ to $b = \sigma\{1 + 2R/MV^2\}^{\frac{1}{2}}$, so that

$$\int \sin^2 \theta\, b\, db = \frac{\sigma^2}{4}(1 + 2R/MV^2).$$

Substituting this value for $\phi(v)$ in Langevin's expression we find

$$k = \frac{3}{8}\frac{e}{\sigma^2}\frac{1}{N'(1 + Rh)}\left\{\frac{h}{\pi M}\right\}^{\frac{1}{2}} \quad \ldots\ldots\ldots\ldots(8).$$

Motion of Ions through Mixed Gases.

39·3. If the ions move through a mixture of gases the momentum communicated to them per unit time will be the sum of the momenta contributed by the constituent gases, thus

$$P = \tfrac{3\cdot 2}{3} N u \left\{ \Sigma N_r \left(\frac{\pi M_r}{h}\right)^{\frac{1}{2}} \left(\frac{K_r h}{n}\right)^{\frac{2}{n}} C_r \int_0^\infty \epsilon^{-z^2} z^{5 - \frac{4}{n}} dz \right\} \quad ..(9),$$

where N_r, M_r, K_r, C_r refer to one of the gases; if the law of force is the same for all the gases C_r will be the same also.

Since

$$\frac{1}{k} = \frac{P}{Nue},$$

we see that the reciprocal of the mobility should be a linear function of the concentrations of the gases.

In the case of elastic collisions, if there are two gases A and B

$$\frac{1}{k} = \frac{8}{3e}\left\{N_1\sigma_1{}^2\left(\frac{\pi M_1}{h}\right)^{\frac{1}{2}} + N_2\sigma_2{}^2\left(\frac{\pi M_2}{h}\right)^{\frac{1}{2}}\right\},$$

where N_1 and N_2 are the numbers of molecules of A and B respectively per unit volume, σ_1 and σ_2 are respectively the sums of the radius of the ion and that of a molecule of A and B, and M_1, M_2 the harmonic means of the mass of an ion and that of a molecule of A and B respectively.

When the mobility is calculated on the theory of forces varying inversely as the $(n + 1)$th power of the distance, the reciprocal of the mobility is proportional to the expression inside the bracket on the right-hand side of equation (9). If the force between an ion and a molecule does not depend on the size of the ion as long as the charge remains the same, K_r the force between the ion and a molecule of the r gas will not be affected by the presence of the other molecules, so that in this case the reciprocal of the mobility should be a linear function of the percentages of the various gases in the mixture. In the case of the elastic collisions this would not necessarily be the case, as the size of an ion in the mixed gases might conceivably differ from that in either gas separately, thus σ_1 and σ_2 the sums of the radii of the ion and the molecules might possibly involve N_1 and N_2; if this were so the relation between $1/k$ and the N's would not be a linear one.

Blanc[1] measured the mobilities of the ions in mixtures of CO_2 and H_2 and of air and CO_2 in all proportions, and he found in the former mixture the relation between the reciprocal of the mobility and the percentage composition was quite accurately a linear one, while in the second case it was very approximately so; in this paper too he showed that the mobility of an ion made in CO_2 and driven through air was the same as that of an ion made in the air itself.

Przibram[2] measured the mobilities in various mixtures of the vapours of alcohol and water and found that in this case the reciprocal of the mobility was not a linear function of the percentage composition.

[1] Blanc, *Comptes Rendus*, cxlvii. p. 39, 1908; *Bull. Soc. Franç. de Phys.* 1908, p. 156. [2] Przibram, *Phys. Zeits.* xiii. p. 545, 1912.

Wellisch has measured the mobilities of the ions in mixtures of methyl iodide, CH_3I, a very heavy gas, and hydrogen the lightest gas known. The following table contains the reciprocal of the mobility of the positive ion as found by Wellisch[1] and the value calculated on the assumption that it is a linear function of the partial pressures of the gases.

Pressure in mm. of mercury.

CH_3I	H_2	Observed	Calculated
70	—	·374	—
70	59	·397	·386
70	183	·458	·410
70	315	·485	·436
70	687	·5	·509
51	—	·253	·272
51	16	·263	·275
51	61	·285	·284
51	334	·333	·337
51	714	·377	·412
25	—	·133	·133
25	85	·169	·150
25	360	·250	·204
25	732	·285	·277
12	373	·168	·137
12	751	·224	·210
6	379	·105	·105
6	757	·192	·179

Though in some cases the differences between the observed and calculated results are considerable, the fact that they are sometimes of one sign and sometimes of the opposite, and that in nearly half the cases the agreement is within the errors of experiment, points I think to the conclusion that even with this mixture, where the molecular weights are so different, the reciprocal of the mobility is probably a linear function of the concentration.

Loeb and Ashley[2] have made measurements of the mobilities in mixtures of air and ammonia. They plot the mobility against the concentration, giving a curve. If however the reciprocal of the mobility is taken, it varies in an almost linear manner with

[1] Wellisch, *Proc. Roy. Soc.* lxxxii. p. 513.
[2] Loeb and Ashley, *Proc. Nat. Acad. Sci.* x. p. 351, 1924.

the concentration. Mayer[1] finds that the reciprocal law holds for mixtures of hydrogen and oxygen, while Loeb and du Sault[2] have proved it for mixtures of hydrogen and acetylene.

Przibram[3] has given another expression for the mobility of the ions in a mixture of two gases A and B. If k_1 and k_2 are the mobilities of the ions in the pure gases A and B respectively, at the pressure p, then

$$k_1 = \frac{A}{p\sigma_1^2 \sqrt{M_1}}, \qquad k_2 = \frac{A}{p\sigma_2^2 \sqrt{M_2}}.$$

He assumes that k_{12} the mobility of the ion in the mixed gases is of the form

$$k_{12} = \frac{A}{(p_1 + p_2)\sigma_{12}^2 \sqrt{M_{12}}},$$

where p_1 and p_2 are the partial pressures of the gases,

$$\sigma_{12}^2 = \frac{p_1\sigma_1^2 + p_2\sigma_2^2}{p_1 + p_2}, \qquad M_{12} = \frac{p_1 M_1 + p_2 M_2}{p_1 + p_2};$$

substituting these values we find

$$k_{12} = \frac{p(p_1 + p_2)^{\frac{1}{2}} k_1 k_2 (M_1 M_2)^{\frac{1}{2}}}{(M_1 p_1 + M_2 p_2)^{\frac{1}{2}} (p_1 k_2 M_2 + p_2 k_1 M_1)^{\frac{1}{2}}}.$$

This formula however does not give quite as good agreement with experiment as the simpler one just discussed and it has not the strong support from dynamical considerations to which the latter is entitled.

The Effect of the Masses of the Ions on their Mobility.

39·4. We see from the expressions that we have obtained for the mobility that whatever may be the law of force between ions and molecules, the connection between the mobility and the masses of the ions and of the molecules through which they move is always the same, and identical with that which exists when the collisions are like those between elastic spheres. In every case the masses only occur in the factor $M^{-\frac{1}{2}}$, where $M = M_1 M_2/(M_1 + M_2)$. When M_1 is large compared with M_2, M is approximately equal to M_2 and is thus independent of M_1. The mass of the ion,

---

[1] Mayer, *Phys. Zeits.* xxviii. p. 637, 1927.
[2] Loeb and du Sault, *Proc. Nat. Acad. Sci.* xiii. p. 510, 1927.
[3] Przibram, *Phys. Zeits.* xiii. p. 545, 1912.

provided it is a considerable multiple of that of a molecule of the gas through which it is moving, will thus not affect its mobility. On the elastic sphere theory of collision, if the increase in mass were accompanied by an increase in the size of the ion, the occurrence of σ^2 in the denominator of the expression for k would make the mobility of the heavy ion less than that of the light one.

In the case of positive ions in an elementary gas, the smallest ion might be an atom, so that if the molecule were diatomic $M_1 = M_2/2$ and therefore $M = M_2/3$.

For the greatest possible positive ion M_1 would be infinite compared with M_2 and therefore $M = M_2$. Since the mobility varies as $M^{-\frac{1}{2}}$, the range in mobility due to the variation in M would be $\sqrt{3}$ to 1. On the collision theory this would have to be multiplied by σ^2/σ_0^2, where σ, σ_0 are respectively the distances between the centre of a molecule and the centres of the largest and smallest ions. For negative ions the range is greater. When the negative ion is an electron M_1 is very small compared with M_2, so that $M = M_1$, while for the most massive negative ion $M = M_2$; so that in this case the range in mobilities is $(M_2/M_1)^{\frac{1}{2}}$ or in the case of air about 225 to 1. In the collision theory this has to be multiplied by σ^2/σ_0^2. If the larger ion were a single molecule this quantity would be equal to four, so that the mobility of an electron might be expected to be at least 1000 times that of the negative ion.

The values actually found by Loeb and Wahlin for electrons in nitrogen for small fields are 30,500 and 18,000 cm./sec. or about 16,000 and 10,000 times the mobility of a heavy negative ion. These results are strong support for the collision theory, which may as we have seen be modified so as to take account of the attractive forces between ions and molecules without altering the dependence on σ^2/σ_0^2. It should be remembered that the experiments of Ramsauer, Townsend and others show that the effective cross-section of the molecule for electron collisions is a complicated function of the velocity of the electron. H. A. Wilson[1] and K. T. Compton[2] have both found theoretical expressions for

[1] H. A. Wilson, *Proc. Roy. Soc.* ciii. p. 53, 1923.
[2] K. T. Compton, *Phys. Rev.* xxi. p. 717, 1923.

the mobilities of free electrons. Wilson considers a gas whose molecules consist of shells of electrons surrounding positive nuclei. Compton brings in the loss of energy due to the finite mass of the molecules; he finds

$$k = \frac{0 \cdot 815 el}{\sqrt{2me}\left[\tfrac{1}{2}\Omega + \left(\tfrac{\Omega^2}{4} + W^2\right)^{\tfrac{1}{2}}\right]^{\tfrac{1}{2}}},$$

where $e\Omega$ is the mean kinetic energy of the gas atoms, eW that of the electrons, and l the mean free path of the electrons. This equation gives a rough agreement with Loeb's values for helium, and Loeb says it agrees well with Townsend's experiments on the same gas (see p. 148), if in both cases l is taken as half the value to be expected on the kinetic theory.

Wahlin has found that Compton's formula gives fair agreement with his experiments on the mobilities of free electrons in nitrogen, but only if l is taken as about 2·7 times the value obtained on the ordinary kinetic theory. It should be remarked that in these experiments the variation with strength of field is very marked. Thus Wahlin found the mobility at 50 volts per cm. only 1/3 of that for vanishingly small fields. This was at atmospheric pressure. A formula due to Loeb fitted Wahlin's experiments on CO if suitable constants were taken, but failed for Loeb's own experiments on helium. The work of § 20·6 shows that the free path of an electron depends greatly on its velocity.

It seems improbable that any really satisfactory formula for the mobility of free electrons will be found until more is known of the nature of the collisions between electrons and molecules at slow speeds. It is to be hoped that the new wave mechanics will lead to advances in this direction, and provide a detailed explanation of Ramsauer's remarkable results.

The Effect of Temperature on Mobility.

39·5. Since the quantity h which occurs in the expressions for the mobility is inversely proportional to T the absolute temperature, we see from the equations in Art. 39·2 that if the actions between ions and molecules are regarded as

(1) like those between elastic spheres, since k contains the factor $h^{\tfrac{1}{2}}$ it will vary as $T^{-\tfrac{1}{2}}$;

(2) like those between elastic spheres, account however being taken of the effect of attractions between ions and molecules in increasing the number of collisions, since k contains the factor $\frac{h^{\frac{1}{2}}}{1+ah}$ it will vary as $\frac{T^{\frac{1}{2}}}{T+a}$;

(3) like those between centres of force varying inversely as the $(n+1)$th power of the distance, since k contains the factor $h^{\frac{1}{2}-\frac{2}{n}}$ it will vary as $T^{\frac{2}{n}-\frac{1}{2}}$.

If the ions are aggregates of molecules, the size and mass of the ions may alter with the temperature; the alteration in size would affect the mobility in cases (1) and (2) where k contains the factor $1/\sigma^2$, in case (3) the size of the ion would not affect the mobility if the force between the ion and the molecule depended only on the charge on the ion and not on its size. Any alteration in the mass of the ion would affect the mobility in the same way in all three cases; this effect however would be comparatively small.

Phillips' experiments show that k cannot vary as T^p if p has a finite value, as according to them k is practically constant over a considerable range of temperature; if $n = 4$, i.e. if the force between the ion and the molecule varied inversely as the fifth power of the distance, p would be zero and the mobility would be independent of the temperature. Phillips however observed considerable diminution in mobility at low temperature.

Sutherland[1] has shown that the law indicated in (2) expresses Phillips' result fairly well over a considerable range of temperature, though not below about 100 absolute. If $a = 333$ for the negative ions and 509 for positive, this would give a maximum mobility for the negative ions at 60° C. and for the positive at 236° C.

We could avoid all the difficulties from the variation of the size of the ion with temperature if we worked with the negative ion in well purified argon or nitrogen, as the ion in this case seems always an electron: experiments of this kind should give decisive evidence as to the nature of the action between electrons and molecules.

[1] Sutherland, *Phil. Mag.* vi. 15, p. 341, 1909.

Erikson's results also show a maximum at a certain temperature, but this is rather lower than for Phillips' results. Kovarik's experiments show no well-marked maximum.

The Formation of Complex Ions.

40·1. On the supposition that the action between ion and molecule is that between elastic spheres, equation (6) enables us to calculate numerically the value of k if the mass of the ion is supposed known. First suppose that the ion is a charged molecule of the gas, and that the diameter σ is that found by the kinetic theory from measurements of viscosity. Putting in the numbers appropriate to oxygen we find for k at atmospheric temperature and pressure, $1\cdot59 \times 10^3$ in E.S.U. or $5\cdot3$ cm./sec. per volt/cm. This is about three times the experimental value for the negative ion and four times that of the positive. A similar discrepancy is found in other cases. This is in accordance with the fact that the coefficient of diffusion of ions is of the order of that of the molecules of a very heavy gas, and much less than that of a gas of the same molecular weight as that in which the ions are formed. It suggests that the ions are heavier than molecules, and consist of a complex group of molecules. We have seen in the discussion of recombination that to obtain agreement between theory and experiment we are obliged to suppose that the ion has several times the mass of a molecule. This was the case whether the coefficient of recombination was calculated from first principles or indirectly from the observed value for the mobility. We also saw that, on certain assumptions, it was more likely that an ion should attach itself to a molecule than that it should remain unchanged till it recombined with one of the opposite sign.

The question of whether the ions remain of molecular dimensions during their whole life or form complex groups as suggested above, has been much discussed. One apparent way of escape from the group theory is to take account of the effect of the probable electrostatic attraction between an ion and a neutral molecule in shortening the mean free path and so decreasing the mobility. But, as was remarked in the discussion on recombination, and as can be seen directly from equation (8) where this effect is allowed for, the condition that it should be important is that the work required to separate an ion from a molecule should be larger than the mean thermal energy at the temperature in question. But this is precisely the condition that formation of complexes should occur in the state of thermodynamic equilibrium. Regarding the process as a reversible reaction $MM' \rightleftarrows M + M'$, where M' represents a charged molecule and M an uncharged one, the extent of the formation of the complex is represented by the equilibrium constant K defined by

$$K = p_M \cdot p_{M'}/p_{MM'},$$

where the quantities p are the partial pressures of the various constituents measured in atmospheres.

Nernst's equation may be written

$$\log_{10} K = -\frac{U}{4 \cdot 571 T} + \frac{\Sigma 3 c_p}{2 \alpha T} \log_{10} T + \Sigma C,$$

where U is the heat set free per gramme molecule by the formation of the complex, c_p is the specific heat per molecule, C is the corresponding chemical constant and the summations refer to the constituents taking part in the reaction, those on the left-hand side of the equation of reaction having a negative sign. At atmospheric pressure this may be written

$$\log_{10} \frac{p_{M'}}{p_{MM'}} = -\frac{R}{1 \cdot 56 \alpha T} + \frac{\Sigma 3 c_p}{2 \alpha T} \log_{10} T + \Sigma C,$$

where R is as before the work required to separate an ion from a molecule. The last two terms are unfortunately difficult to calculate. For a monatomic substance a theoretical expression has been found for C, namely $-1 \cdot 6 + \frac{3}{2} \log_{10} M$, where M is the atomic weight, but even for a gas like helium the complex is bimolecular. One may perhaps, however, get a rough idea of the magnitudes of the terms in this case by assuming the expression for C to hold throughout and supposing that the formation of the complex reduces the number of degrees of freedom of ion and molecule from 6 to 5. In the case of a positive ion of helium this would give

$$\log_{10} \frac{p_{M'}}{p_{MM'}} = -\frac{R}{1 \cdot 56 \alpha T} + 2 \cdot 54.$$

Thus the numbers of free and complex ions would be equal when $R/\alpha T = 4\cdot 0$ or $Rh = 3\cdot 0$, and if Rh is markedly greater than this the number of free ions will be small. But equation (8) shows that the mobility is inversely as $1 + Rh$, thus a value of Rh large enough to account for the mobility will cause the formation of complexes. While it is possible that an accurate knowledge of the chemical constants might modify this conclusion[1], it seems improbable that the attractive forces can be large enough to account for the observed mobility without leading to the formation of complexes.

A further argument in favour of the existence of complex ions is the independence of mobility on the mass of the ion when foreign ions are introduced into a gas. This could have been explained otherwise by supposing that the charge could shift readily from the ion to a molecule of the gas, so that however the ions started, they became after a short time charged molecules of the gas through which they were moving, but this explanation is excluded by the results of experiments with radioactive recoil atoms.

The marked difference between the mobilities of the positive and negative ions in the simple gases might be explained by supposing that the negative ion is a free electron for part of the time. Owing to the enormous mobility of free electrons, a very short period of freedom would account for the observed difference in mobility. A strong argument against this view is that the ratio of the mobilities is much the same for gases which, according to Loeb's results, show great differences in the ease with which they combine with electrons. If we accept the 'cluster' view, according to which the masses of both positive and negative ions would be infinite as far as equation (1) is concerned, it might

[1] Ehrenfert and Trkal, *Ann. der Phys.* lxv. p. 609, 1921, have given as an expression for diatomic molecules

$$i = \log_{10}\left\{\frac{4\pi}{\sigma} m^{\frac{3}{2}} J \frac{(2\pi k)^{\frac{5}{2}} k}{h^5}\right\},$$

where m is the mass and J the moment of inertia of the molecule, σ is a 'symmetry factor' and k is Boltzmann's constant. Wismewski, *Zeits. f. Phys.* xliv. p. 392, 1927, finds that this fits a number of gases if σ is put equal to one. If this is done, and J taken for a complex helium ion as $m \times 10^{-16}$, the constant comes out as $3\cdot 0$ instead of $2\cdot 54$ as above.

still be possible to account for the difference by supposing that the positive ion had the power of attaching itself to a large polar molecule of some impurity which would increase both σ and Rh. As Grindley has pointed out, the forces between a polar molecule and a positive and a negative ion may be quite different.

The polar molecule will tend to present a different portion of its surface to the ion in the two cases, and thus the distortion produced in the molecule by the ionic charge will be different for the two ions. Again the polar molecule may have one charge, e.g. an electron or proton, at a point on its surface and the neutralising charge deep inside, so that the attraction of an ion will depend on its sign. Loeb and Cravath[1] have shown that in H_2O and H_2S, where the molecules have probably protons near the surface, $\bar{u}$ is less than $\overset{+}{u}$. The same holds for HCl, probably for a similar reason.

This view of the cause of the difference in mobility would fit in well with Erikson's discovery of ageing, as it might well be some time before the ion found a suitable polar impurity.

Large Ions.

41·1. Besides the ions we have been considering above, with a mobility of the order of 1 cm./sec. per volt/cm., other much slower ions are known. These ions are also characterised by a small coefficient of diffusion, so that the usual process of filtering does not remove them. Ions of this kind exist in the air especially near big cities, they are also formed when air is bubbled through water. The ions found in electrolytic gas immediately after its formation are of this type. McClelland, in measurements of the mobilities of ions from flames, showed the gradual change of the mobility with temperature as the ions left the flame from ·23 to ·04 cm./sec., and Bloch showed that on allowing the gas to stand for 15 to 20 minutes the mobility fell still further to ·001.

Important work has been done on the slow ions of the atmosphere by Langevin, who devised a method by which the presence of ions of different mobilities can be detected. The principle of the method is as follows. Suppose we have a stream

[1] Loeb and Cravath, *Phys. Rev.* xxvii. p. 811, 1926.

of ionised gas flowing with velocity u between two parallel plates A and B, and apply an electric force X so to drive the positive ions towards the plate A, where the number i received in unit time is measured by an electrometer connected with A. Unless the mobility of the ions exceeds a certain amount depending on the velocity of the stream, d, the distance between the plates and the strength of the electric field, some of the ions will escape from the field. If the mobility of an ion is k and if T be the time the gas takes to pass between the plates, X the electric force, then the ions which reach the plate are those whose distance from A when they first came under the electric field is not greater than kXT, and hence of the ions of this mobility in the stream of gas only the fraction $\dfrac{kXT}{d}$ give up their charge to the plate; when k is greater than d/XT, all the ions with this mobility will give up their charges. Hence if U be the number of cubic centimetres of gas passing between the plates in unit time, $f(k)\,dk$ the number of ions per cubic centimetre of the gas which have mobilities between k and $k+dk$, e the charge on an ion, $\kappa = d/XT$, and i the charge given to the electrometer per second,

$$i = eU \int_{\kappa}^{\infty} f(k)\,dk + \frac{eUT}{d} \int_0^{\kappa} Xkf(k)\,dk,$$

$$\frac{di}{dX} = \left(-eUf(\kappa) + \frac{eUT}{d} X\kappa f(\kappa) \right) \frac{d\kappa}{dX} + \frac{eUT}{d} \int_0^{\kappa} kf(k)\,dk$$

$$= \frac{eUT}{d} \int_0^{\kappa} kf(k)\,dk,$$

$$\frac{d^2 i}{dX^2} = -\frac{eUT}{d} \kappa f(\kappa) \frac{d}{X^2 T}$$

$$= -\frac{eU\kappa f(\kappa)}{X^2}.$$

Hence if we plot the curve representing the relation between i and X we shall be able to determine $f(\kappa)$, the density of the ions which have a mobility κ.

Thus suppose all the ions have the same mobility, so that $f(\kappa)$ vanishes except for one particular value of κ, say κ_1, then the i and X curve would be straight except at one point where

$$X = d/\kappa_1 T,$$

where there would be a nick in the curve. The presence of curvature in the i and X curve at any point shows that there are in the gas ions which in the time T can under the electric force corresponding to this point travel through the distance d. If the ions are of distinct types with mobilities $\kappa_1, \kappa_2, \kappa_3$, the i-X curve will be a series of straight lines with a nick for each type of ion. If the mobility of an ion depends only on the gas through which it moves, there will be only one nick in the i-X curves obtained by this method.

Mobilities of Slowly Moving Ions.

41·2. Langevin[1] applied this method to the air from the top of the Eiffel Tower and found two nicks in the i-X curves, one corresponding to the ordinary ions with a mobility of about 1·5 cm./sec. volt/cm., the other corresponding to a mobility of only $\frac{1}{3000}$ cm./sec., *i.e.* several thousand times less than the mobility of the ordinary ions. This mobility is of the same order as that of the ions produced by bubbling air through water, of those liberated by chemical actions, which probably are produced by the bubbling of the gas liberated by the chemical action through the liquid reagents, and of the ions produced when air is passed over phosphorus; they are probably charged particles of dust of various kinds and the number of them varies very widely in different localities, being much greater in large towns than in the open country. In air taken from near the ground in Paris, Langevin found that the quantity of electricity carried by these slowly moving ions was fifty times that carried by the ions of the ordinary type.

These slow ions are particles of dust held in suspension in a gas which is exposed to some ionising agent which produces a supply of the ordinary ions, the latter settle on the particles of dust and form the slow ions. The number of these slow ions present when the gas is in a steady state will only depend on the number of dust particles in the gas and will not be affected by the number of ions of the ordinary type produced by the ionising agent. For when there is equilibrium the number of particles which in unit time acquire a positive charge must equal the number which lose such a charge. A positively electrified dust particle might lose

[1] Langevin, *Comptes Rendus*, cxl. p. 232, 1905.

its charge by meeting and coalescing with a negative small ion, or by coalescing with a negatively electrified dust particle; these dust particles are however so sluggish in their movements that unless the dust particles are thousands of times more numerous than the ions we may neglect the second source of loss in comparison with the first. Let A be the number of uncharged dust particles in a cubic centimetre of gas, P and N the numbers of those with positive and negative charges respectively, and p, n the numbers of positive and negative small ions. The number of dust particles which acquire a positive charge per second will be $\alpha A p$, the number losing such a charge by coalescing with a negative ion is $\beta P n$, where α and β are constants; hence for equilibrium

$$\alpha A p = \beta P n.$$

Similarly by considering the negatively charged particles we get

$$\alpha' A n = \beta' N p,$$

hence we see that the proportion between the charged and the uncharged particles of dust depends only upon the ratio of p to n and not upon the absolute magnitude of either of these quantities. Thus though it would take much longer to reach the equilibrium state with a feeble source of ionisation than with a strong one, there would be as much dust electrified in one case as in the other, when that state was reached; de Broglie estimates that in this state about one-tenth of the particles would be electrified[1].

The mobility of a good many different kinds of dust particles is of the order $\frac{1}{1000}$ cm./sec. It would be a matter of some interest to see how the mobility varied with the pressure; we should expect the variation to be between that of ordinary ions when the mobility varies inversely as the pressure and drops big enough to be easily visible when the mobility is independent of the pressure. McClelland showed many years ago[2] that the ions in flame gases gradually diminished in mobility as well as in number with the time which had elapsed since the gas left the flame, so that in this case we apparently get a continuous gradation between ordinary ions and those which have a mobility of the order of 10^{-3}. Aselmann[3] found in the negative ions produced by the splashing

[1] de Broglie, *Ann. de Chimie et de Physique*, viii. 16, p. 1, 1909.
[2] McClelland, *Phil. Mag.* v. 46, p. 29, 1898.
[3] Aselmann, *Ann. der Phys.* xix. p. 960, 1906.

of pure water mobilities ranging from 4 to 2.7×10^{-4}, the greater number being between 4 and 1.6×10^{-2}; and for positive ions obtained by adding a little sodium chloride to the water the mobilities ranged from 8.8×10^{-2} to 3.43×10^{-4}, the greater number being between 8.8×10^{-2} and 6×10^{-4}. In these cases we seem to have a nearly continuous series of mobilities. In many cases however there is a gap between the ordinary ions and those having the mobilities of the order 10^{-3}. A convenient method, due to de Broglie, of testing for the presence of fine dust particles is to ionise the gas by X-rays or radium and then test for the presence of these large ions; the electrical method will enable us to determine the number of particles which get charged and their mobilities will give some information as to their size. In this way de Broglie showed the presence of these particles in flames such as the oxy-hydrogen flame where one of the products of combustion is liquid, while they could not be detected in the carbon monoxide flame. He detected them in gas through which an electric spark had passed, and in gas which had been bubbled through liquids. He observed some of these particles under the ultra-microscope and noticed that, if the particles were under electric forces, as soon as some radium was brought near some of them moved in one direction, others in the opposite, while some did not move at all.

Becker[1] has shown that these large ions are often produced by chemical action set up by the agent used to ionise the gas, such as X-rays or cathode rays. These produce ozone in gases containing oxygen, and the action of ozone, if suitable substances are present, may give rise to liquid or solid products which act as nuclei for the large ions.

Pollock[2] finds that the mobility of the large natural ions in air depends on the amount of moisture present, being roughly constant when the relative humidity is constant. He also found a group of ions of intermediate mobility, about 1/15 to 1/150 cm./sec. These disappeared when the pressure of water vapour exceeded 17 mm., and Pollock considers that they can grow by condensation into the larger 'Langevin' ions, of about 1/3000 mobility. If

[1] Becker, *Ann. der Phys.* xxxvi. p. 209, 1911.
[2] Pollock, *Phil. Mag.* xxix. pp. 514, 636, 1915.

the dust nucleus was of soluble material one would expect it to attract water vapour and form a liquid drop which would grow until the solution became so dilute that its vapour pressure was the same as the pressure of water vapour in the atmosphere, apart from a correction due to the curvature of the surface.

Besides the 'intermediate' ions described by Pollock, McClelland[1] and Nolan have found a number of discrete groups of ions of various mobilities among the ions formed by splashing and bubbling air through water and alcohol, and by the oxidation of phosphorus. Both these experiments and Pollock's were done with an air blast apparatus of the Zeleny type, more or less modified (Method III or IV), the 'groups' appearing as breaks in the slope of the ionisation-voltage curves. McClelland and Nolan's results have been criticised by Blackwood[2] using at first the same design of apparatus, and then one almost identical with Zeleny's original pattern which gives much greater power of 'resolving' different groups. He used ions formed by spraying and from a hot wire, but in no case found any sign of groups with definite mobility.

The average mobility decreased with age. Nolan criticised these results on the ground that the ions remained too long in the apparatus, but Busse[3], in the course of a very careful investigation on the ions formed by splashing water, could find no trace of groups even when the ions had a life as short as ·05 sec. His experiments seem conclusive so far as ions of this kind are concerned, but there does seem to be a tendency for natural ions to have a mobility round about $\frac{1}{3000}$, and perhaps also Pollock's intermediate ions should be regarded as a definite group. In some of these experiments sufficient care does not seem to have been taken to ensure the absence of eddies in the air flow.

Slow ions formed by ultra-violet light are an important possible source of error in investigations of the ionisation produced in gases by this light. They are due to a photoelectric emission of

[1] McClelland and P. J. Nolan, *Proc. Roy. Irish Acad.* xxxiii. p. 24, 1916; xxxiv. p. 51, 1918; xxxv. p. 1, 1919. J. T. Nolan, *Proc. Roy. Irish Acad.* xxxiii. p. 10, 1916; *Proc. Roy. Soc.* xciv. p. 126, 1917.

[2] Blackwood, *Phys. Rev.* xvi. p. 85, 1920.

[3] Busse, *Ann. der Phys.* lxxvi. p. 493, 1925.

electrons from small dust particles or other solid or liquid nuclei, and for this reason great care must be taken to filter gases before experiments of this kind.

Theory of Mobility of Large Ions.

41·3. The case of an ion of the size of a dust particle will approximate to that of a solid sphere moving through a viscous gas. Stokes has shown that the force which must be applied to the sphere to make it move with uniform velocity V through a gas whose coefficient of viscosity is μ is $6\pi\mu a V$, where a is the radius of the sphere. Thus if an electric force X acts on a sphere carrying a charge e of electricity

$$6\pi\mu a V = Xe,$$

or

$$V = \frac{Xe}{6\pi\mu a}.$$

Thus the mobility of the charged sphere is $e/6\pi\mu a$, and as μ is independent of the pressure of the gas the mobility of ions of this type would not vary with the pressure; the mobility of the smaller ions, as we have seen, varies inversely as the pressure. A formula for the mobility which covers both cases may be deduced by a method due to E. Cunningham[1]. It is based on the principle that when a body A is moving through a gas B, with a velocity V relative to the gas at a great distance from A, the gas in its neighbourhood will be carried along with it to some extent and will have an average velocity βV in the same direction; β will depend on the size of the body and also on the pressure of the gas. In Stokes' investigation β was assumed to be unity, in the ordinary investigation of ionic mobility it is assumed to be zero. We can however calculate β by equating the expressions for the momentum communicated to the body in unit time by the impact of the molecules through which it is moving, calculated (1) by Stokes' method, (2) by Maxwell's. If the velocity of the gas next the sphere is βV, the momentum calculated by Stokes' method is

$$6\pi\mu a\beta V,$$

[1] Cunningham, *Proc. Roy. Soc.* A. lxxxiii. p. 357, 1910.

while by Maxwell's method, assuming the collisions are like those between elastic spheres, it is (equation (5), p. 170) equal to

$$\tfrac{8}{3} n \left(\frac{\pi M}{h} \right)^{\frac{1}{2}} \sigma^2 V (1 - \beta),$$

since $V(1-\beta)$ is the relative velocity of the gas and ion. When the mass of the ion and its size are large compared with a molecule, $M = M_2$ the mass of a molecule of the gas and $\sigma = a$ the radius of an ion. Now

$$\mu = \frac{n}{3} M_2 \lambda \bar{c},$$

where λ is the free path of a molecule of the gas and $\bar{c} = 2/\sqrt{M_2 \pi h}$; hence

$$\mu = \tfrac{2}{3} n \left(\frac{M_2}{\pi h} \right)^{\frac{1}{2}} \lambda \quad \ldots\ldots\ldots\ldots\ldots\ldots (1),$$

and Maxwell's expression may be written as

$$4\pi\mu \frac{a^2}{\lambda} V (1 - \beta).$$

Equating this to Stokes' expression we have

$$6\pi\mu\beta a V = 4\pi\mu \frac{a^2}{\lambda} V (1 - \beta),$$

and
$$\beta = \frac{1}{1 + 3\lambda/2a},$$

and both Stokes' and Maxwell's expressions reduce to

$$\frac{6\pi\mu a}{1 + 3\lambda/2a}.$$

The mobility of the ion is equal to

$$e (1 + 3\lambda/2a)/6\pi\mu a,$$

and V the speed of the sphere falling under gravity is given by

$$\frac{6\pi\mu a}{1 + 3\lambda/2a} V = \tfrac{4}{3} \pi a^3 (\rho - \sigma) g,$$

whence
$$V = \frac{2a^2}{9\mu} (\rho - \sigma)(1 + 3\lambda/2a) g \quad \ldots\ldots\ldots\ldots (2),$$

where ρ is the density of the sphere and σ that of the gas through which it falls.

This formula is important because the size of small drops is frequently determined by the speed with which they fall through a gas under the action of gravity. The formula has been tested by Millikan[1], Zeleny[2] and McKeehan[3]. The agreement is satisfactory if the numerical factor 3/2 is replaced by ·864 according to Millikan or by unity according to McKeehan. It must be remembered that the magnitude of the free path is to some extent a matter of definition. The free path in equation (2) is defined by, and is to be calculated from, equation (1).[4] The coefficient of λ/a will depend upon the assumptions made as to the nature of a collision, so it is preferable to leave it to be determined by experiment. If we take McKeehan's value unity for this coefficient, the term λ/a may be written as B/pa, where p is the pressure of the gas in millimetres of mercury and $B = \cdot 0075$ for air.

For a further account of the application of this formula to small charged particles, see pp. 305, 306.

For air $\mu = 1\cdot 83 \times 10^{-4}$; substituting this value we find that the force required to drive an ion at a speed V through a gas at atmospheric pressure is

$$\frac{3\cdot 44 a \times 10^{-3} V}{1 + 10^{-5}/a}.$$

If V is the speed acquired by the ion under a force of a volt per centimetre, since this force is $10^8 \times 1\cdot 6 \times 10^{-20}$ or $1\cdot 6 \times 10^{-12}$ dynes, we have

$$\frac{2\cdot 15 \times 10^9 \, aV}{1 + 10^{-5}/a} = 1,$$

an equation which may be used to determine a. Thus for the slow ions investigated by Langevin $V = 1/3000$ and hence

$$\frac{\cdot 717 \times 10^6 \, a}{1 + 10^{-5}/a} = 1,$$

or $a = 4\cdot 5 \times 10^{-6}$ cm. approximately.

[1] Millikan, *Phil. Mag.* vi. 34, p. 10, 1917.
[2] Zeleny and McKeehan, *Phys. Rev.* xxx. p. 535, 1910.
[3] McKeehan, *Phys. Zeits.* xii. p. 707, 1911.
[4] But see note, p. 306, for Millikan's method of calculation.

Potential Gradient between two Parallel Plates immersed in an Ionised Gas and maintained at different Potentials.

42. It was shown first by Zeleny[1], and then independently by Child[2], that when electricity is passing between two plates immersed in ionised gas, the potential gradient between the plates is not uniform, but is greatest in the neighbourhood of the electrodes. The difference of potential between one of the plates and any point in the gas may be measured by having a water or mercury dropper at the point; the most convenient way, however, is to place at the point a fine wire, which will ultimately assume the potential of the point. When the wire is used it is necessary however to take several precautions: in the first place, if the number of ions in the gas is small, the wire will only take up the potential very slowly, and it is important that the instrument used for measuring the potential of the wire should have very small capacity. This consideration often makes it desirable to use a small gold-leaf electroscope to measure the potential of the wire instead of a quadrant electrometer, which though more sensitive to differences of potential has yet a very much greater capacity. Another point to be remembered is that if a wire is placed in a region where the ions are all of one sign, its potential can only change one way. Thus if it is a region where there are only positive ions, its potential can increase but cannot decrease, and thus if the potential of the wire gets by some accident too high, it cannot sink to its true value.

A characteristic curve for the distribution of potential between the plates, due to Zeleny, is given in Fig. 47. It will be seen that the gradient near the centre of the field is uniform, but that near the plates the gradients get much steeper and that they are steeper at the negative than at the positive plate.

From the equation $\dfrac{d^2V}{dx^2} = -4\pi\rho$, where V is the potential at a distance x from the plate and ρ the density of the electrification, we can, if we know the distribution of potential, calculate the density of the electrification at any point between the plates.

[1] Zeleny, *Phil. Mag.* v. 46, p. 120, 1898.
[2] Child, *Wied. Ann.* lxv. p. 152, 1898.

The density corresponding to the potential curve in Fig. 47 is shown in Fig. 48.

We see that near the positive plate there is an excess of negative electricity while near the negative plate the positive electricity is in excess. With the small potential differences used in this experiment the regions where there is an excess of one kind of electricity over the other are in the immediate neighbourhood of the plates, the density of the free electricity being exceeding small in the

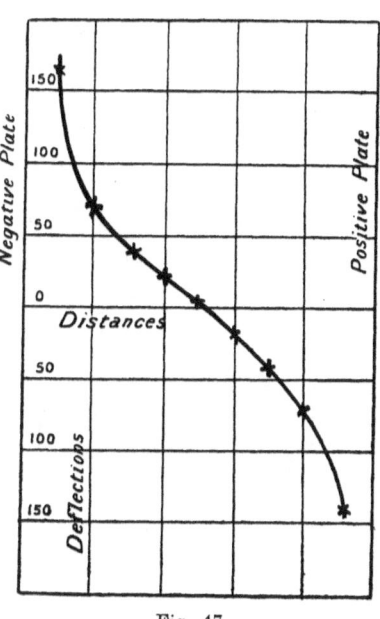

Fig. 47.

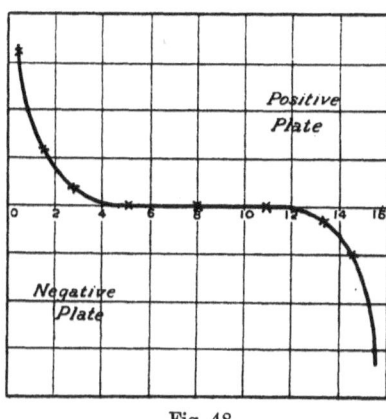

Fig. 48.

central portion of the field. If a larger potential difference had been applied to the plates, the regions with free electricity would have expanded, and with very large potential differences these regions would fill the whole of the space between the plates. In the example given, the greatest density of the electrification is about 2×10^{-4} electrostatic units; as the charge on an ion is about $4 \cdot 77 \times 10^{-10}$ such units, the number of positive ions in a cubic centimetre would exceed that of negative by about 4×10^5. Taking the number of molecules in a cubic centimetre of the gas as $2·7 \times 10^{19}$, the ratio of the excess of ions of one sign to the number of molecules is only $1·5 \times 10^{-14}$. As most of the negative ions would be driven away from the neighbourhood of the negative plate, this will approximately represent the ratio of the number of free ions to the number of molecules. This example is an illustration of the very small amount of ionisation which is sufficient to account for many of the phenomena of the conduction of electricity through gases.

CHAPTER IV

MATHEMATICAL THEORY OF THE CONDUCTION OF ELECTRICITY THROUGH A GAS CONTAINING IONS

43. WE shall now proceed to develop the theory of electric conduction through an ionised gas on the basis that the velocities of the ions are proportional to the electric force acting upon them. We shall take the case of two infinite parallel metal plates maintained at different potentials and immersed in an ionised gas; the lines of electric force are everywhere at right angles to the plates; they are thus always parallel to a line which we shall take as the axis of x.

Let n_1, n_2 be respectively the numbers of positive and negative ions per unit volume at a place fixed by the coordinate x, let q be the number of positive or negative ions produced in unit time per unit volume at this point by the ionising agent; let X be the electric intensity at this point, k_1, k_2 the velocities of the positive and negative ions under unit electric intensity, so that the velocities of these ions at this point are respectively $k_1 X$, $k_2 X$; let e be the charge on an ion. The volume density of the electrification, supposed due entirely to the presence of the ions, is $(n_1 - n_2) e$; hence we have

$$\frac{dX}{dx} = 4\pi (n_1 - n_2) e \dots \dots \dots \dots (1).$$

If i is the current through unit area of the gas, and if we neglect any motion of the ions except that caused by the electric field, we have

$$n_1 e k_1 X + n_2 e k_2 X = i \quad \dots \dots \dots \dots (2).$$

From equations (1) and (2) we get

$$n_1 e = \frac{1}{k_1 + k_2} \left\{ \frac{i}{X} + \frac{k_2}{4\pi} \frac{dX}{dx} \right\} \dots \dots \dots \dots (3),$$

$$n_2 e = \frac{1}{k_1 + k_2} \left\{ \frac{i}{X} - \frac{k_1}{4\pi} \frac{dX}{dx} \right\} \dots \dots \dots \dots (4).$$

If we measure the distribution of electric force between the plates, we can from these equations, if we know k_1 and k_2, determine n_1 and n_2; or if in addition to the distribution of electric force, we measure, by the methods previously given, n_1, n_2 at various points in the field, we can use these equations to determine k_1 and k_2, the velocities of the ions.

When the gas is in a steady state, the number of negative and of positive ions in each unit of volume must remain constant with respect to the time, thus the losses of these ions must be balanced by the gains. Now ions are lost in consequence of the recombination of the positive and negative ions: these ions will come into collision with each other, and in a certain fraction of the whole number of collisions the positive and negative ions will combine to form a single system which is electrically neutral and which no longer acts as an ion; the number of collisions in unit volume in unit time is proportional to $n_1 n_2$. We shall suppose that the number of positive or negative ions which recombine in unit volume in unit time is $\alpha n_1 n_2$: this is the rate at which unit volume is losing positive and negative ions in consequence of recombination; in consequence of ionisation it is gaining them at the rate q, and in consequence of the motion of the ions under the electric force it is losing positive ions at the rate $\dfrac{d}{dx}(n_1 k_1 X)$ and negative ones at the rate $-\dfrac{d}{dx}(n_2 k_2 X)$. The diffusion of the ions causes unit volume to lose positive and negative ions at the rates

$$-D_1 \frac{d^2 n_1}{dx^2}, \quad -D_2 \frac{d^2 n_2}{dx^2},$$

where D_1 and D_2 are the coefficients of diffusion of the positive and negative ions. Unless the electric field is very weak the motion of the ions by diffusion is, except in quite exceptional cases, insignificant in comparison with that under the electric field. We shall therefore for the present leave diffusion out of account. Hence when the gas is in a steady state we have

$$\frac{d}{dx}(n_1 k_1 X) = q - \alpha n_1 n_2 \quad \ldots\ldots\ldots\ldots (5),$$

$$-\frac{d}{dx}(n_2 k_2 X) = q - \alpha n_1 n_2 \quad \ldots\ldots\ldots\ldots (6).$$

If k_1 and k_2 are constant at all parts of the field, we have from (1), (5) and (6)

$$\frac{d^2 X^2}{dx^2} = 8\pi e\,(q - an_1 n_2)\left(\frac{1}{k_1} + \frac{1}{k_2}\right) \quad \ldots\ldots\ldots(7).$$

From this equation, if we know the distribution of X^2 between the plates, we can determine whether ionisation or recombination is in excess at any point, for from (7) $q - an_1 n_2$ and $d^2 X^2/dx^2$ have the same sign; hence when ionisation is in excess of recombination, i.e. when $q - an_1 n_2$ is positive, $d^2 X^2/dx^2$ is positive and the curve whose ordinate is X^2 is convex to the axis of x; when recombination is in excess of ionisation the curve for X^2 is concave to the axis of x.

Substituting in equation (7) the values of n_1, n_2 given by equations (3) and (4), we get

$$\frac{d^2 X^2}{dx^2} = 8\pi e \left(\frac{1}{k_1} + \frac{1}{k_2}\right) \left\{ q - \frac{\alpha}{e^2 X^2 (k_1 + k_2)^2} \right.$$
$$\left. \times \left(i + \frac{k_2}{8\pi}\frac{dX^2}{dx}\right)\left(i - \frac{k_1}{8\pi}\frac{dX^2}{dx}\right) \right\} \ldots(8).$$

I have not been able to get a general solution of this differential equation except when q is constant and $k_1 = k_2$; in that case putting $X^2 = y$ and $\frac{dy}{dx} = p$ we get, writing k for either k_1 or k_2,

$$p \frac{dp}{dy} = \frac{16\pi e}{k} \left\{ q - \frac{\alpha}{4e^2 k^2 y}\left(i^2 - \frac{k^2 p^2}{64\pi^2}\right) \right\} \quad \ldots\ldots(9).$$

Integrating this we get

$$\frac{k^2 p^2}{64\pi^2} - i^2 = \frac{qek}{2\pi\left(1 - \dfrac{\alpha}{8\pi ek}\right)} y + Cy^{\frac{\alpha}{8\pi ek}} \quad \ldots\ldots(10),$$

where C is a constant of integration. From this equation we can find the ratio of X_0, the electric intensity midway between the plates, to X_1, the electric intensity close to a plate. For when $k_1 = k_2$ the distribution of electric force is symmetrical and midway between the plates dX/dx and $p = 0$; let us further assume that we are dealing with a case like that in Fig. 48, where there is no free electricity for some distance from the plane midway between the plates, so that here d^2X/dx^2 also vanishes; hence from (9) and (10) we have

$$X_0{}^2 = \frac{ai^2}{4e^2k^2q},$$

$$\frac{-X_0{}^2 \dfrac{4k^2e^2q}{a}}{1 - \dfrac{a}{8\pi ek}} = CX_0^{\frac{a}{4\pi ek}} \ldots\ldots\ldots\ldots (11).$$

Now at the positive plate $n_1 = 0$ and at the negative plate $n_2 = 0$; hence at either plate $n_1 n_2 = 0$, but

$$n_1 n_2 = \frac{1}{4k^2e^2X^2}\left(i^2 - \frac{k^2p^2}{64\pi^2}\right);$$

hence if X_1 is the value of X at either plate, we have

$$\frac{-X_1{}^2 \dfrac{qek}{2\pi}}{1 - \dfrac{a}{8\pi ek}} = CX_1^{\frac{a}{4\pi ek}} \ldots\ldots\ldots\ldots (12).$$

Hence by (11) and (12)

$$\frac{8\pi ke}{a} = \left(\frac{X_0}{X_1}\right)^{\frac{a}{4\pi ek} - 2},$$

or writing β for $8\pi ke/a$ we get

$$\frac{X_0{}^2}{X_1{}^2} = \beta^{\frac{\beta}{1-\beta}}.$$

We see from this equation that X_0/X_1 is never greater than unity, for $\beta^{\frac{\beta}{1-\beta}}$ diminishes from unity to zero as β increases from $\beta = 0$ to $\beta =$ infinity. Since β does not involve either q or i, the ratio of the electric intensities does not depend upon either the intensity of the ionisation or the current between the plates. For air at atmospheric pressure $k = 520$ (since unit electrostatic force is 300 volts per centimetre), a is about $1\cdot 6 \times 10^{-6}$, and $e = 4\cdot 77 \times 10^{-10}$; substituting these values we find $\beta = 3\cdot 9$ for air at atmospheric pressure. Since k is inversely and a directly proportional to the pressure, β is inversely proportional to the square of the pressure, and thus is very large at the pressure of a few millimetres of mercury. Putting $\beta = 4$ we find

$$\frac{X_1}{X_0} = 4^{\frac{2}{3}} = 2\cdot 51 \text{ approximately.}$$

At low pressures β is large, in this case $X_1/X_0 = \beta^{\frac{1}{2}}$ approximately, and thus the ratio of X_1 to X_0 varies inversely as the pressure.

The experiments we have described on the distribution of electric force between the plates show that when the current is small, the regions where X differs appreciably from X_0 are confined to two layers near the plates, the distribution of X between the plates being represented by a curve like that shown in Fig. 49. We can very easily find an inferior limit to λ, the thickness of one of these layers. For let P be a point on the boundary of the

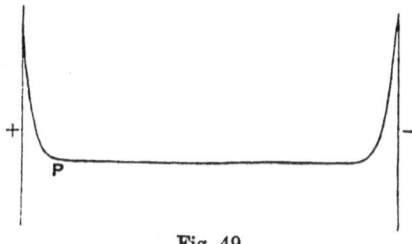

Fig. 49.

layer next the electrode, then since X becomes constant at P, there are at this point as many positive as negative ions per unit volume, and if the velocities of the ions are the same, half the current must be carried by the positive and half by the negative ions. Thus if i is the current through unit area, and e the charge on an ion, $i/2e$ positive ions must cross unit area of a plane through P in unit time; and all these positive ions must be produced in the region between P and the positive plate. But if λ is the thickness of the layer, the number of positive ions produced in unit time corresponding to each unit area of the plate is $q\lambda$, the number that cross unit area at P cannot therefore be greater than $q\lambda$, and can only be as great when there is no recombination of the ions between P and the positive plate, hence

$$q\lambda > \frac{i}{2e},$$

or $\lambda > i/2eq$; thus $i/2eq$ is an inferior limit to λ. If I is the maximum current, l the distance between the plates, $I = qle$; hence $i/2I$ is an inferior limit to λ/l.

44. Though we cannot find a general solution of the equations (1), (2), (5), (6) when k_1 is not equal to k_2 we see at once that a particular solution of these equations is given by the relations

$$n_1 = n_2 = (q/\alpha)^{\frac{1}{2}},$$

$$k_1 n_1 Xe = \frac{k_1}{k_1 + k_2} i,$$

$$k_2 n_2 Xe = \frac{k_2}{k_1 + k_2} i,$$

$$X = \left(\frac{\alpha}{q}\right)^{\frac{1}{2}} \frac{i}{e(k_1 + k_2)}.$$

This solution corresponds to a constant value of the electric force between the plates, and indicates that the proportion of the current carried by the positive and negative ions respectively is the same as the ratio of the velocities of these ions. This solution, though it may apply to the central portion of the field, cannot however hold right up to the plates. For suppose P is a point between the plates at which this solution applies. Then across unit area at P, $ik_1/(k_1 + k_2) e$ positive ions pass in unit time, and these must come from the region between P and the positive plate; if the distance of P from this plate is λ this region cannot furnish more than $q\lambda$ positive ions in unit time, and can only do this when there is no recombination; hence the preceding solution cannot hold at a distance from the positive plate less than

$$\frac{k_1}{k_1 + k_2} \frac{i}{qe}.$$

Similarly it cannot hold at a distance from the negative plate less than

$$\frac{k_2}{k_1 + k_2} \frac{i}{qe}.$$

We shall assume that the preceding solution does hold at distances from the plates greater than the preceding values: and further that in the layers in which the solution does not hold there is no recombination of the ions.

Let us consider the state of things at the positive plate between $x = 0$ and $x = \lambda_1$, where

$$\lambda_1 = \frac{k_1}{k_1 + k_2} \frac{i}{qe}.$$

Then, since in this region there is no recombination, equations (1), (5), (6) become

$$\frac{dX}{dx} = 4\pi (n_1 - n_2) e,$$

$$\frac{d}{dx}(k_1 n_1 X) = q,$$

$$\frac{d}{dx}(k_2 n_2 X) = -q.$$

If q is constant we have

$$k_1 n_1 X = qx,$$

$$k_2 n_2 X = \frac{i}{e} - qx,$$

where the constant of integration has been chosen so as to make $n_1 = 0$ when $x = 0$: substituting these values for n_1, n_2 in the equation giving dX/dx we get

$$X \frac{dX}{dx} = 4\pi e \left\{ qx \left(\frac{1}{k_1} + \frac{1}{k_2} \right) - \frac{i}{ek_2} \right\},$$

or

$$X^2 = 8\pi e \left\{ \frac{1}{2} qx^2 \left(\frac{1}{k_1} + \frac{1}{k_2} \right) - \frac{ix}{ek_2} \right\} + C \ldots (13),$$

where C is a constant which may be determined from the condition that when $x = \lambda_1$

$$X^2 = \frac{\alpha}{q} \frac{i^2}{e^2 (k_1 + k_2)^2};$$

from this we find

$$C = \frac{\alpha}{q} \frac{i^2}{e^2 (k_1 + k_2)^2} \left\{ 1 + \frac{4\pi e}{\alpha} \frac{k_1}{k_2} (k_1 + k_2) \right\}.$$

C is the value of X^2 when $x = 0$, *i.e.* at the positive plate; if we call this value X_1, and if X_0 is the constant value of X between the layers, we have

$$X_1 = X_0 \left\{ 1 + \frac{4\pi e}{\alpha} \frac{k_1}{k_2} (k_1 + k_2) \right\}^{\frac{1}{2}};$$

thus X_1 is always greater than X_0 and the ratio X_1/X_0 does not depend upon the amount of ionisation or the strength of the current between the plates.

If X_2 is the value of X at the negative plate, we can prove in a similar way that

$$X_2 = X_0 \left\{ 1 + \frac{4\pi e}{\alpha} \frac{k_2}{k_1} (k_1 + k_2) \right\}^{\frac{1}{2}}.$$

Thus if k_2, the velocity of the negative ion, is very large compared with k_1, the velocity of the positive ion, the value of X at the negative plate is large compared with its value at the positive, and the thickness of the layer in which X is variable, is greater at the negative than it is at the positive plate. A curve representing the distribution of electric intensity between the plates in this case is represented in Fig. 50.

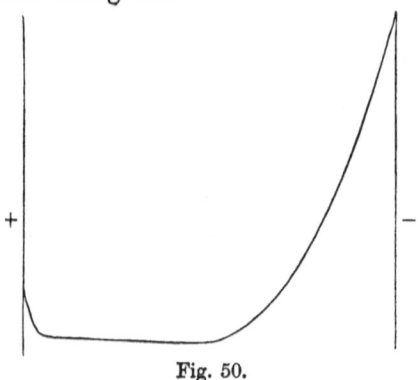

Fig. 50.

If we put
$$\beta_1 = \frac{4\pi e}{\alpha} \frac{k_1}{k_2}(k_1 + k_2); \quad \beta_2 = \frac{4\pi e}{\alpha} \frac{k_2}{k_1}(k_1 + k_2),$$
we have $\quad X_1 = X_0(1 + \beta_1)^{\frac{1}{2}}; \quad X_2 = X_0(1 + \beta_2)^{\frac{1}{2}};$
when β_1 and β_2 are large we have approximately
$$X_1 = X_0 \beta_1^{\frac{1}{2}}; \quad X_2 = X_0 \beta_2^{\frac{1}{2}}.$$

In the special case when the velocities of the positive and negative ions are equal $\beta_1 = \beta_2$ and $X_1/X_0 = (8\pi e k_1/\alpha)^{\frac{1}{2}}$; this agrees when β is large with the result found by the independent investigation of this case given on p. 196.

The fall of potential V_1 across the layer next the positive plate whose thickness is λ_1 is equal to
$$\int_0^{\lambda_1} X \, dx;$$
substituting the value of X given by equation (13) and integrating we find
$$V_1 = \frac{1}{2} X_1 \lambda_1 + \frac{1}{2} \frac{X_0 \lambda_1}{\sqrt{\beta_1}} \log(\sqrt{\beta_1} + \sqrt{1 + \beta_1})$$
$$= \frac{1}{2} X_0 \lambda_1 \left\{ (1 + \beta_1)^{\frac{1}{2}} + \frac{1}{\sqrt{\beta_1}} \log(\sqrt{\beta_1} + \sqrt{1 + \beta_1}) \right\}.$$

Since $X_0 = \left\{\dfrac{\alpha}{q}\right\}^{\frac{1}{2}} \dfrac{i}{e(k_1+k_2)}$, and $\lambda_1 = \dfrac{k_1}{k_1+k_2} \dfrac{i}{qe}$,

$$V_1 = \frac{1}{2} \frac{\alpha^{\frac{1}{2}}}{q^{\frac{3}{2}}} \frac{i^2 k_1}{e^2 (k_1+k_2)^2} \left\{(1+\beta_1)^{\frac{1}{2}} + \frac{1}{\sqrt{\beta_1}} \log(\sqrt{\beta_1} + \sqrt{1+\beta_1})\right\} \ldots (14).$$

Thus the fall of potential across this layer is proportional to the square of the current.

If V_2 is the change in potential in crossing the layer next the negative electrode, we find similarly

$$V_2 = \frac{1}{2} \frac{\alpha^{\frac{1}{2}}}{q^{\frac{3}{2}}} \frac{i^2 k_2}{e^2 (k_1+k_2)^2} \left\{(1+\beta_2)^{\frac{1}{2}} + \frac{1}{\sqrt{\beta_2}} \log(\sqrt{\beta_2} + \sqrt{1+\beta_2})\right\}.$$

If β_1 and β_2 are very large we have approximately

$$V_1 = \frac{1}{2} \frac{\alpha^{\frac{1}{2}}}{q^{\frac{3}{2}}} \frac{i^2}{e^2} \frac{k_1}{(k_1+k_2)^2} \beta_1^{\frac{1}{2}} = \frac{\sqrt{\pi} i^2}{k_2^2} \left(\frac{k_1 k_2}{qe(k_1+k_2)}\right)^{\frac{3}{2}},$$

$$V_2 = \frac{1}{2} \frac{\alpha^{\frac{1}{2}}}{q^{\frac{3}{2}}} \frac{i^2}{e^2} \frac{k_2}{(k_1+k_2)^2} \beta_2^{\frac{1}{2}} = \frac{\sqrt{\pi} i^2}{k_1^2} \left(\frac{k_1 k_2}{qe(k_1+k_2)}\right)^{\frac{3}{2}}.$$

Thus

$$\frac{V_1}{V_2} = \frac{k_1^2}{k_2^2},$$

or the falls of potential at the positive and negative plates are proportional to the squares of the velocities of the positive and negative ions.

Let us consider how the fall of potential varies with the pressure of the gas: if p is the pressure, k_1 and k_2 are inversely proportional to p, and q is directly proportional to p, hence we see that for a given current V_1 and V_2 vary inversely as p.

The Relation between the Potential Difference between the Plates and the Current.

45. The fall of potential between the plates is made up of the fall of potential at the layers which we have already calculated and the fall of potential in the space between the layers where the electric intensity is uniform and equal to X_0; the breadth of this space is $l - (\lambda_1 + \lambda_2)$, where l is the distance between the plates, and since $\lambda_1 + \lambda_2$ is equal to i/qe, the fall of potential in this space is equal to

$$X_0 \left(l - \frac{i}{qe}\right),$$

or to

$$\left(\frac{a}{q}\right)^{\frac{1}{2}} \frac{i}{e(k_1 + k_2)} \left(l - \frac{i}{qe}\right);$$

adding to this the values for the fall of potentials across the layers we get, if V is the potential difference between the plates,

$$V = \frac{1}{2} \frac{a^{\frac{1}{2}}}{q^{\frac{3}{2}}} \frac{i^2}{e^2 (k_1 + k_2)^2} \Big\{ k_1 (1 + \beta_1)^{\frac{1}{2}} + \frac{k_1}{\sqrt{\beta_1}} \log(\sqrt{\beta_1} + \sqrt{1 + \beta_1})$$

$$+ k_2 (1 + \beta_2)^{\frac{1}{2}} + \frac{k_2}{\sqrt{\beta_2}} \log(\sqrt{\beta_2} + \sqrt{1 + \beta_2}) \Big\}$$

$$+ \frac{a^{\frac{1}{2}}}{q^{\frac{1}{2}}} \frac{i}{e(k_1 + k_2)} \left(l - \frac{i}{qe}\right) \quad \ldots\ldots\ldots(15).$$

This equation is of the form

$$V = Ai^2 + Bi,$$

thus the curve whose ordinate is i and abscissa V is a parabola. This equation ceases to be an approximation to the truth when the two layers touch, i.e. when $\lambda_1 + \lambda_2 = l$ or $i = qel$; in this case the current is the greatest that can be carried by the ionised gas. The minimum value of the potential difference required to produce this current is got by putting $i = qel$ in equation (15); we see that the potential difference required to produce saturation is proportional to the square of the distance between the plates and to the square root of the intensity of ionisation.

46. The study of the distribution of electric intensity between the plates when the maximum current is passing leads to an easy way of finding the ratio of the velocities of the positive and negative ions, for as in this case there is no recombination, equations (5) and (6), p. 194, give

$$k_1 n_1 X = qx \quad \ldots\ldots\ldots\ldots\ldots\ldots(16),$$
$$k_2 n_2 X = q(l - x) \quad \ldots\ldots\ldots\ldots\ldots(17),$$

where x is measured from the positive plate. At the point between the plates where the force is a minimum

$$\frac{dX}{dx} = 0 = 4\pi (n_1 - n_2) e,$$

hence at this point $n_1 = n_2$, so that if x is the distance of the point P where X is a minimum from the positive plate we have by equations (16) and (17)
$$\frac{k_1}{k_2} = \frac{x}{l-x};$$
thus the ratio of the velocities of the positive and negative ions is equal to the ratio of the distances of P from the positive and negative plates, so that if we have determined P by measuring the distribution of potential between the plates we can at once deduce the ratio of the velocities.

47. Mie[1] has by successive approximations obtained solutions of equation (8), p. 195, (1) when the current is only a small fraction of the saturation current, (2) when the current is nearly saturated. The results of his investigation are shown in Fig. 51. Fig. 51

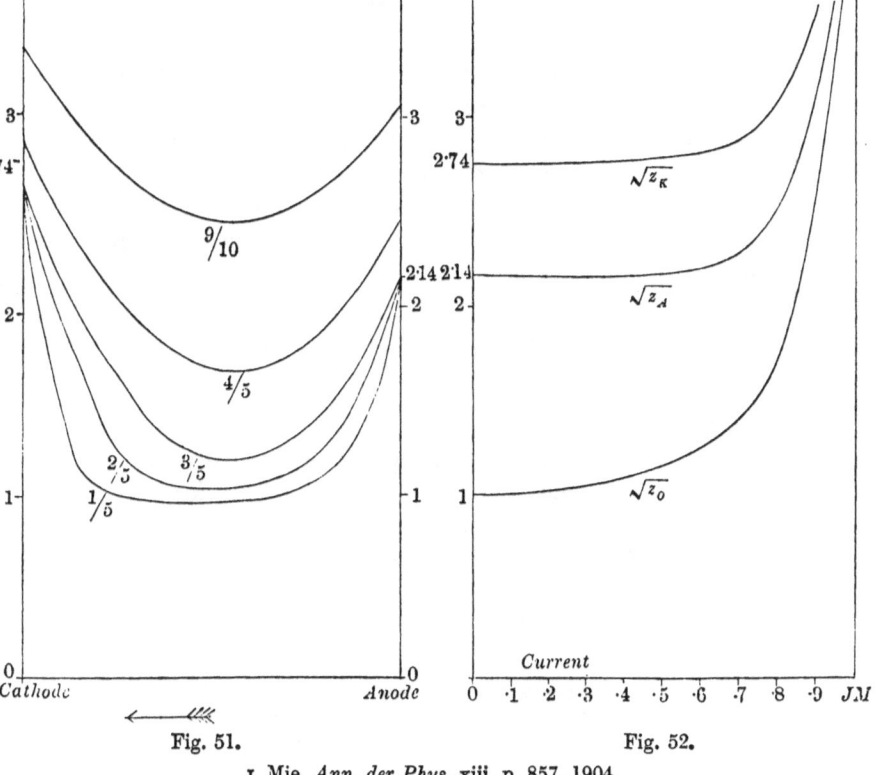

Fig. 51. Fig. 52.

[1] Mie, *Ann. der Phys.* xiii. p. 857, 1904.

represents the distribution of the electric force along the current for various values of the current expressed as fractions of the saturation current. It will be seen that until the current amounts to about 1/5 of the maximum current the type of solution is that indicated on p. 197, *i.e.* the electric force is constant except in the neighbourhood of the electrodes, where it increases rapidly. In Fig. 52 the quantities denoted by $\sqrt{z_0}$, $\sqrt{z_A}$, $\sqrt{z_K}$ are proportional respectively to the ratios of the minimum electric force, the force at the anode, and the force at the cathode to the current. The result indicated by the solution on p. 196, that the ratios of the forces close to the electrodes to the minimum electric force are independent of the current, is seen to hold for a wide range of currents.

Solutions of equation (7), p. 195, at pressures chosen so as to make $\alpha/4\pi ek$ have special values, have been given by G. W. Walker[1] and Robb[2].

Seeliger[3] has also investigated the form of the saturation curve for an ionised gas between plane parallel electrodes. In an elaborate research Seemann[4] has determined this experimentally, using X-rays as the source of ionisation. He finds excellent agreement with the theories of Mie and Seeliger, which in his case give practically identical results. There is tolerable agreement with the approximate theory of § 44.

Case when the Carriers of Negative Electricity are Electrons.

48·1. In this case the velocity of the negative carrier is very much greater than that of the positive, *i.e.* with the notation of § 44, k_2 is very large compared with k_1; when this is so we see from the results given in that article that except near the cathode the electric force X_0 is given by the equation

$$X_0 = \sqrt{\frac{\alpha}{q} \frac{i}{ek_2}} \quad \ldots\ldots\ldots\ldots\ldots\ldots(18).$$

[1] Walker, *Phil. Mag.* Nov. 1904.
[2] Robb, *Phil. Mag.* Aug. and Dec. 1905.
[3] Seeliger, *Ann. der Phys.* xxxiii. p. 319, 1910.
[4] Seemann, *Ann. der Phys.* xxxviii. p. 781, 1912.

The force at the cathode is approximately
$$2\left(\frac{\pi e}{qk_1}\right)^{\frac{1}{2}}\frac{i}{e} \quad \ldots\ldots\ldots\ldots\ldots\ldots(19);$$
the thickness of the space within which the force falls from this value to X_0 is
$$\frac{i}{qe} \quad \ldots\ldots\ldots\ldots\ldots\ldots\ldots\ldots(20),$$
and the fall of potential in crossing this space is
$$\left(\frac{\pi e}{q^3 k_1}\right)^{\frac{1}{2}}\frac{i^2}{e^2} \quad \ldots\ldots\ldots\ldots\ldots\ldots(21);$$
if l is the distance between the electrodes, V the whole fall of potential is given by the equation
$$V = \left(\frac{\pi e}{q^3 k_1}\right)^{\frac{1}{2}}\frac{i^2}{e^2} + \left(l - \frac{i}{qe}\right)\left(\frac{\alpha}{qk_2^2}\right)^{\frac{1}{2}}\frac{i}{e} \quad \ldots\ldots\ldots(22).$$
Thus the relation between the potential difference and current is of the form
$$V = Ai^2 + Bli,$$
and in general the term Ai^2 is the more important.

The force at a distance x from the cathode is by equation (13) given by
$$X^2 = \frac{4\pi eq}{k_1}(x - \lambda)^2 + \frac{\alpha}{qe^2}\frac{i^2}{k_2^2},$$
where
$$\lambda = \frac{i}{qe}.$$

Unless x is very nearly equal to λ, the first term on the right-hand side of this equation is much greater than the second, so that approximately
$$X = \sqrt{\frac{4\pi eq}{k_1}}(\lambda - x);$$
hence the force diminishes linearly as the distance from the cathode increases; and the density of the electrification is from this equation equal to
$$\sqrt{\frac{eq}{4\pi k_1}},$$
and is thus approximately constant at small distances from the cathode.

49·1. The expressions we have obtained throw light on the very interesting effects which are observed when a bead of salt is put in a flame through which an electric current is passing; the salt vaporises and the ionisation in the salt vapour is much more intense than in the other parts of the flame. Hittorf found, however, that unless the bead of salt is introduced near the cathode it produces but little effect upon the current; when however the salt vapour reaches the cathode the salt enormously increases the current. This is what we should expect from the preceding theory. When the salt is not near the cathode it will only affect the value of X_0, it will diminish this for a constant current. Since the introduction of the salt will increase the value of q, the introduction of the salt in this region will diminish the potential fall in the part of the flame away from the cathode. The potential fall in this region is but a small fraction of the potential difference between the electrodes, so that changes in it will produce but a comparatively small effect. When however the salt vapour extends up to the cathode and produces great ionisation, then since the fall of potential is proportional to $q^{-\frac{2}{3}}$ an increase in q will produce a very large diminution in the potential fall at the cathode; as this potential is by far the greater part of the potential difference between the electrodes it will require far less potential difference to produce a given current in a salted flame than in an unsalted one, and with the same potential difference the current will be increased greatly by the salt.

On the other hand anything which reduced the ionisation near the cathode, such as a reduction of temperature, would lead to a great diminution in the current.

Case when the Ionisation is confined to a Thin Layer.

50. In the preceding investigation we have supposed that the ionisation is uniformly distributed between the plates; there are however many very important cases when ionisation only takes place in a thin layer of the gas, the rest of the space between the plates being free from the action of the ionising agent. We proceed now to the consideration of this case, beginning with the one where the ionised layer is close to one of the plates A. Let us suppose that A is the positive plate, then all the ions in the space between the plates must have been dragged by the action of the electric field from the layer, hence these ions must be all positive, so that the current is carried entirely by positive ions. Let there be n_1 of these ions per cubic centimetre and let X be the electric force, i the current, then using the same notation as before our equations are now

$$\frac{dX}{dx} = 4\pi n_1 e,$$

$$k_1 n_1 X e = i;$$

from these equations we get

$$\frac{XdX}{dx} = \frac{4\pi i}{k_1},$$

or
$$X^2 = \frac{8\pi i x}{k_1} + C \quad \ldots\ldots\ldots\ldots (23),$$

where C is the constant of integration; it is evidently the value of X^2 close to the positive plate.

If V is the potential difference between the plates, and l their distance apart, we have

$$V = \int_0^l X dx = \frac{k_1}{12\pi i} \left[\left(\frac{8\pi i l}{k_1} + C\right)^{\frac{3}{2}} - C^{\frac{3}{2}} \right] \quad \ldots (24).$$

To find an expression for C we must turn our attention to the layer of ionised gas; let us suppose that the current is small compared with that required to saturate this layer, then the number of free positive or negative ions in unit volume of the layer $= (q/\alpha)^{\frac{1}{2}}$, if q as before measures the intensity of ionisation; if there is no great change in the electric force as we pass from the gas into the layer, the sum of the velocities of the positive and negative ions will be of the order $(k_1 + k_2) C^{\frac{1}{2}}$, and as i the current equals the number of ions multiplied by the sum of the velocities of the ions, $e(k_1 + k_2) C^{\frac{1}{2}} (q/\alpha)^{\frac{1}{2}}$ will be of the same order as i; hence C is comparable with

$$\frac{i^2 \alpha}{q e^2 (k_1 + k_2)^2}.$$

Hence C will be small compared with $8\pi i l/k_1$ if

$$\frac{i \alpha k_1}{8\pi q e^2 (k_1 + k_2)^2 l}$$

is a small quantity.

If δ is the thickness of the ionised layer, I the saturation current,
$$I = q e \delta;$$
thus the preceding quantity will be small if
$$\frac{1}{2} \frac{i}{I} \frac{\delta}{l} \frac{k_2}{(k_1 + k_2)} \frac{1}{\beta_2} \text{ is small,}$$
where
$$\beta_2 = \frac{4\pi e}{\alpha} \frac{k_2}{k_1} (k_1 + k_2).$$

If δ/l, i/I are small, then since β_2 is greater and $k_2/(k_1 + k_2)$ less than unity, we see that the quantity under consideration will be small. When this is the case we can, in equation (24), neglect C in comparison with $\dfrac{8\pi i l}{k_1}$, and the equation becomes
$$V^2 = \frac{32\pi}{9 k_1} i l^3 \quad \ldots\ldots\ldots\ldots\ldots (25).$$

We see that the current is proportional to V^2, and thus increases more rapidly with increasing potential difference than if the conduction followed Ohm's law. We shall see examples of this when we consider the passage of electricity from hot metals immersed in gases. In this case by far the greater part of the ionisation occurs in the layer next the metal and, as Pringsheim[1] has shown, the current increases more rapidly than the potential difference. The current is proportional to k_1, the velocity of the ion which carries it; thus since the velocity of the negative ion is greater than that of the positive, the current for the same difference of potential between the plates is greater when the ionisation takes place next the negative plate than when next the positive, in other words the current is greater in one direction than in the opposite; this unipolar conductivity as it is called is very marked indeed in conduction through hot gases and flames containing salts. Rutherford[2] has observed it when the ionisation was due to X-rays or radium radiation. We see from (25) that for a given potential difference the current is independent of q, the intensity of ionisation; the maximum current between the plates will of course depend upon the intensity of the ionisation,

[1] Pringsheim, *Wied. Ann.* lv. p. 507, 1895.
[2] Rutherford, *Phil. Mag.* vi. 2, p. 210, 1901.

but as long as the currents are only a small fraction of the maximum corresponding to the ionisation, they will be independent of the amount of ionisation next the plate; we see too that the current does not depend on the charge carried by the ion.

The current for a given difference of potential varies inversely as the cube of the distance between the plates; as the current varies as the square of the potential difference, if the average electric intensity between the plates remains constant as the distance diminishes, the current will vary inversely as the distance between the plates.

When the ionisation is confined to a layer next the plate A, we can stop the flow of ions and therefore of electricity to the plate B by interposing between A and B a third plate, and the passage of electricity will be stopped just as well by a plate of metal as by a non-conductor; thus we get the somewhat paradoxical effect of completely stopping a current between two plates by interposing between them an excellent conductor of electricity. An example of this effect will be considered when we discuss the passage of electricity through very hot gases.

If the layer of ionised gas is situated between the plates at a distance l_1 from the positive and l_2 from the negative plate, then if V is the potential difference between the plates, we can easily prove by the preceding method that

$$V = \tfrac{2}{3}\sqrt{8\pi i}\left\{\frac{l_1^{\frac{3}{2}}}{k_2^{\frac{1}{2}}} + \frac{l_2^{\frac{3}{2}}}{k_1^{\frac{1}{2}}}\right\},$$

where k_1 and k_2 are respectively the mobilities of the positive and negative ions. We see that if k_1 is not equal to k_2 the current for the same potential difference will not, unless $l_1 = l_2$, be the same in one direction as in the opposite. If the velocity of the negative ion is greater than that of the positive, the current will be greatest when its direction is such that the negative plate is nearer to the ionised layer than the positive. From this we conclude that want of symmetry in the distribution of ionisation will give rise to unipolar conductivity. The distribution of electric intensity when the ionised layer is between the plates is represented in Fig. 53.

The preceding results are only true when the electric intensity close to the ionised layer is small compared with its value some distance away. We shall now consider from another point of view the condition for this to be the case. If the saturation current, I, passed between the plates the maximum electric force would be greater than $\sqrt{8\pi Il/k_1}$, and the potential difference greater than $l^{\frac{3}{2}}\sqrt{32\pi I/9k_1}$: unless then the potential difference exceeds this value we cannot have saturation and therefore the force near the

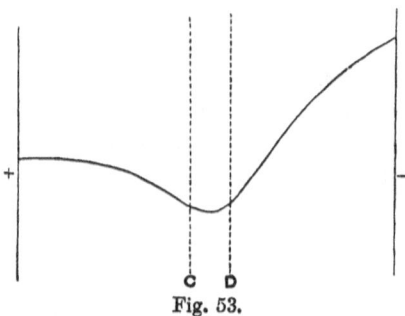

Fig. 53.

plate cannot be great enough to drag all the ions produced near the plate away into the field. In order to get these ions away we have not only to contend against the recombination of the ions, but also against the tendency of the ions to diffuse back into the plate; indeed in many important cases when only ions of one sign come into the gas, as for example when a metal plate emits negative ions by exposure to ultra-violet light, or when a red-hot plate emits positive or negative ions, diffusion is the only thing the field has to overcome in order to saturate the current. In such cases the ions unless removed by the field accumulate round the plate, until the number striking against the plate in unit time is equal to the number emitted by the plate in that time. These ions, like the molecules of any gas, have an average velocity of translation U depending on their absolute temperature, and the electric field will not be strong enough to remove these ions from the plate unless it is able to impart a velocity U to them. If the field required to do this is small compared with $\sqrt{8\pi Il/k_1}$ then we are justified in putting C in equation (23), p. 207, equal to zero. If however the force required to produce the velocity is comparable with $\sqrt{8\pi Il/k_1}$ we cannot assume that C is zero and the force near the plate, as is generally the case when the currents are produced by ultra-violet light, may be comparable with that in other parts of the field. The greater the value of I the more likely are the conditions justifying the zero value of C to be satisfied.

50·1. If we take into account the diffusion of the electrons, then if D is the coefficient of diffusion, and k the mobility, the rest of the notation being the same as before,

$$\frac{k}{4\pi} X \frac{dX}{dx} - De\frac{dn}{dx} = i,$$

or

$$\frac{k}{4\pi} X \frac{dX}{dx} - \frac{D}{4\pi}\frac{d^2 X}{dx^2} = i,$$

hence

$$\frac{kX^2}{8\pi} - \frac{D}{4\pi}\frac{dX}{dx} = i(x+\delta) \quad \ldots \ldots \ldots \ldots (26),$$

where δ is a constant of integration. This is a form of Riccati's equation. Introduce a new variable w, defined by the equation

$$X = -\frac{2D}{k}\frac{d\log w}{d\xi},$$

where

$$\xi = x + \delta.$$

Equation (26) may be written as

$$\frac{d^2 w}{d\xi^2} = \frac{2\pi i k}{D^2} \xi w;$$

or putting

$$\left(\frac{2\pi i k}{D^2}\right)^{\frac{1}{3}} \xi = \eta \quad \ldots \ldots \ldots \ldots \ldots \ldots (27),$$

$$\frac{d^2 w}{d\eta^2} = \eta w.$$

The solution of this equation is (see Watson's *Bessel's Functions*, p. 88),

$$w = \eta^{\frac{1}{2}} \{AJ_{\frac{1}{3}}(i\eta^{\frac{3}{2}}/\frac{3}{2}) + BJ_{-\frac{1}{3}}(i\eta^{\frac{3}{2}}/\frac{3}{2})\} \quad \ldots \ldots (28),$$

where A and B are arbitrary constants, and the J's denote Bessel's Functions. It is only the ratio of A to B that is of any significance; we can determine the value of this ratio and of δ by the condition that when $x = 0$, since $nc/\sqrt{6\pi}$ is, if c is the velocity due to thermal agitation, the number of ions passing through unit area in unit time,

$$ecn = (I - i)\sqrt{6\pi},$$

and when $x = l$,

$$i = kXne + \frac{nc}{\sqrt{6\pi}} e,$$

or

$$D\frac{dn}{dx} + \frac{nc}{\sqrt{6\pi}} = 0,$$

$$n = \frac{1}{4\pi}\frac{dX}{d\xi}$$

$$= -\frac{D}{2\pi k}\frac{d^2}{d\xi^2}\log w$$

$$= -\frac{i^{\frac{2}{3}}}{(2\pi kD)^{\frac{1}{3}}}\frac{d^2}{d\eta^2}\log w.$$

The expressions in the general case when no supposition is made about the magnitude of η are very complicated; we shall consider two cases: (1) when η is small, (2) when η is large. The condition that η should be small in the region between the plates is from (27) that $2\pi ikl^3/D^2$ should be small. When η is small, if we expand w in ascending powers of η, we see from the form of the differential equation that

$$w = C + D\eta^3 + \text{higher powers of } \eta;$$

and from the equation

$$\frac{d^2w}{d\eta^2} = w\eta$$

we see

$$D = C/6;$$

hence approximately

$$w = C\left(1 + \frac{\eta^3}{6}\right)$$

$$= C\left(1 + \frac{2\pi ik}{6D^2}\xi^3\right);$$

hence

$$X = -\frac{2D}{k}\frac{\pi ik}{D^2}\xi^2$$

$$= -\frac{2\pi i}{D}\xi^2,$$

and
$$ne = \frac{1}{4\pi}\frac{dX}{d\xi}$$
$$= -\frac{i}{D}\xi = -\frac{i}{D}(x+\delta).$$

When $x = 0$,
$$\frac{nce}{\sqrt{6\pi}} = I - i,$$

hence
$$I - i = -\frac{c}{\sqrt{6\pi}}\frac{\delta}{D}i;$$

and when $x = l$,
$$D\frac{dn}{dx} + \frac{nc}{\sqrt{6\pi}} = 0,$$

hence
$$i = -\frac{ci}{\sqrt{6\pi}D}(l+\delta),$$

and
$$\frac{i}{I} = \frac{1}{2 + \frac{c}{\sqrt{6\pi}}\frac{l}{D}}.$$

Since D is of the order $c\lambda$, where λ is the mean free path of the electron, which for the equations to apply must be small compared with l,
$$\frac{i}{I} = \frac{\sqrt{6\pi}D}{cl} \text{ approximately.}$$

It should be noticed that the condition that η should be small can only be satisfied when the currents passing through the gas are excessively minute. For since all the quantities are to be measured in electrostatic units, $k/D = 1.3 \times 10^4$. Hence for η to be small
$$2\pi i \times 1.3 \times 10^4 l^3/D$$
must be small; D will be inversely proportional to the pressure, so that *ceteris paribus* the chance that η should be small is greater at low pressures than at high; if the pressure is so low that the free path of an electron is one millimetre, D will be about 10^6, so that η will be about
$$2\pi i \times 1.3 \times 10^{-2} l^3;$$
hence if l is one centimetre, i, measured in electrostatic units, must not be much greater than unity, *i.e.* must not be greater than about $.3 \times 10^{-9}$ amperes.

When η is very large, since for large values of x
$$J_n(ix) = \frac{\epsilon^{-x}}{\sqrt{x}},$$
$$J_{-n}(ix) = \frac{\epsilon^{x}}{\sqrt{x}}$$
approximately, we see from equation (28) that
$$w = A\eta^{-\frac{1}{4}}\epsilon^{-\frac{\eta^{\frac{3}{2}}}{\frac{3}{2}}} + B\eta^{-\frac{1}{4}}\epsilon^{\frac{\eta^{\frac{3}{2}}}{\frac{3}{2}}}.$$

Since $X = -\dfrac{2D}{k}\dfrac{1}{w}\dfrac{dw}{dx}$, we see that if X is to be positive, B must vanish; hence
$$w = A\eta^{-\frac{1}{4}}\epsilon^{-\frac{\eta^{\frac{3}{2}}}{\frac{3}{2}}},$$
$$\log w = -\tfrac{1}{4}\log\eta - \frac{\eta^{\frac{3}{2}}}{\frac{3}{2}}.$$

Since
$$\eta = \left(\frac{2\pi i k}{D^2}\right)^{\frac{1}{3}}\xi,$$
and
$$X = -\frac{2D}{k}\frac{d}{d\xi}\log w,$$
$$X = \frac{2D}{k}\left(\frac{2\pi i k}{D^2}\right)^{\frac{1}{2}}\xi^{\frac{1}{2}} + \frac{2D}{4k\xi},$$

where, since η is large, the first term is large compared with the second and $\xi = x + \delta$; hence
$$X = \sqrt{\frac{8\pi i}{k}}(x+\delta)^{\frac{1}{2}} + \frac{D}{2k(x+\delta)}.$$

The first term on the right-hand side, which is large compared with the second, is the value of X obtained in § 50. Thus even when diffusion is taken into account it produces but little effect except in the unusual case when $2\pi i k l^3/D^2$ is small. From the condition that, when $x = 0$, $\dfrac{nce}{\sqrt{6\pi}} = I - i$, we find $\delta = \dfrac{1}{48\pi^2}\dfrac{i}{(I-i)^2}\dfrac{c^2}{k}$, and the condition that η should be large requires that
$$\frac{c^2}{48\pi^2}\left(\frac{i}{I-i}\right)^2\left(\frac{2\pi}{k^2 D^2 c^2}\right)^{\frac{1}{3}}$$
should also be large.

CHAPTER V

EFFECT PRODUCED BY A MAGNETIC FIELD ON THE MOTION OF THE IONS

51. WHEN a charged ion is moving in a magnetic field it experiences a mechanical force whose direction is at right angles to the direction of motion of the ion, at right angles also to the magnetic force and equal in magnitude to $HeV \sin \theta$, where H is the magnetic force, V the velocity of the ion, e its charge, and θ the angle between H and V; H and e are to be expressed in the electromagnetic system of units. The relation between the direction of this force F, V and H, for a positively charged ion, is shown in Fig. 54.

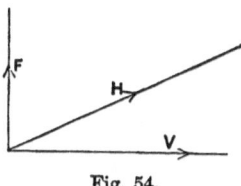

Fig. 54.

Now suppose that we have an ion moving through a gas, the viscosity of the gas causing the velocity of the ion to be proportional to the force acting upon it. Then if X, Y, Z are the components of the electric intensity, α, β, γ those of the magnetic force, u, v, w those of the velocity, the mechanical force exerted on the ion by the magnetic field has for components

$$e(\beta w - \gamma v), \quad e(\gamma u - \alpha w), \quad e(\alpha v - \beta u),$$

while the components of the mechanical force due to the electric field are Xe, Ye, Ze. Thus as the velocity of the ion is proportional to the mechanical force acting upon it, we have

$$\left. \begin{aligned} u &= R(X + \beta w - \gamma v) \\ v &= R(Y + \gamma u - \alpha w) \\ w &= R(Z + \alpha v - \beta u) \end{aligned} \right\} \quad \ldots\ldots\ldots\ldots(1);$$

R is evidently the velocity of the ion under unit electric intensity when there is no magnetic field. Solving equations (1) we find

$$u = \frac{RX + R^2(\beta Z - \gamma Y) + R^3\alpha(\alpha X + \beta Y + \gamma Z)}{1 + R^2(\alpha^2 + \beta^2 + \gamma^2)}$$

$$v = \frac{RY + R^2(\gamma X - \alpha Z) + R^3\beta(\alpha X + \beta Y + \gamma Z)}{1 + R^2(\alpha^2 + \beta^2 + \gamma^2)} \quad \ldots(2).$$

$$w = \frac{RZ + R^2(\alpha Y - \beta X) + R^3\gamma(\alpha X + \beta Y + \gamma Z)}{1 + R^2(\alpha^2 + \beta^2 + \gamma^2)}$$

The first term in the numerator of these expressions represents a velocity parallel and proportional to the electric force; the second term a velocity at right angles both to the electric and magnetic forces and proportional to $R^2 HF \sin\phi$, where H, F, and ϕ represent respectively the magnetic and electric forces and the angle between them; the third term represents a velocity parallel to the magnetic force and proportional to $R^3 H^2 F \cos\phi$. The relative importance of these terms depends upon the value of RH: if this quantity is small, the first term is the most important and the ion moves parallel to the electric force; if on the other hand RH is large, the last term is the most important and the ion moves parallel to the magnetic force. Since R is the velocity of the ion under unit electric force, and the unit force on the electromagnetic system is 10^{-8} of a volt per cm., the value of R for an ion moving through air at atmospheric pressure would be 1.5×10^{-8}, since the velocity of the ion under a volt per cm. is about 1·5 cm./sec. Thus at atmospheric pressure it would not be feasible to get a magnetic field strong enough to make RH large. As R varies inversely as the pressure of a gas through a considerable range of pressures it might at very low pressures be possible to make RH large and thus make the ions travel along the lines of magnetic force.

Let us take the case of an ion placed in a field in which both the electric and magnetic forces are uniform; let the electric force be parallel to the axis of x and let the magnetic force be in the plane of xz, then $Y = 0$, $Z = 0$, $\beta = 0$, and equations (2) become

$$u = \frac{RX(1 + R^2\alpha^2)}{1 + R^2(\alpha^2 + \gamma^2)} = RX \text{ approximately, if } R^2(\alpha^2 + \gamma^2) \text{ is small,}$$

$$v = \frac{R^2\gamma X}{1 + R^2(\alpha^2 + \gamma^2)} = R^2\gamma X,$$

$$w = \frac{R^3\alpha\gamma X}{1 + R^2(\alpha^2 + \gamma^2)} = R^3\alpha\gamma X.$$

Thus the effect of the magnetic force is to give the ion a velocity $R\gamma u$ at right angles to both the electric and magnetic forces and a velocity $R^2 \gamma (a^2 + \gamma^2)^{\frac{1}{2}} u$ in the plane of xz at right angles to the magnetic force.

If both positive and negative ions are present and if R_1 is the value of R for the positive and R_2 that for the negative ion, and if $u_1, v_1, w_1; u_2, v_2, w_2$ are respectively the velocities of the positive and negative ions, then if there are n positive and negative ions per unit volume the current parallel to y will be equal to $ne(v_1 - v_2)$ or, substituting the values of v_1 and v_2, to

$$ne(R_1^2 - R_2^2)\gamma X = (R_1 - R_2)\gamma I,$$

if I is the main current parallel to x; thus if the velocities of the positive and negative currents are unequal the magnetic field will give rise to a side current proportional to the main one and the direction of the current will be deflected through an angle whose tangent is $(R_1 - R_2)\gamma$. If we retain terms proportional to $(RH)^2$, where H is the magnetic force, we see that there will be an additional current proportional to $(R_1^2 + R_2^2 - R_1 R_2)\gamma(a^2 + \gamma^2)^{\frac{1}{2}} I$ in the plane of xz at right angles to the magnetic force.

When the electric field is not uniform but, like that due to a charged particle, radiates from a point, we can prove without difficulty that an ion in a uniform magnetic field will describe a spiral traced on a cone of revolution, the axis of the cone being parallel to the magnetic force.

Motion of a Free Ion under the action of Magnetic Force only.

52·1. When the pressure of the gas through which the particle is moving is so low that the effects of collisions may be neglected, the equations of motion when a, β, γ are the components of the magnetic force, m the mass, e the charge on the particle, are

$$m\frac{d^2x}{dt^2} = e\left(\beta\frac{dz}{dt} - \gamma\frac{dy}{dt}\right),$$

$$m\frac{d^2y}{dt^2} = e\left(\gamma\frac{dx}{dt} - a\frac{dz}{dt}\right),$$

$$m\frac{d^2z}{dt^2} = e\left(a\frac{dy}{dt} - \beta\frac{dx}{dt}\right).$$

Since the force is at right angles to the direction of motion the velocity is constant.

Motion under Constant Magnetic Force.

52·2. Take the axis of z parallel to the direction of the magnetic force, then $\alpha = 0$, $\beta = 0$ and γ is constant, hence

$$\frac{d^2 z}{dt^2} = 0,$$

so that
$$\frac{dz}{dt} = \text{constant}$$
$$= V \cos \alpha,$$

if V is the velocity of the particle and α the direction of projection it makes with the magnetic force,

$$m \frac{d^2 x}{dt^2} = - e\gamma \frac{dy}{dt}, \quad m \frac{d^2 y}{dt^2} = e\gamma \frac{dx}{dt}.$$

Thus
$$m \frac{dx}{dt} = - e\gamma (y + p),$$

$$m \frac{dy}{dt} = e\gamma (x + q),$$

where p and q are constants.

Since
$$\left(\frac{dx}{dt}\right)^2 + \left(\frac{dy}{dt}\right)^2 + \left(\frac{dz}{dt}\right)^2 = V^2,$$

and
$$\frac{dz}{dt} = V \cos \alpha,$$

$$\left(\frac{dx}{dt}\right)^2 + \left(\frac{dy}{dt}\right)^2 = V^2 \sin^2 \alpha,$$

so that
$$m^2 V^2 \sin^2 \alpha = e^2 \gamma^2 (\overline{x + p}^2 + \overline{y + q}^2);$$

hence the projection of the path of the particle on the plane of xy is a circle whose radius is

$$\frac{m}{e} \frac{V \sin \alpha}{\gamma}.$$

The time taken for the particle to make a revolution round this circle is the circumference of the circle divided by the tangential velocity $V \sin \alpha$, and is thus equal to

$$\frac{2\pi m}{e\gamma}.$$

It is thus independent of the velocity of the particle or the direction of projection. Choosing the axis of z so that it passes through the centre of this circle $p = q = 0$, and we have

$$x = \frac{m}{e} \frac{V \sin \alpha}{\gamma} \cos \frac{He}{m} (t + \epsilon),$$

$$y = \frac{m}{e} \frac{V \sin \alpha}{\gamma} \sin \frac{He}{m} (t + \epsilon),$$

$$z = Vt \cos \alpha.$$

The path of the particle is thus a spiral on a cylinder of radius $mV \sin \alpha / e\gamma$, whose axis is parallel to the magnetic force.

Magnetic Force due to a Single Pole.

53·1. Choosing the pole as the origin of coordinates, we have

$$\alpha = \frac{Mx}{r^3}, \quad \beta = \frac{My}{r^3}, \quad \gamma = \frac{Mz}{r^3},$$

where M is the strength of the pole and r the distance between the pole and the particle.

We have

$$\left. \begin{aligned} m \frac{d^2 x}{dt^2} &= \frac{eM}{r^3} \left(y \frac{dz}{dt} - z \frac{dy}{dt} \right) \\ m \frac{d^2 y}{dt^2} &= \frac{eM}{r^3} \left(z \frac{dx}{dt} - x \frac{dz}{dt} \right) \\ m \frac{d^2 z}{dt^2} &= \frac{eM}{r^3} \left(x \frac{dy}{dt} - y \frac{dx}{dt} \right) \end{aligned} \right\} \quad \ldots\ldots\ldots (3),$$

hence
$$m \left(y \frac{d^2 x}{dt^2} - x \frac{d^2 y}{dt^2} \right) = \frac{eM}{r} \left(\frac{dz}{dt} - \frac{z}{r} \frac{dr}{dt} \right)$$

$$= eM \frac{d}{dt} \frac{z}{r};$$

thus
$$m \left(y \frac{dx}{dt} - x \frac{dy}{dt} \right) = eM \frac{z}{r} + h_3 \quad \ldots\ldots\ldots\ldots (4),$$

where h_3 is a constant.

Similarly
$$m \left(x \frac{dz}{dt} - z \frac{dx}{dt} \right) = eM \frac{y}{r} + h_2 \quad \ldots\ldots\ldots\ldots (5),$$

$$m \left(z \frac{dy}{dt} - y \frac{dz}{dt} \right) = eM \frac{x}{r} + h_1 \quad \ldots\ldots\ldots\ldots (6).$$

Multiplying the first of these equations by z, the second by y, the third by x, and adding we get

$$eMr + h_3 z + h_2 y + h_1 x = 0 \quad \ldots\ldots\ldots\ldots (7).$$

If θ is the angle between the line joining the pole and the particle, and the line whose direction cosines are proportional to h_1, h_2, h_3,

$$\sqrt{h_1^2 + h_2^2 + h_3^2} \cos \theta = \frac{x}{r} h_1 + \frac{y}{r} h_2 + \frac{z}{r} h_3;$$

hence equation (7) shows that the radius vector from the pole to the particle makes a constant angle θ with the line whose direction cosines are proportional to h_1, h_2, h_3.

The expressions on the left-hand side of equations (3), (4), (5), are the z, y, x components of the moment of momentum of the particle about O; denoting these components by W, V, U we see that

$$U^2 + V^2 + W^2 = e^2 M^2 + \frac{2eM}{r}(h_3 z + h_2 y + h_1 x) + h_1^2 + h_2^2 + h_3^2$$

$$= -e^2 M^2 + h_1^2 + h_2^2 + h_3^2 \ldots\ldots\ldots\ldots\ldots (8);$$

hence $U^2 + V^2 + W^2$ is constant, *i.e.* the magnitude of the moment of momentum is constant throughout the motion; denoting its value by Γ we have from (8)

$$h_1^2 + h_2^2 + h_3^2 = \Gamma^2 + e^2 M^2;$$

hence
$$\cos \theta = \frac{eM}{\sqrt{\Gamma^2 + e^2 M^2}}.$$

We see from the equations for U, V, W that they are the components of the resultant of two vectors, one OP' along OP, the line joining the pole to the particle, such that $OP' = eM$, the other a vector, constant both in magnitude and direction, whose components are h_1, h_2, h_3. Hence we can find h_1, h_2, h_3 by the following construction. If P is the point from which the particle is projected, v the velocity of projection, the vector representing the moment of momentum will be the line OQ perpendicular to the plane containing OP and the direction of projection, and equal in length to vp, where p is the perpendicular from O on the direction of projection. This is the resultant of a vector OP' along OP where $OP' = eM$, and the vector parallel to the axis of the cone whose components are h_1, h_2, h_3; thus the direction of the vector is along $P'Q$ and its magnitude is $\sqrt{v^2p^2 + e^2M^2}$.

The moment of momentum vp is constant, and therefore since v is constant, p must be constant. Hence the path of the particle is such that the perpendicular from O on to its tangent is of constant length; the path of the particle is on the surface of the cone, hence if the cone is unwrapped into a plane the paths must become straight lines on the plane, *i.e.* the paths are geodesics on the cone.

Again, from (3),

$$x\frac{d^2x}{dt^2} + y\frac{d^2y}{dt^2} + z\frac{d^2z}{dt^2} = 0,$$

or $$\frac{d}{dt}\left(x\frac{dx}{dt} + y\frac{dy}{dt} + z\frac{dz}{dt}\right) = \left(\frac{dx}{dt}\right)^2 + \left(\frac{dy}{dt}\right)^2 + \left(\frac{dz}{dt}\right)^2$$
$$= v^2;$$

hence $$r\frac{dr}{dt} = v^2 t + c_1,$$

or $$r^2 = v^2 t^2 + 2c_1 t + c_0,$$

where c_0 and c_1 are constants which can be determined from the initial velocity and position. We see that the expression for r^2 is the same as if there was no magnetic force, so that the distance of the particle from the pole at any time will be the distance which would have separated them if no magnetic force had been present.

Motion of an Electrified Particle under the action of the Magnetic Force due to a Uniformly Magnetised Sphere.

54·1. This case is of importance in connection with the theory that the aurora is due to high speed electrified particles coming from the sun.

Taking the axis of magnetisation as the axis of z, then if M is the magnetic moment of the sphere,

$$\gamma = \frac{3z^2 - r^2}{r^5}M, \quad \beta = \frac{3yz}{r^5}M, \quad \alpha = \frac{3xz}{r^5}M.$$

Hence as in § 53·1 we get

$$m \frac{d}{dt}\left(y \frac{dx}{dt} - x \frac{dy}{dt}\right) = eM \frac{d}{dt}\left(\frac{z^2}{r^3} - \frac{1}{r}\right),$$

so that

$$m \left(y \frac{dx}{dt} - x \frac{dy}{dt}\right) = eM (z^2 - r^2) \frac{1}{r^3} + C \quad \ldots \ldots (9),$$

where C is the constant of integration.

Again
$$m \frac{d^2z}{dt^2} = \frac{3eMz}{r^5}\left(x \frac{dy}{dt} - y \frac{dx}{dt}\right)$$

$$= \frac{3e^2M^2z}{mr^8}(r^2 - z^2) - \frac{3eMz}{mr^5} C \quad \ldots \ldots (10).$$

Again
$$\frac{d}{dt}\left(r \frac{dr}{dt}\right) = x \frac{d^2x}{dt^2} + y \frac{d^2y}{dt^2} + z \frac{d^2z}{dt^2} + v^2,$$

where v is the velocity of the particle; hence

$$\frac{d}{dt}\left(r \frac{dr}{dt}\right) = \frac{e}{m} \frac{M}{r^3}\left(x \frac{dy}{dt} - y \frac{dx}{dt}\right) + v^2$$

$$= \frac{3e^2M^2}{m^2} \frac{1}{r^6}(r^2 - z^2) - \frac{eM}{m^2r^3} C + v^2 \quad \ldots \ldots (11).$$

Equations (10) and (11) determine the path of the particle; they are discussed very fully by Störmer[1].

When $C = 0$, *i.e.* when the particle is projected from an infinite distance with no moment about the axis of z,

$$m \left(y \frac{dx}{dt} - x \frac{dy}{dt}\right) = - \frac{eM\rho^2}{r^3},$$

where ρ is the distance of the particle from the axis of z.

But
$$y \frac{dx}{dt} - x \frac{dy}{dt} = \rho^2 \frac{d\theta}{dt},$$

where $\frac{d\theta}{dt}$ is the angular velocity of a plane passing through the particle and the axis of z; hence

$$\frac{d\theta}{dt} = - \frac{eM}{mr^3},$$

so that the angular velocity varies inversely as the cube of the distance. From this result we can easily show that when the

---

[1] Störmer, *Videnskabs-Selskabets Skrifter, Math. Nat. Klasse*, Stockholm, No. 5, 1909; No. 4, 1913; No. 14, 1913.

particle is projected radially from infinity in the equatorial plane of the sphere the path of the particle is a rectangular hyperbola.

If the perpendicular from the centre of the magnet on the direction of motion at infinity of the particle is p_0, the equation to the path of the particle is from (9), since z is always zero,

$$\frac{mv}{Me}(p - p_0)\, r = 1,$$

where p is the perpendicular from the centre of the sphere on the direction of motion of the particle when it is at a distance r from the centre. It follows from this equation that the minimum distance from the centre is the root of the equation

$$\frac{mv}{Me}(r - p_0)\, r = 1,$$

or

$$r = \frac{p_0}{2} + \sqrt{\frac{p_0^2}{4} + \frac{Me}{mv}}.$$

Motion of an Ion under the joint action of Electric and Magnetic Forces.

55. We shall now investigate the motion of an ion when it is acted on simultaneously by both electric and magnetic forces; we shall take the case when both these forces are constant. Let the axis of z be parallel to the direction of the magnetic force, and the plane of xz parallel to the direction of the electric force. Let H be the magnetic force, $X, 0, Z$ the components of the electric force, then if m is the mass of an ion, e its charge, and x, y, z its coordinates the equations of motion are

$$m\frac{d^2x}{dt^2} = Xe - He\frac{dy}{dt} \quad \ldots\ldots\ldots\ldots\ldots(12),$$

$$m\frac{d^2y}{dt^2} = He\frac{dx}{dt} \quad \ldots\ldots\ldots\ldots\ldots\ldots(13),$$

$$m\frac{d^2z}{dt^2} = Ze \quad \ldots\ldots\ldots\ldots\ldots\ldots\ldots(14).$$

From equation (14) we have

$$z = \frac{1}{2}\frac{Ze}{m}t^2 + w_0 t \quad \ldots\ldots\ldots\ldots(15),$$

where w_0 is the velocity of projection parallel to z, the origin of coordinates being supposed to be taken at the point of projection.

From equations (12) and (13) we have

$$y = c + \frac{X}{H} t + A \cos \omega t + B \sin \omega t \dots \dots (16),$$

$$x = a - A \sin \omega t + B \cos \omega t \dots \dots \dots (17),$$

where a, c, A and B are arbitrary constants and $\omega = He/m$. Writing equations (16) and (17) in the form

$$y - c = \frac{X}{\omega H} \omega t + A' \cos (\omega t - \alpha),$$

$$x - a = - A' \sin (\omega t - \alpha),$$

we see that the projection of the path of the ion on the plane of xy is a trochoid, generated by a circle whose radius is $X/\omega H$ rolling on a line perpendicular to the electric force, the distance of the tracing point from the centre of the rolling circle being A'. Since the average value of the periodic terms tends to vanish when the time over which the average extends is large compared with $1/\omega$, we see, from equations (16) and (17), that the equations

$$y = c + \frac{X}{H} t,$$

$$x = a,$$

give the average positions of the ion, and that the average velocity parallel to y is X/H while that parallel to x vanishes. As the velocity parallel to z at the time t is $\frac{Ze}{m} t + w_0$ we see that if Z is finite the velocity parallel to z will ultimately become infinite compared with the components parallel to the other axes, thus in this case the ions will ultimately move along the lines of magnetic force; we must remember however that this reasoning only applies when the electric field has a finite component in this direction.

If we determine the constants in (16) and (17) in terms of v_0, u_0, the initial values of the components of the velocity of projection of the ion parallel to the axes of y and x respectively, we have, the origin being taken at the point of projection,

$$y = \frac{u_0}{\omega} (1 - \cos \omega t) + \frac{X}{H} t + \left(v_0 - \frac{X}{H}\right) \frac{1}{\omega} \sin \omega t \dots (18),$$

$$x = \left(\frac{X}{H} - v_0\right) \frac{1}{\omega} (1 - \cos \omega t) + \frac{u_0}{\omega} \sin \omega t \dots \dots (19).$$

If $X = 0$, *i.e.* if the directions of the electric and magnetic forces coincide, we have

$$\left(x + \frac{v_0}{\omega}\right)^2 + \left(y - \frac{u_0}{\omega}\right)^2 = \frac{v_0^2 + u_0^2}{\omega^2};$$

thus the projection of the path of the ion on the plane of xy is a circle, and the path of the ion is a helix of gradually increasing pitch with its axis parallel to the lines of magnetic force.

If $Z = 0$, *i.e.* if the electric force is at right angles to the magnetic, and if in addition u_0, v_0, w_0 all vanish, we have

$$y = \frac{X}{\omega H}(\omega t - \sin \omega t),$$

$$x = \frac{X}{\omega H}(1 - \cos \omega t).$$

This is the equation to a cycloid, the radius of the generating circle being $X/\omega H$ or Xm/eH^2, the line on which it rolls is perpendicular to the electric force. The greatest distance measured in the direction of the electric force which the particle can get from its point of projection is $2Xm/eH^2$; the average velocity in this direction is zero while the average velocity parallel to y, *i.e.* in the direction at right angles both to the electric and magnetic forces, is finite and equal to X/H. If the ion were projected with the velocity w parallel to the axis of z it would retain this velocity unaltered, and the average direction of motion of the ion would be at right angles to the electric force and along a line making an angle $\tan^{-1} X/wH$ with the direction of the magnetic force.

56. If $u_0 = 0$ and $v_0 = X/H$ we have by equations (18) and (19)

$$y = v_0 t,$$

$$x = 0.$$

Thus in this case the path of the ion in the plane of xy is the same as if there were neither electric nor magnetic forces acting upon it: the force Xe acting on the particle due to the electric field is in this case just balanced by the force Hev_0 due to the magnetic field.

57. Returning to the general case represented by equations (17) and (18) we easily deduce that the maximum velocity V parallel to the plane of xy attained by the ion is given by the equation

$$V = \frac{X}{H} + \left\{ u_0^2 + \left(\frac{X}{H} - v_0 \right)^2 \right\}^{\frac{1}{2}};$$

thus until u_0 and v_0 are comparable with X/H, the maximum velocity attained is very approximately $2X/H$ and is independent of the velocity of projection, and the charge and the mass of the ion.

The maximum displacement ξ measured parallel to the direction of the electric force is given by the equation

$$\xi = \left\{ \frac{X}{H} - v_0 \right\} \frac{1}{\omega} + \frac{1}{\omega} \left\{ u_0^2 + \left(\frac{X}{H} - v_0 \right)^2 \right\}^{\frac{1}{2}},$$

and thus until u_0 and v_0 become comparable with X/H, the distance travelled by the ion parallel to the lines of electric force will be very approximately independent of the velocity of projection of the ion.

If ξ, ξ_1, ξ_2 are the maximum displacements corresponding to the electric forces X, X_1, X_2, the magnetic force being the same throughout, we find from the preceding equation that

$$\omega = \frac{2}{H(\xi - \xi_1)(\xi_1 - \xi_2)(\xi_2 - \xi)} \{X\xi(\xi_2 - \xi_1) + X_1\xi_1(\xi - \xi_2) + X_2\xi_2(\xi_1 - \xi)\},$$

$$u_0^2 = \frac{-2\omega(\xi\xi_1\xi_2)}{H(\xi - \xi_1)(\xi_1 - \xi_2)(\xi_2 - \xi)} \{X(\xi_2 - \xi_1) + X_1(\xi - \xi_2) + X_2(\xi_1 - \xi)\},$$

$$v_0 = \frac{1}{H(\xi - \xi_1)(\xi_1 - \xi_2)(\xi_2 - \xi)} \{X\xi(\xi_2^2 - \xi_1^2) + X_1\xi_1(\xi^2 - \xi_2^2) + X_2\xi_2(\xi_1^2 - \xi^2)\}.$$

These equations supply a method for finding e/m and the velocity of projection of the ion.

If the charged particles are projected from the plane $x = 0$ in all directions with a constant velocity v the distances they will travel from this plane will vary between

$$\frac{2}{\omega}\left(\frac{X}{H} - v\right) \text{ and } \frac{2}{\omega}\left(\frac{X}{H} + v\right).$$

The smaller value is the distance reached by those particles which are projected parallel to the plane in the direction for which the force on the moving particle due to the magnetic field tends towards the plane, the larger value is the distance travelled by particles projected parallel to the plane in the opposite direction.

If there is a second plane at a distance d from the plane of projection, then with a constant magnetic field all the particles will reach the second plane when the electric force X_1 is equal to or greater than the value given by the equation

$$d = \frac{2}{\omega}\left(\frac{X_1}{H} - v\right),$$

while none of them will reach the plane if the electric force X_2 is equal to or less than the value given by the equation

$$d = \frac{2}{\omega}\left(\frac{X_2}{H} + v\right);$$

from these equations we get

$$v = \frac{1}{2}\frac{(X_1 - X_2)}{H}, \quad \frac{e}{m} = \frac{(X_1 + X_2)}{H^2 d}.$$

With a constant electric force, all the particles will be prevented from reaching the second plane if the magnetic force H_1 is equal to or greater than the value given by the equation

$$d = \frac{2m}{eH_1}\left(\frac{X}{H_1} + v\right),$$

while none of them will be stopped by a magnetic force H_2 equal to or less than the value given by the equation

$$d = \frac{2m}{eH_2}\left(\frac{X}{H_2} - v\right);$$

from these equations we get

$$v = \frac{X(H_1 - H_2)}{H_1 H_2}; \quad \frac{e}{m} = \frac{2X}{H_1 H_2 d}.$$

These results may be applied to find v and e/m.

58. The case when the electric and magnetic forces are at right angles to each other is a very important one as it includes the fields produced by electric waves. In these waves the electric and magnetic forces are not constant, but in the case of a simple harmonic wave may be taken as proportional to $\cos pt$. When the waves are all divergent the electric force is equal to V times the magnetic force, where V is the velocity with which the electric waves travel through the medium. Thus if the direction of propagation of the wave is parallel to the axis of y and if the magnetic force is parallel to the axis of z and equal to $H_0 \cos \theta$, the electric force will be parallel to the axis of x and equal to $VH_0 \cos \theta$, where $\theta = p\left(t - \dfrac{y}{V}\right)$. The equations of motion of a charged particle acted on by this electric wave are

$$m \frac{d^2 x}{dt^2} = e\left(V - \frac{dy}{dt}\right) H_0 \cos \theta,$$

$$m \frac{d^2 y}{dt^2} = e \frac{dx}{dt} H_0 \cos \theta.$$

From these equations we have, if dx/dt and θ vanish simultaneously,

$$\frac{dx}{dt} = \frac{e}{m} \frac{V}{p} H_0 \sin \theta,$$

$$\frac{d^2 \theta}{dt^2} + \frac{H_0^2 e^2}{m^2} \sin 2\theta = 0.$$

From the analogy of the second of these equations with that representing the motion of a simple pendulum we see that the character of the motion of the ions will depend upon the value of $H_0 e/pm$; if this quantity is large, θ varies between definite limits and the average velocity of the ions parallel to x will vanish while that parallel to y will be equal to V: thus the wave will in this case carry the charged particles along with it. When however $H_0 e/pm$ is a small quantity, the effect of the wave will be to superpose on the undisturbed motion a small vibratory motion proportional to H_0 and parallel to the electric force in the wave and thus at right angles to its direction of propagation. It can be shown[1] that the maximum velocity in the direction of propagation is proportional to H_0^2.

[1] J. J. Thomson, *Phil. Mag.* vi. 4, p. 253, 1902.

CHAPTER VI

DETERMINATION OF THE RATIO OF THE CHARGE TO THE MASS OF AN ION

59. THE value of e/m—the charge on an ion divided by its mass—has been determined by the application of some of the results discussed in the preceding chapter. The first case we shall consider is that of the ion in cathode rays.

Cathode rays are the streams which start from the cathode in a highly exhausted tube through which an electric discharge is passing; they produce a vivid phosphorescence when they strike against the glass of the tube. In the chapter on cathode rays (vol. II) we shall give the evidence which leads us to the conclusion

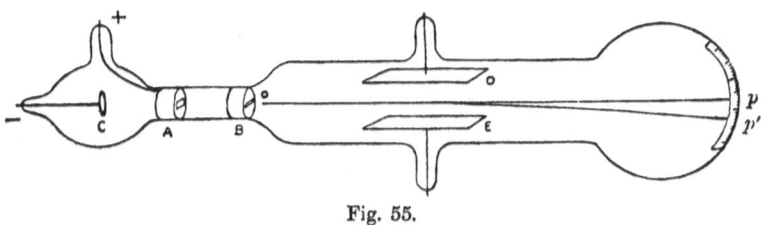

Fig. 55.

that they consist of negatively electrified particles starting from the neighbourhood of the cathode and moving with a very high velocity along straight lines. Assuming that this is the nature of the cathode rays we shall show here how to determine the velocity of the particles and the value of e/m. Suppose that we have a highly exhausted tube of the pattern shown in Fig. 55.

In this tube C is the cathode, A the anode, B is a thick metal disc connected with the earth, slits a millimetre or so in diameter being bored through the middle of the disc and through the anode; some of the cathode rays starting from the neighbourhood of the cathode pass through these holes, thus in the part of the tube to the right of the disc we have a pencil of negatively electrified particles travelling along straight lines parallel to the line joining the holes in the discs; the place where these particles strike the glass is marked by a patch of bright phosphorescence p. Suppose now that the tube is placed in a uniform magnetic field, the lines of force being at right angles to the path of the ions; the paths of the ions will now be circles, the radii of the circles being (see p. 218) mv/eH, where m is the mass of the ion, e its charge, v its velocity, and H the strength of the magnetic field. The place at which these particles strike the tube will no longer be at p but at some other point p', the direction of pp' being at right angles to the magnetic force. Since op' is an arc of a circle of which op is a tangent, we have

$$pp'(2R + pp') = op^2,$$

where R is the radius of the circle; hence

$$2R = \frac{op^2}{pp'} - pp',$$

or, since $R = mv/eH$, we have

$$2\frac{mv}{eH} = \frac{op^2}{pp'} - pp'.$$

If the magnetic field is not uniform we may proceed as follows. Since ρ the radius of curvature at any point of the path of the ion is given by the equation

$$\frac{1}{\rho} = \frac{He}{vm},$$

and since, when the path of the ion is fairly flat, $1/\rho$ is very approximately equal to d^2y/dx^2, where y and x are the coordinates of the ion, x being measured along the undisturbed path, and y at right angles to it, we have

$$\frac{d^2y}{dx^2} = \frac{He}{vm},$$

so that
$$pp' = \frac{e}{vm}\int_0^{op}[\int_0^x H dx]\,dx \quad \ldots\ldots\ldots\ldots(1).$$

Hence if we measure pp' and know the distribution of the magnetic force H along the tube we can from this equation determine the value of e/vm. This gives us a relation between v and m/e. We can determine v in the following way: two parallel metal plates D and E are placed in the tube, the plates being parallel to the lines of magnetic force and parallel also to the undisturbed path of the rays; these plates are maintained at a known difference of potential by connecting them to the terminals of a battery. Thus we have an electric field between the plates the lines of force of which are at right angles to the lines of magnetic force and to the direction of motion of the ions; this electrostatic force Y tends to deflect the ions, the force acting on an ion being Ye; the force due to the magnetic field acts in the same straight line and is equal to Hev. Adjust the sign of the difference of potential so that the electric and magnetic forces tend to oppose each other, then keeping one of the forces fixed, say the electric force, alter the value of the other until the two forces just balance; this stage can be ascertained by observing when the phosphorescent patch p' is restored to its undisturbed position. When this stage is reached we have

$$Ye = Hev,$$

or

$$v = \frac{Y}{H} \quad \ldots\ldots\ldots\ldots\ldots\ldots(2).$$

Thus by measuring Y/H we can determine the velocity of the ions composing the cathode rays. As we know e/vm from the experiments on the magnetic deflection we can deduce the values of both e/m and v. Equation (2) depends upon the assumption that the magnetic and electric fields are both uniform and coincident; if this condition is not fulfilled we must proceed as follows. Suppose that p'' is the displaced position of p when the electric field alone is acting on the rays, then we can prove without difficulty that

$$pp'' = \frac{e}{v^2 m} \int_0^{op} [\int_0^x Y dx] \, dx \quad \ldots\ldots\ldots\ldots(3);$$

hence if we know the distribution of the electric field and the value of pp'' we can by equation (3) find the value of $e/v^2 m$, and since by equation (2) we can determine e/vm we have the data for determining both v and e/m.

In order to apply this method it is necessary that the pressure of gas in the discharge tube in which the rays are produced should be very low; the passage of cathode rays through a gas makes it a conductor and thus, as the rays are shielded from the electrostatic field by the gas through which they move, the electrostatic repulsion is hardly appreciable; if, however, the pressure of the gas is very low the conductivity of the gas is so small that there is hardly any appreciable shielding effect and the deflection produced by the electric field is easily observed.

If the pressure of the gas through which the charged particles pass is not exceedingly low the velocity of the particles will be reduced by collision with the molecules of the gas; to investigate this effect let us suppose that the velocity of the particle after traversing a distance x is equal to $v_0 \epsilon^{-kx}$. The equation giving y, the magnetic deflection of the particle, will be

$$\frac{d^2y}{dx^2} = \frac{He}{mv_0} \epsilon^{kx}.$$

If the magnetic field is uniform we have

$$\frac{dy}{dx} = \frac{He}{mv_0 k} (\epsilon^{kx} - 1)$$

and

$$y = \frac{He}{mv_0 k} \left(\frac{\epsilon^{kx} - 1}{k} - x \right).$$

If the particle travels through the uniform field for a distance l and then through a distance d under no magnetic force, before reaching the screen, the deflection at the screen will be

$$y_{(x=l)} + d \left(\frac{dy}{dx} \right)_{x=l}$$

$$= \frac{He}{mv_0 k} \left(\frac{\epsilon^{kl} - 1}{k} - l \right) + \frac{dHe}{mv_0 k} (\epsilon^{kl} - 1),$$

or if kl be small

$$= \frac{He}{mv_0} l \left\{ \frac{l}{2} \left(1 + \frac{kl}{3} \right) + d \left(1 + \frac{kl}{2} \right) \right\}.$$

If z be the deflection due to the electrostatic force X we have

$$\frac{d^2z}{dt^2} = \frac{Xe}{m};$$

thus

$$\frac{dz}{dt} = \frac{Xe}{m} t; \quad z = \frac{1}{2} \frac{Xe}{m} t^2.$$

Again,

$$\frac{dx}{dt} = v_0 \epsilon^{-kx},$$

so that

$$t = \frac{1}{kv_0} (\epsilon^{kx} - 1).$$

Hence if as before the particle travels through a distance l under a constant electric force X and then for a distance d under no force,

z, the deflection on reaching the screen, will be given by the equation

$$z = \frac{Xe}{mv_0^2}\frac{l^2}{2}(1+kl) + \frac{Xe}{mv_0^2}\,dl\left(1+\frac{3kl}{2}\right),$$

hence we have

$$\frac{y^2}{z} = \frac{H^2 e}{m X}\, l\left(\frac{l}{2}+d\right)\left(1 - kl\,\frac{\left(\frac{l}{6}+\frac{d}{2}\right)}{\frac{l}{2}+d}\right).$$

Thus the effect of the retardation of the rays by the gas will be to make the value of e/m given by the uncorrected expression

$$\frac{y^2}{z} = \frac{H^2 e}{Xm}\, l\left(\frac{l}{2}+d\right)$$

too small. In a fairly good vacuum the correction is small; thus if the gas in the tube is air at the pressure ·01 mm., k for particles moving with a velocity of 3×10^9 cm./sec. is about ·0085. On the other hand, the residual gas will tend to make the electrical force in the part of the space between the plates traversed by the rays less than V/D, where V is the potential difference and D the distance between the plates; thus if we put $X = V/D$ in the above equation the value of e/m would, if any residual gas were present, tend to be too large.

Using this method the author in 1897[1] obtained the values for v and e/m given in the following table: the first column contains the name of the gas filling the tube: the different numbers given under one gas relate to experiments made at different pressures.

Gas	v	m/e	Gas	v	m/e
Air.....	$2 \cdot 8 \times 10^9$	$1 \cdot 3 \times 10^{-7}$	Air*.....	$2 \cdot 8 \times 10^9$	$1 \cdot 1 \times 10^{-7}$
Air.....	$2 \cdot 8 \times 10^9$	$1 \cdot 1 \times 10^{-7}$	Hydrogen	$2 \cdot 5 \times 10^9$	$1 \cdot 5 \times 10^{-7}$
Air.....	$2 \cdot 3 \times 10^9$	$1 \cdot 2 \times 10^{-7}$	Carbonic acid	$2 \cdot 2 \times 10^9$	$1 \cdot 5 \times 10^{-7}$
Air*.....	$3 \cdot 6 \times 10^9$	$1 \cdot 3 \times 10^{-7}$			

The mean of the values of m/e is $1 \cdot 3 \times 10^{-7}$ or $e/m = 7 \cdot 7 \times 10^6$. We see too that within the limits of the errors of the experiments

[1] J. J. Thomson, *Phil. Mag.* v. 44, p. 293, 1897.

the value of e/m is the same whether the tube be filled with air, hydrogen or carbonic acid, so that it does not depend upon the nature of the gas. This result was first obtained by the writer[1] by another method; the pressure in the discharge tube was adjusted so that the potential difference between the electrodes in the discharge tube was the same for all the gases tried; photographs were taken of the rays when deflected by a constant magnetic field and from these it was found that the deflected rays occupied the same position whether the gas in the tube was hydrogen, air, carbonic acid or methyl iodide; these gases give a wide range of densities, as the density of methyl iodide is about 70 times that of hydrogen. The constancy of the value of e/m for the ions which constitute the cathode rays is in striking contrast with the variability of the corresponding quantity in the ions which carry the current through liquid electrolytes. Experiments were made on the effect of altering the metal of which the cathode was made; the experiments marked with an asterisk in the preceding table were made with platinum electrodes, all the others were made with aluminium electrodes; it will be seen that the values of e/m are the same in the two cases. A further series of experiments on this point has been made by H. A. Wilson[2], who used cathodes made of aluminium, copper, iron, lead, platinum, silver, tin and zinc, and found the same value for e/m in all cases.

If we compare the value of e/m, viz. $7 \cdot 7 \times 10^6$, for the ions in the cathode rays, with the value of the corresponding quantity for the ions which carry the current through liquid electrolytes, we are led to some very interesting conclusions; the greatest value of e/m in the case of liquid electrolysis is when the ion is the hydrogen ion, in this case e/m is about 10^4. When we discuss the electric charge carried by the ion in the cathode rays we shall find that it is equal in magnitude to the charge carried by the hydrogen ion, in liquid electrolysis; it follows then that the mass of the hydrogen ion must be 770 times[3] that of the ion in the cathode rays; hence the carrier of the negative electricity in these rays must be very small compared with the mass of the hydrogen atom.

[1] J. J. Thomson, *Proc. Camb. Phil. Soc.* ix. p. 243, 1897.
[2] H. A. Wilson, *Proc. Camb. Phil. Soc.* xi. p. 179, 1901.
[3] More recent measurements give a ratio of 1844 : 1; see below, p. 265.

We shall return to this point when we have studied other phenomena involving gaseous ions.

60. Kaufmann[1] has introduced a method by which the deflections due to the electric and magnetic forces take place simultaneously and can be measured with great accuracy. A photographic plate is placed at right angles to the path of the undeflected rays and the electric and magnetic fields are arranged so as to produce deflections in directions at right angles to each other. Suppose the magnetic field produces a vertical, the electric a horizontal deflection. Let O be the point where the undeflected rays strike the plate, then if all the rays have the same velocity the deflected rays will strike the plates at a point O', the vertical distance between O, O' will be

$$\frac{1}{2}\frac{He}{mv}l\left(\frac{l}{2}+d\right),$$

and the horizontal distance

$$\frac{1}{2}\frac{Xe}{mv^2}l\left(\frac{l}{2}+d\right);$$

hence if we measure these distances on the plates we have the means of determining v and e/m. If the particles have different velocities, then when the rays are deflected we shall have instead of one point O', a number of points O_1, O_2, O_3; by measuring the displacements of these points we can determine the various velocities of the particles and the values of e/m. We shall consider later how Kaufmann applied this method to find the effect of the velocity on the value of e/m for particles whose velocity is comparable with that of light.

Ions in Lenard Rays.

61. Lenard[2] has determined by the method described in Art. 59 the velocity and the value of e/m for the Lenard rays; these rays are cathode rays which have escaped from the discharge tube through a window of very thin aluminium foil. In his experiments the rays, after escaping from the discharge tube, entered a highly exhausted vessel where they were deflected by

[1] Kaufmann, *Göttingen Nach.* Nov. 8, 1901.
[2] Lenard, *Wied. Ann.* xliv. p. 279, 1898.

electric and magnetic forces in the way described in the preceding article; the results of these experiments are given in the following table:

v cm./sec.	e/m
$6 \cdot 7 \times 10^9$	$6 \cdot 49 \times 10^6$
7×10^9	$6 \cdot 32 \times 10^6$
$8 \cdot 1 \times 10^9$	$6 \cdot 36 \times 10^6$

The mean of the values of e/m is $6 \cdot 39 \times 10^6$, which agrees well with the value $7 \cdot 7 \times 10^6$ found above. It will be noticed that the velocities of the ions in this case are much greater than in the preceding; taking the two sets together we have velocities of the ions ranging from $2 \cdot 2 \times 10^9$ to $8 \cdot 1 \times 10^9$ cm./sec. without any indication of a change in the value of e/m.

Lenard[1] has also made some very interesting experiments on the effect of an external electric field in accelerating or retarding the motion of the ions. The apparatus used for this purpose is shown in Fig. 56.

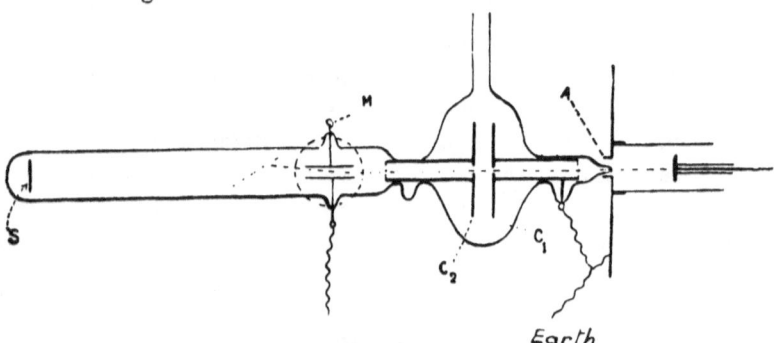

Fig. 56.

The rays after coming through the window A pass through small holes in two parallel circular metallic plates C_1 and C_2; of these C_1 is always kept connected with the earth, while C_2 is charged positively or negatively by means of an electrical machine; after leaving this condenser the rays pass between two plates M, used for producing the electrostatic deflection, on to a screen S;

[1] Lenard, *Wied. Ann.* xlv. p. 504, 1898.

the dotted circle round M represents the coil used for producing the magnetic deflection. The velocities of the ions were measured (1) when the plates of the condenser C_1C_2 were at the same potential, (2) when they were maintained at different potentials; it was found that when the plate C_2 was negatively electrified the velocity in case (2) was less than that in (1), while when the plate C_2 was positively electrified it was greater; if v_1 is the velocity of the ions in case (1), v_2 that in case (2), then assuming that the whole change in the energy is due to the action of the electric field we have

$$\tfrac{1}{2}m\,(v_2^2 - v_1^2) = eV \qquad \ldots\ldots\ldots\ldots(1),$$

where V is the potential difference between the plates, V being taken positive when C_2 is at a higher potential than C_1. The results of Lenard's experiments are given in the following table; the fourth column contains the value of e/m calculated by equation (1).

v_1 (cm./sec.)	v_2 (cm./sec.)	V (electromagnetic units)	e/m
$\cdot 7 \times 10^{10}$	$\cdot 35 \times 10^{10}$	-291×10^{10}	$6\cdot 2 \times 10^6$
$\cdot 68 \times 10^{10}$	$\cdot 34 \times 10^{10}$	-210×10^{10}	$8\cdot 1 \times 10^6$
$\cdot 62 \times 10^{10}$	$\cdot 89 \times 10^{10}$	$+291 \times 10^{10}$	$6\cdot 9 \times 10^6$
$\cdot 77 \times 10^{10}$	$\cdot 47 \times 10^{10}$	-291×10^{10}	$6\cdot 4 \times 10^6$
$\cdot 79 \times 10^{10}$	$1\cdot 0 \times 10^{10}$	$+291 \times 10^{10}$	$6\cdot 6 \times 10^6$
$\cdot 88 \times 10^{10}$	$1\cdot 07 \times 10^{10}$	$+291 \times 10^{10}$	$6\cdot 5 \times 10^6$

The constancy of the value of e/m is a strong confirmation of the truth of the theory that the rays are charged particles in rapid motion.

Method of determining the value of e/m and v by measuring the Energy carried by the Cathode Rays.

62. Many other methods have been employed to measure e/m. One, used by the writer[1], was to measure the energy carried by the rays. To do this a narrow pencil of rays passed through a small hole in a metal cylinder and fell upon a thermo-couple, the couple was heated by the impact of the rays, and by measuring

[1] J. J. Thomson, *Electrician*, May, 1897; *Phil. Mag.* v. 44, p. 293, 1897.

by means of a galvanometer the rate at which the temperature of the junction increased, the amount of heat communicated to the junction in unit time was determined. Let us call this amount Q; then if we assume that all the energy possessed by the cathode rays is converted into heat we have

$$\tfrac{1}{2} N m v^2 = Q,$$

where N is the number of particles which enter the cylinder through the hole in unit time, m is the mass and v the velocity of a particle.

If e is the charge of the ion, then in each unit of time Ne units of negative electricity will enter the cylinder; the rate at which the negative charge increases can easily be measured if the cylinder is insulated and connected with an electrometer; let E be the rate of increase of the negative electricity inside the cylinder, then we have

$$Ne = E.$$

Eliminating N from these equations we get

$$\frac{1}{2}\frac{m}{e} v^2 = \frac{Q}{E}.$$

If we observe the magnetic deflection produced by a known magnetic field we determine mv/e, hence since we have just seen how to determine mv^2/e we can deduce the values of v and m/e.

The results of experiments made in this way are shown below:

Gas	v	e/m
Air	$2 \cdot 4 \times 10^9$	$1 \cdot 1 \times 10^7$
Air	$3 \cdot 2 \times 10^9$	$1 \cdot 4 \times 10^7$
Hydrogen	$2 \cdot 5 \times 10^9$	$1 \cdot 0 \times 10^7$

We may remark in passing that the charges of negative electricity carried by the rays are very large, thus with quite a small hole (about 1 mm. in radius) in the cylinder the potential of the cylinder would change sometimes as much as 5 volts per second when exposed to the rays, even though it was connected with a condenser having a capacity about $\cdot 15$ microfarad.

Methods of determining v and e/m from the Magnetic Deflection and Potential Difference between the electrodes of the discharge tube.

63. These methods, which were first used by Schuster[1] in 1890, are based on the following principles. If V is the potential difference between the terminals of the tube, then the work done on an ion in passing from one end of the tube to the other is Ve; hence the kinetic energy acquired by the ion cannot be greater than Ve, so that

$$\tfrac{1}{2} mv^2 \not> Ve.$$

From the observation of the effect of the magnet on the discharge (Schuster measured the radii of the circles which are the path of the ions in a strong magnetic field) we know the value of mv/e; let us call this quantity q, then from the preceding equation we have

$$e/m \not> \frac{2V}{q^2}.$$

To find an inferior limit for e/m, Schuster took v equal to the velocity of mean square of the atoms of the gas in the tube; calling this velocity U we have

$$e/m \not< \frac{U}{q}.$$

Schuster found for air by this method

$$e/m \not> 11 \times 10^5,$$
$$e/m \not< 10^3.$$

If we assume that the charge on the nitrogen atom is three times that on the atom of hydrogen in the electrolysis of liquids and if m is the mass of the nitrogen atom, then e/m is equal to 2×10^3; as this is within the limits for e/m previously found, Schuster concluded that the negatively electrified particles in the cathode rays in a tube filled with nitrogen are atoms of nitrogen. We have seen that more recent investigations have led to quite a different conclusion.

64. Several determinations of the values of e/m and v have been made on the assumption that the kinetic energy possessed by the ion is equal to the energy that would be acquired by the

[1] Schuster, *Proc. Roy. Soc.* xlvii. p. 526.

ion in falling through the potential difference V between the anode and the cathode; on this assumption we have

$$\tfrac{1}{2}mv^2 = Ve \quad \ldots\ldots\ldots\ldots\ldots\ldots(1),$$

and if q or mv/e is determined by the magnetic deflection we have

$$\frac{e}{m} = \frac{2V}{q^2}.$$

Determinations of e/m on this principle have been made by Kaufmann[1] and subsequently by Simon[2]. Kaufmann found by this method that

$$\frac{e}{m} = 1\cdot 86 \times 10^7.$$

And Simon, who made a very large number of experiments in which the potential difference between the cathode and anode ranged from 4860 to 11840 volts, found that

$$\frac{e}{m} = 1\cdot 865 \times 10^7.$$

The value of e/m was found to be independent of the potential difference. A Wimshurst machine was used to produce the discharge, as this maintains a very much more uniform potential difference than an induction coil.

This method is open to objection, for it assumes that the kinetic energy of the ion is equal to the work done on an ion starting in the cathode itself and thus experiencing the maximum fall of potential possible in the tube, and also that all the work done by the electric field is spent in increasing the kinetic energy of the ion while none of this energy is lost by the collisions of the ion with the molecules of the gas through which it passes.

If these assumptions are incorrect the use of equation (1) will lead to an over-estimate of the kinetic energy of the ion and therefore, since $e/m = mv^2/eq^2$, the value of e/m calculated by this method will tend to be too large.

Experiments made by Seitz[3] show, however, that when the pressure of the gas is very low the kinetic energy acquired by the electrons is equal to Ve. The deflection of the electrons by an

[1] Kaufmann, *Wied. Ann.* v. 61, p. 544; 62, p. 596, 1897; 65, p. 431, 1898.
[2] Simon, *Wied. Ann.* v. 69, p. 589, 1899.
[3] Seitz, *Ann. der Phys.* viii. p. 233, 1902.

electric field, the heat produced by them, and, on the preceding assumption, the fall of potential at the cathode all involve the same quantity, the kinetic energy of the electron. Seitz determined the kinetic energy by each of these methods and found that they gave the same value.

The method used by Lenard, and described on p. 236, though it depends upon the same equations is not open to these objections, as in this method the potential difference which enters into the equations is applied to the ions after they have been produced and started on their path, and in this case the increase in the kinetic energy must equal the work done if we can neglect the loss of kinetic energy of the ions produced by collisions with the molecules of the gas; this effect can be eliminated by working at very low pressures and varying the length of path traversed by the ion under the electric field.

Becker[1] has determined e/m for cathode rays by this method; he finds $e/m = 1\cdot 8 \times 10^7$; the velocity of the rays was about 10^{10} cm./sec.

Malassez[2] in a similar way finds $e/m = 1\cdot 76 \times 10^7$ for rather slower rays.

65. In January, 1897, Wiechert[3] published a determination of the values between which e/m must lie. The principles on which this determination is based are as follows: by measuring the magnetic deflection in a field of known strength we can determine $\dfrac{m}{e} v$; to get a second relation between m/e and v, Wiechert put

$$\frac{1}{2}\frac{m}{e} v^2 = kV \quad \ldots\ldots\ldots\ldots\ldots\ldots (1),$$

where V is the difference of potential between the electrodes in the discharge tube and k an unknown quantity which cannot be greater than unity. To get the maximum value of v, and therefore the maximum of e/m, k in equation (1) was put equal to unity. To get minimum values for v and e/m Wiechert assumed that

[1] Becker, *Ann. der Phys.* xvii. p. 381, 1905.
[2] Malassez, *Ann. Chim. Phys.* xxiii. pp. 231, 397, 491, 1911.
[3] Wiechert, *Sitzungsber. d. Physikal.-ökonom. Gesellsch. zu Königsberg i. Pr.* xxxviii. p. 1, 1897.

the kinetic energy of the ions in the cathode rays was greater than that due to a fall through a potential difference equal to the 'cathode fall of potential.' The cathode fall of potential is the difference between the potential of the cathode and that of a point on the outer boundary of that dark space in the discharge which adjoins the cathode. Warburg has shown that this cathode fall of potential is independent of the magnitude of the current through the gas, of the pressure of the gas and, within certain limitations, of the nature of the electrodes. As its value in air is about 270 volts, Wiechert assumed that a minimum value for kV was 200 volts. The grounds for this assumption do not seem obvious; *a priori* it would seem more probable that the minimum value to take for kV should have been the potential difference, not between the cathode and the outer boundary of this dark space, but between this boundary and the place where the magnetic deflection of the rays was determined, for we know that the rays are fully developed at this boundary, and it is by no means so certain that at moderate pressures they all exist close to the cathode. Using these assumptions, however, Wiechert found for the maximum value of e/m the value 4×10^7 and for the minimum value 4×10^6.

66. Wiechert[1] has also determined by direct measurement the velocity of the ions in the cathode rays, using a method first applied by Des Coudres[2] for this purpose. The principle of the method is as follows: suppose that $ABCD$, $A'B'C'D'$ are two circuits traversed by very rapidly alternating currents, such as those produced by the discharge of a Leyden jar, let us suppose that the currents in the two circuits are in the same phase, and that these circuits are placed close to a tube along which cathode rays are passing. The currents in the circuits will give rise to electric and magnetic forces which will deflect the rays as they pass by the circuits. If the velocity of the rays were infinite, then the deflections produced by the two circuits on the rays would be equal and in the same direction; if however the rays take a finite time to travel from one circuit to the other, and if the distance between the circuits is adjusted so that this time is equal

[1] Wiechert, *Wied. Ann.* lxix. p. 739, 1899.
[2] Des Coudres, *Verhandl. d. physikal. Gesellsch. zu Berlin*, xiv. p. 86, 1895.

to half the period of vibration of the current, then the deflection produced by the first circuit will be equal and opposite to that produced by the second; or if the distance between the circuits is such that the time taken by the rays to pass from one circuit to the other is equal to one-quarter of the period of the currents, then when the effect produced by the circuit $ABCD$ is a maximum that produced by $A'B'C'D'$ will be zero.

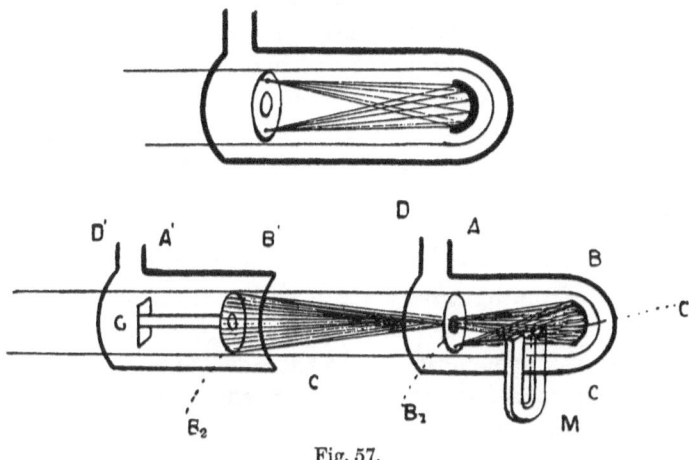

Fig. 57.

The arrangement used to apply these principles to determine the velocity of the cathode rays is represented in Fig. 57; $ABCD$, $A'B'C'D'$ are the circuits carrying the currents produced by the discharge of the jars, C is a concave cathode, B_1, B_2 metal diaphragms perforated at the centre, G a screen covered with some material which becomes phosphorescent when bombarded by the cathode rays. M is a horse-shoe magnet which deflects the rays from the hole in the diaphragm B_1, so that when no currents are passing through $ABCD$, $A'B'C'D'$ the cathode rays are stopped by the diaphragm and the phosphorescent screen remains dark. When a current passes through $ABCD$ the pencil of cathode rays is deflected and swings backwards and forwards like a pendulum; if during the swing the pencil strikes the hole in B_1 some of the rays will get through B_1 and B_2, and the screen G will be illuminated. The brightness of the illumination will be greatest when the hole in B_1 is just at the extremity of the swing caused by the current in $ABCD$, for in this case the pencil is momentarily at rest, and the time the pencil remains on the opening is therefore a maximum. If there is no current in $A'B'C'D'$ the position of the phosphorescent spot on the screen will be on the line joining the holes in the two diaphragms; if a current in the same phase as that through $ABCD$ is passing through $A'B'C'D'$, then since the cathode rays that reach the diaphragm are displaced upwards by the current in $ABCD$, they will be similarly displaced by that in $A'B'C'D'$, and the phosphorescent patch will be above the line joining the holes in the diaphragm, while if the current in $A'B'C'D'$ is in the opposite phase the patch will be displaced downwards, the direction of the displacement of the patch will be reversed by reversing the poles of the magnet. If however the phases of the currents in $ABCD$, $A'B'C'D'$ differ by a quarter of a period, then when the vertical displacement due to $ABCD$ is a maximum that due to $A'B'C'D'$ will be zero, and the vertical distribution of the light on the screen G will not be affected by reversing the magnet M. We can ensure that the rays which get through the opening in B_1 are those which are passing when the vertical displacement due to the current in $ABCD$ is greatest, by gradually increasing the deflection of the rays by moving the magnet M; when we have got M into such a position that any further increase in the deflection prevents any rays from reaching the screen, we know that only those which suffer the maximum deflection come under the action of $A'B'C'D'$; if then we move $A'B'C'D'$ into such a position that the vertical distribution of phosphorescence on the screen is not affected by reversing M, we know that when the rays are passing $A'B'C'D'$ the current in this circuit differs in phase by a quarter-period from the phase of the current in $ABCD$ when the rays were passing that circuit. If the circuits $ABCD$, $A'B'C'D'$ are arranged so that the currents in them are simultaneously in the same phase, we know that the rays must have taken a time equal to one-quarter of a period of the currents to pass from $ABCD$ to $A'B'C'D'$. The period of the currents can be determined by Lecher's method[1], hence knowing the distance between the circuits we can determine the velocity of the rays.

[1] Lecher, *Wied. Ann.* xci. p. 850, 1890.

The arrangement used to carry out this method is represented in Fig. 58. CC are two pairs of parallel plates; the upper pair of plates are connected with the spark gap F, which is also connected with the terminals of an induction coil, the lower pair of plates are connected symmetrically with the circuits $ABCD$, $A'B'C'D'$. The cathode rays are produced by a system in electrical connection with that producing the alternating currents. L and L

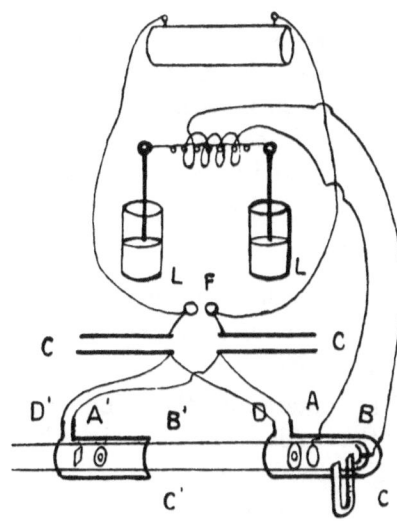

Fig. 58.

are two Leyden jars whose outer coatings are connected with the extremities of the spark gap F, the inner coatings of the jars are connected with the primary coil of a high tension transformer, the secondary coil of which is connected with the anode and cathode of the discharge tube. In order to prevent the rays being scattered to the walls of the tube during their passage from one circuit to another a magnetising spiral was wound round the tube producing a magnetic force parallel to the length of the tube; this concentrated the rays along the axis of the tube and made the observations easier. With this contrivance it was found possible not merely to find a position of $A'B'C'D'$, when the currents differed by a quarter of a period, when the rays passed through them, but to find the second position when they differed by three-quarters of a period.

If λ is the distance between the circuits when they differ by a quarter-period, L the wave-length of the electrical waves passing through these circuits, v the velocity of the rays, and V the velocity of light, then

$$\frac{v}{V} = \frac{\lambda}{L/4}.$$

Thus, in one experiment, $L = 940$ cm., $\lambda = 39$, hence v is about 5×10^9. The pressure was between $\frac{1}{2}$ and $\frac{1}{4}$ of a millimetre. v being determined, we get e/m from the value of mv/e, which is found by measuring the magnetic deflection of the rays. The determination of v by this method is difficult and we cannot expect a high degree of accuracy. As the result of his experiments, Wiechert came to the conclusion that the value of e/m is between $1{\cdot}55 \times 10^7$ and $1{\cdot}01 \times 10^7$. The most probable value he gives as $1{\cdot}26 \times 10^7$.

Determination of e/m for the Negative Ions produced when ultra-violet light falls on a metal plate, the gas through which the ions pass being at a very low pressure.

67. The writer[1] determined the values of e/m for the negative ions produced by the incidence of ultra-violet light on a metal plate by the following method. It is proved on p. 225 that when an ion starts from rest from the plane $x = 0$, at the time $t = 0$, and is acted on by a uniform electric field of strength X, parallel to the axis of x, and by a uniform magnetic force H, parallel to z, the position of the particle at the time t is given by the equations

$$x = \frac{m}{e}\frac{X}{H^2}\left\{1 - \cos\left(\frac{e}{m}Ht\right)\right\},$$

$$y = \frac{m}{e}\frac{X}{H^2}\left\{\frac{e}{m}Ht - \sin\left(\frac{e}{m}Ht\right)\right\},$$

where x and y are the coordinates of the ion. The path of the ion is thus a cycloid and the greatest distance the ion can get from the plane $x = 0$ is equal to $2mX/eH^2$.

Suppose now that we have a number of ions starting from the plane $x = 0$, and moving towards the parallel plane $x = a$, supposed to be unlimited in extent; if a is less than $2mX/eH^2$

[1] J. J. Thomson, *Phil. Mag.* v. 48, p. 547, 1899.

all the ions which start from $x = 0$ will reach the plane $x = a$, while if a is greater than $2mX/eH^2$ none of the ions will reach this plane. If $x = 0$ is a zinc plate illuminated by ultra-violet light, and thus the seat of a supply of negative ions, and $x = a$ a metal plate connected with an electrometer, then when a definite electric intensity is established between the plates, so that the number of ions which leave the plate in unit time is fixed, and if a is less than $2Xm/eH^2$, all the ions which start from $x = 0$ will reach the plane $x = a$. Thus the rate at which the plate connected with the electrometer receives a negative charge will be the same when there is a magnetic force acting across the plate as when there is no such force. If however a is greater than $2Xm/eH^2$, then no ion which starts from $x = 0$ will reach the plane $x = a$, and this plate will not receive any negative charge: so that in this case the magnetic field entirely stops the supply of negative electricity to the plate connected with the electrometer. Thus, on this theory, if the distance between the plates is less than a certain value, the magnetic force produces no effect on the rate at which the plate connected with the electrometer receives a negative charge, while when the distance is greater than this value the magnetic force entirely stops the supply of negative electricity to the plate. The actual phenomena are not so abrupt as this theory indicates. We find in practice that when the plates are near together the magnetic force produces only an exceedingly small effect, and this an increase in the rate of charging of the plate. On increasing the distance between the plates, we come to a stage where the magnetic force produces a very great diminution in the rate of charging; it does not, however, stop it abruptly, as there is a considerable range in which the magnetic field diminishes but does not entirely stop the supply of negative electricity to the plate. At still greater distances the current to the plate under the magnetic force is quite insignificant compared with the current when there is no magnetic field. We should get this gradual instead of abrupt decay of the current if the ions were projected with finite velocity, or if, instead of all starting from the plane $x = 0$, they started from a layer of finite thickness t; in this case the first ions which failed to reach the plate would be those which started from $x = 0$, this would occur when $a = 2mX/eH^2$; some ions would however continue to reach the plate until $a = t + 2mX/eH^2$. Thus if we measure the distance between the plates when the magnetic force first begins to retard the current, we can, if we know the values of X and H, determine the value of e/m. The finite thickness of the layer from which the ions start may be explained by the use of a principle which we shall find of great importance in many other phenomena connected with the discharge of electricity through gases: it is that when ions move through a gas with a velocity exceeding a certain limit, the ions by their collisions with the molecules of the gas through which they move produce fresh ions. Thus when the negative ions which start from the metal surface acquire under the electric field a certain velocity they will produce new ions, and thus the ionisation will not be confined to the metal plate but will extend through a layer of finite thickness.

In using this method of determining e/m it is necessary to have the gas between the plates at a very low pressure, so low that the mean free path of the ion is at least comparable with the distance between the plates; if this is not the case the resistance offered to the motions of the ions by the viscosity of the gas prevents the preceding investigation from being applicable.

The mean value of e/m found in these experiments was $7 \cdot 3 \times 10^6$. It thus agrees very well with the value $7 \cdot 6 \times 10^6$ found for the same quantity for the carriers of the negative electricity in the cathode rays: and proves that the carriers of electricity in the two cases are the same, or, as we may express it, that a metal plate emits cathode rays when illuminated by ultra-violet light.

68. Lenard[1] in 1900 also measured the value of e/m in the case of the discharge of negative electricity through gas at a very low pressure from a cathode illuminated by ultra-violet light. The arrangement he used is represented in Fig. 59. A is an aluminium plate on which the ultra-violet light shines: this light comes from a spark between zinc electrodes and enters the tube through the quartz window B. E is another metal electrode perforated in the middle and connected with the earth; it shields

[1] Lenard, *Ann. d. Phys.* ii. p. 359, 1900.

the right-hand part of the apparatus from the electrostatic action of the charged electrode A. D and C are electrodes which can be connected with an electrometer. When A is charged up a stream

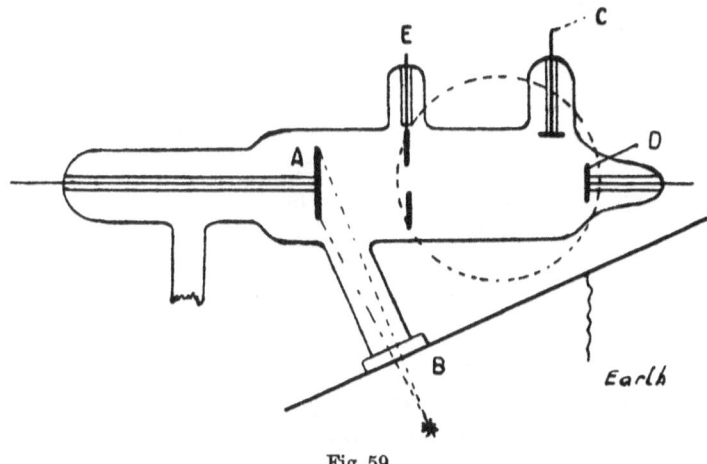

Fig. 59.

of negative electricity goes through the opening in E, and striking against the plate D, charges up the electrometer with negative electricity. If the electrometer be connected with C instead of with D, it will not receive any charge. We can however give C a charge by deflecting the stream of negative ions by a magnet

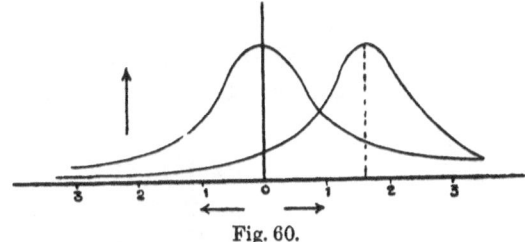

Fig. 60.

until they strike against C. As we still further increase the magnetic field the ions will be deflected by the field past C, and the charge communicated to C will fall off rapidly. The amount of negative electricity received by the electrodes D and C respectively, as the magnetic force is increased, was in Lenard's experiments represented by the curves in Fig. 60. The ordinates are the charges received by the electrodes and the abscissæ the values of the magnetic force. The curve to the left is for the electrode D, that to the right for C. Since the negative ions are not exposed to any electric field in the part of the tube to the right of E, their paths in this region under a constant magnetic field will be circles whose radii are equal to mv/eH. Now C will receive the maximum charge when the circle with this radius passing through the middle of the hole in E, and having its tangent at this point horizontal, passes also through the middle of the electrode C. The radius R of this circle is fixed by the relative positions of E and C. Hence, if we measure H when C receives its maximum charge, we have

$$R = \frac{mv}{eH} \quad \ldots\ldots\ldots\ldots\ldots\ldots(1).$$

The velocity is determined by the assumption that the work done by the electric field, when the ion passes from A to E, is spent in increasing the kinetic energy of the ion (we have already considered on p. 240 the objections which may be raised against this assumption): this leads to the equation

$$\tfrac{1}{2}mv^2 = Ve \quad \ldots\ldots\ldots\ldots\ldots\ldots(2),$$

where V is the potential difference between A and E. From equations (1) and (2) the values of e/m and v can be determined. In this way Lenard found that e/m for the negative ions produced by the action of ultra-violet light in a gas at a very low pressure is equal to $1 \cdot 15 \times 10^7$.

Reiger[1] found for the negative ions emitted by glass when exposed to ultra-violet light values of e/m ranging from $9 \cdot 6 \times 10^6$ to $1 \cdot 2 \times 10^7$.

68·1. A careful series of experiments has been made by Alberti[2] on these ions. Light from a mercury vapour lamp was directed on to the oxidised copper plate K (Fig. 61). The electrons fell through an accelerating field of 15,000–20,000 volts and passed through a fine slit in the anode A. The whole was placed in a uniform magnetic field produced by currents in the three coils shown in cross-section. The rays after leaving A thus described an arc of a circle and struck the phosphorescent screen P. In

[1] Reiger, *Ann. der Phys.* xvii. p. 947, 1905.
[2] Alberti, *Ann. der Phys.* xxxix. p. 1133, 1912.

a plane parallel to that of A were a number of fine wires F parallel to the slit; if the beam struck one of these it cast a sharp shadow on P whose position could be measured with accuracy. From this the curvature of the path was calculated. Since

$$\frac{1}{R} = \frac{He}{vm} \text{ and } \tfrac{1}{2} mv^2 = eV,$$

where V is the potential between K and A, e/m can be found. One

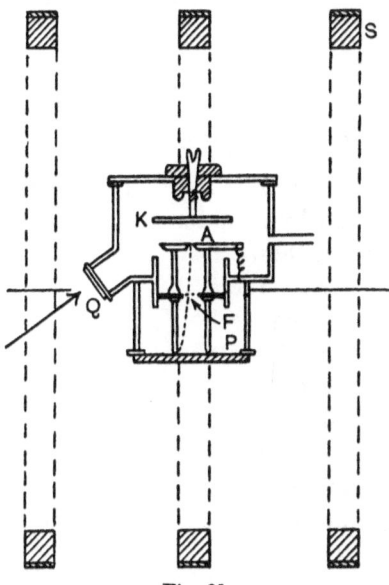

Fig. 61.

of the difficulties in an accurate determination of this kind is to find H. It can be done either by calculation from the dimensions of the coils or by comparing the field with that caused by a standard coil of known dimensions. In this case the two methods gave appreciably different results leading to values of e/m, $1 \cdot 756 \times 10^7$ and $1 \cdot 766 \times 10^7$ respectively.

Value of e/m for the Negative Ions produced by an Incandescent Wire.

69. A metal wire when raised to a white heat in a gas at a very low pressure gives out negative ions; the writer[1] has

[1] J. J. Thomson, *Phil. Mag.* v. 48, p. 547, 1899.

determined the value of e/m for the negative ions given out by an incandescent carbon filament in hydrogen at a very low pressure. The method used was the same as that used by him to determine the value of e/m for the ions produced by the action of ultra-violet light, and which has already been described on p. 246. The value of e/m found in this way was $8\cdot7 \times 10^6$, which agrees within the errors of experiment with the values found for e/m for the ions in the cathode rays, and for those produced by the action of ultra-violet light.

For the particles emitted by a glowing Nernst filament Owen[1] found $e/m = 5\cdot65 \times 10^6$, and for those emitted by glowing lime Wehnelt[2] found $e/m = 1\cdot4 \times 10^7$.

69·1. More recent determinations of e/m for these ions have been made by Classen[3] and by Bestelmeyer[4]. The former used a very neat form of apparatus shown in Fig. 62. The ions were produced from the electrically heated speck of oxide at K. Close to this was a platinum plate A with a fine hole in its centre, through which the ions were drawn by an electric field of 1000 or 4000 volts between A and K. The whole was in a uniform magnetic field due to two Helmholtz coils Sp which bent the rays as shown by the dotted line, so that after passing round in a semicircle they struck the underside of the photographic plate P, which was specially cut into a circle with

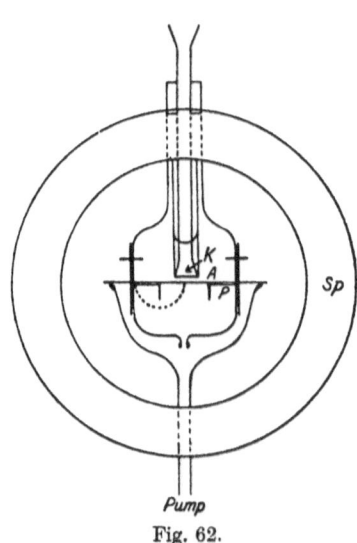

Fig. 62.

a hole through its centre. By reversing the magnetic field the rays could be deflected on to the opposite part of the plate, and the distance between the two impressions so formed gave twice

---

[1] Owen, *Phil. Mag.* vi. 8, p. 230, 1904.
[2] Wehnelt, *Ann. der Phys.* xiv. p. 425, 1904.
[3] Classen, *Phys. Zeits.* ix. p. 762, 1908.
[4] Bestelmeyer, *Ann. der Phys.* xxxv. p. 909, 1911.

the diameter of the path, after making a small correction for the fact that the plane of the plate lay slightly below the centres of the circles described by the rays. The apparatus was exhausted, and the coils were arranged so that, without disturbing the rest of the apparatus, they could be rotated about a vertical axis, thus changing the direction of the magnetic field and the plane of the semicircles described by the rays. For each of six settings of the magnetic field exposures were taken with five slightly varying values of the magnetic field, so that in all thirty diameters could be measured on the plate, each of which gives a value of e/m. An additional advantage of rotating the magnetic field in this way is that it enables the earth's magnetic field to be eliminated. The mean value found after correction to zero velocity (see below, § 70·4) was $1·775 \times 10^7$.

Bestelmeyer[1] used a method which enabled him to follow the path of the rays over a long distance. The rays, produced from a heated speck of lime and accelerated by about 870 volts, were bent round in a circle of about 17·5 cm. diameter, almost returning to the starting-point. The circle was marked out by six diaphragms, and the path of the rays observed by the light they produced in the residual gas of the apparatus, the distances at which they passed from the edges of the diaphragms being measured with microscopes. The magnetic field, about 11 gauss, required to produce the curvature was determined from the current and dimensions of the solenoid which produced it. It was found that the curvature was slightly greater at the end of the path than at the beginning, owing to the slowing down of the rays by collision with the gas molecules. The value $1·767 \times 10^7$ was found for e/m, which would be somewhat reduced if allowance were made for the reduction in velocity.

Value of e/m for the Negative Ions emitted by Radioactive Substances.

70. It has been shown by M. and Madame Curie[2] that the radioactive substance radium emits negative ions. The velocity of these ions and the value of e/m have been determined

[1] Bestelmeyer, *Ann. der Phys.* xxxv. p. 909, 1911.
[2] M. et Mme Curie, *Comptes Rendus* t. 130, p. 647.

by Becquerel[1]. The method he employed was to measure the deflections of the rays produced by an electrostatic and also by a magnetic field. The experiments were made at atmospheric pressure, and the resistance offered to the motion of the ions by the gas through which they pass was neglected: this would not be justifiable in the case of the ions we have hitherto been considering, but as the ions emitted by radium are very much more penetrating than those we have hitherto considered, and are able to travel as far through a gas at atmospheric pressure as other kinds of ions travel through a gas at a very low pressure, we shall probably get approximately the right values for e/m and v for the radium ions even if we neglect the resistance of the gas. The radium was placed below two parallel vertical metal plates, about 3·5 cm. wide and 1 cm. apart; above these metal plates was a horizontal photographic plate protected by a covering of black paper from the action of light; a thin slip of mica, symmetrically situated with respect to the metal plates, was placed over the radium, this cast a shadow on the photographic plate which when the metal plates were at the same potential was at the middle of the field; when a great difference of potential, 10,200 volts, was maintained between the plates the position of this shadow was displaced towards the positive plate. Consider an ion passing between the plates, then if l is the length of its path between the plates, F the electric force acting upon it, the displacement of the ion parallel to the lines of electric force when it leaves the region between the plates is $\frac{1}{2}\frac{Fe}{m}\frac{l^2}{v^2}$, and its direction of motion is displaced through an angle $\tan^{-1}\frac{Fe}{m}\frac{l}{v^2}$; hence if h is the vertical distance of the photographic plate above the upper edge of the parallel metal plate, the point where the ion strikes the plate will be deflected through a space δ parallel to the line of electric force, where δ is given by the equation

$$\delta = \frac{1}{2}\frac{Fe}{m}\frac{l^2}{v^2} + h\frac{Fe}{m}\frac{l}{v^2}$$
$$= \frac{Fe}{m}\frac{l}{v^2}\left\{\frac{l}{2} + h\right\}.$$

[1] Becquerel, *Rapports présentés au Congrès International de Physique à Paris*, t. iii. p. 47, 1900.

The magnetic deflection was found in the following way: a small quantity of radium was placed in a little lead saucer on a photographic plate; as none of the rays from the radium reach the plate the latter is not affected; if however a strong magnetic field, with the lines of force parallel to the plate, acts on the negative ions coming from the radium, these will be bent round and will strike the plate, producing a photograph.

To find the boundary of this photograph, let us take the plane of the photographic plate as the plane of xy, the magnetic force H being parallel to x; the equations of motion of an ion are

$$m \frac{d^2x}{dt^2} = 0, \quad m \frac{d^2y}{dt^2} = He \frac{dz}{dt}, \quad m \frac{d^2z}{dt^2} = -He \frac{dy}{dt};$$

the solutions of these equations are, if $\omega = He/m$, and u, A, B are constants,

$$x = ut,$$
$$y = A(1 - \cos \omega t) + B \sin \omega t,$$
$$z = A \sin \omega t + B (\cos \omega t - 1).$$

If v and w are the values of dy/dt, dz/dt when $t = 0$, we have

$$y = \frac{w}{\omega}(1 - \cos \omega t) + \frac{v}{\omega} \sin \omega t,$$

$$z = \frac{w}{\omega} \sin \omega t + \frac{v}{\omega}(\cos \omega t - 1);$$

when the ion strikes the plane we have $z = 0$, hence

$$\tan \tfrac{1}{2}\omega t = \frac{w}{v}.$$

Now if the ion is projected so as to make an angle θ with the direction of the magnetic force, and if the plane through the direction of projection and the axis of x makes an angle ϕ with the plane of xz, we have, if V is the velocity of projection,

$$u = V \cos \theta, \quad v = V \sin \theta \sin \phi, \quad w = V \sin \theta \cos \phi;$$

hence
$$\tan \tfrac{1}{2}\omega t = \cot \phi$$
$$= \tan \left(\frac{\pi}{2} - \phi\right),$$

thus
$$\omega t = \pi - 2\phi.$$

Substituting this value for t, we find, if ξ and η are the co-ordinates of the point where the ion strikes the photographic plate,

$$\xi = \frac{V \cos \theta}{\omega} (\pi - 2\phi),$$

$$\eta = \frac{2V \sin \theta \cos \phi}{\omega}.$$

Thus, for the particles projected in a plane through the axis of x, the locus of the points where they strike the plate will be an ellipse whose semi-axes are $\frac{2V \cos \phi}{\omega}$ and $\frac{V(\pi - 2\phi)}{\omega}$. For the particle projected in the plane of xz, the semi-axes of the ellipse are $2V/\omega$ and $\pi V/\omega$. An example of such an ellipse is shown in Fig. 63, which is copied from a photograph by Becquerel.

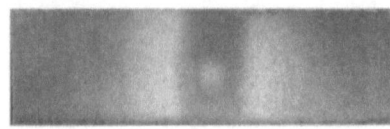

Fig. 63.

By the measurement of the axes of the ellipse we can determine V/ω, i.e. Vm/eH. As the radium emits ions having velocities extending over a considerable range, the impression on the plate is not the arc of a single ellipse, but a band bounded by the ellipses corresponding to the smallest and greatest velocities of the ions. Becquerel took photographs when the ions from the radium went (1) through the air at atmospheric pressure, and (2) through air at very low pressure; the photographs were found to be identical, in fact one-half of the photograph represented in Fig. 63 is produced by ions going through air at atmospheric pressure, and the other half by ions going through air at a very low pressure. The identity of the results in the two cases justifies us in our neglect of the resistance of the air.

As the result of his experiments Becquerel found for one set of rays given out by the radium

$$v = 1·6 \times 10^{10}, \quad e/m = 10^7,$$

thus the value of e/m is the same for these negatively charged ions from radium as for the ions in the cathode and Lenard rays, as well as for those produced by ultra-violet light or by incandescent metals. The velocity of the ions is much greater than any we have met with in the case of ions arising in other ways, amounting as it does to more than half the velocity of light; the ions chosen by Becquerel for this experiment were by no means the fastest given out by the radium. Becquerel detected the existence of others whose velocity was at least half as much again as the velocity of those he measured.

70·1. The constancy of the value of e/m for the ions derived from these varied sources justifies the assumption that we are dealing with identical bodies in all cases, which must therefore be present in very varied kinds of matter. Further, the value of e/m found is the same, within the errors of experiment, as that found by Zeeman and later workers for the electrified particles whose vibrations they assumed to cause the emission of spectral lines. Johnstone Stoney had suggested the term 'electron' for these hypothetical particles, and this name is now generally adopted for the various classes of ions above mentioned which show the same value of e/m. It must be realised that while these ions start as electrons they, in most cases, readily attach themselves to molecules of gas, if these are present, forming heavier ions with a much smaller e/m. A discussion of the optical methods of determining e/m would take us beyond the scope of this book, but they are of two kinds. The original method of Zeeman depended on the splitting up of spectral lines in a magnetic field, the second method depends on the relation between the wave-lengths of lines in the spectra of hydrogen and ionised helium, using Bohr's theory of the atomic structures. While these methods are capable of great numerical accuracy owing to the accuracy attainable in wave-length measurements, they involve questions of atomic structure and dynamics which are not yet settled, and the very direct methods, described above, are of great importance as a check on the results deduced from theoretical considerations.

70·2. Since a moving electrified particle is surrounded by a magnetic field which possesses energy, more work will have to be done to start or stop the particle than if it were uncharged.

It will thus behave as though it had an extra mass due to its charge, a result pointed out by one of the authors in 1881. While the increase in mass for a particle of ordinary size is negligible for such degrees of electrification as are practically possible, it becomes more and more important as the size of the particle decreases, and it was early suggested that much or all of the mass of the electron might be of this electrical character. A detailed calculation shows (*Recent Researches*, p. 21) that the electrical mass is not constant but increases with the velocity, or put in another way, that the energy of the magnetic field increases more rapidly than the square of the velocity, but this effect is only appreciable when the velocity is a considerable fraction of that of light. Since it is possible to obtain β particles from radioactive substances with velocities up to within a few per cent. of that of light, measurements of their e/m should show a marked variation with velocity, and the law of variation thus found could be compared with theory. Actually several laws were arrived at on theoretical grounds corresponding to different assumptions as to the behaviour of the electron in motion, but all agreeing in that the mass increases indefinitely for velocities very near that of light and is practically constant for velocities less than a few per cent. of this. The two formulae which received most support were that due to Abraham

$$m = \frac{3m_0}{4\beta^2} \left[\frac{1+\beta^2}{2\beta} \log \frac{1+\beta}{1-\beta} - 1 \right],$$

where m_0 is the mass for slow speeds and β is the ratio of the velocity to that of light, and that due to Lorentz

$$m = \frac{m_0}{\sqrt{1-\beta^2}},$$

the former being obtained on the assumption that the electron behaves as a rigid sphere, and the latter supposing that a contraction occurs in the direction of motion in the ratio $\sqrt{1-\beta^2} : 1$. If only part of the mass is electrical then, on these theories, a constant term would have to be added to represent the non-electrical mass. The first experiments were made by Kaufmann[1].

[1] Kaufmann, *Göttingen Nach.* Nov. 8, 1901.

70·3. The method used by Kaufmann is illustrated in Fig. 64.

A small piece of radium was placed at C in a vessel from which the air was extracted, the radiations from the radium passed through a strong electric field in the space between the parallel plates P_1, P_2 which were ·1525 cm. apart and maintained at a potential difference of 6750 volts, they then passed through a small hole D in a diaphragm and then on to a photographic plate E; during the whole of their journey from C to E the rays were under the influence of a magnetic field produced by the

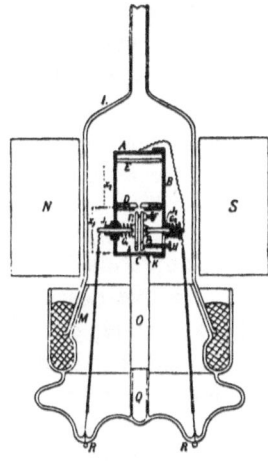

Fig. 64.

electromagnet NS; the deflection due to the magnetic field was at right angles to that due to the electric. If the electric and magnetic fields were not in action all the rays from the radium would strike the photographic plate at the same point; when however the rays are exposed to the electric and magnetic fields the deflection will depend upon the velocity, so that the rays of different velocities will now strike the plate at different points and the impression produced by the radium on a plate will be a curved line; by measuring the photograph the deflection due to the magnetic field and also that due to the electric field can be found, and from these deflections (see p. 235) the values of v the velocity of the particles and the corresponding values of e/m can be found. Kaufmann found that when his plates were exposed for several days he got a clearly defined curve from which he deduced the following values of e/m and v.

$v \times 10^{-10}$	$e/m \times 10^{-7}$
2·83	·63
2·72	·77
2·59	·975
2·48	1·17
2·36	1·31

These experiments were not sufficiently accurate to distinguish with certainty between the various formulae proposed, but they show a very marked change of e/m with velocity, and are consistent with the view that all the mass is electrical. More careful experiments led Kaufmann to the conclusion that the formula of Abraham was more nearly correct than that of Lorentz.

70·4. Further experiments were made by Bestelmeyer[1] and by Bucherer[2]. These experimenters used very similar methods which were based on the principle described in § 59, namely determining the velocity by arranging electric and magnetic fields so that their deflections just cancelled and then finding e/m from the magnetic deflection alone. Bestelmeyer used electrons caused by X-rays falling on a platinum plate; these did not give a large enough range of velocities to enable a discrimination between the formulae to be made with certainty. Bucherer used β-rays from radium fluoride. His arrangement was as follows. The source of rays was placed in the centre of a parallel plate condenser formed of two circular silvered glass plates of 8 cm. diameter, optically flat and 0·25 mm. apart. The lines of magnetic force were parallel to the plates. Suppose a ray to start at an angle θ to H and in the plane of the plates. The electrostatic and magnetic deflections will balance, and so allow the ray to escape from the condenser if

$$eX = ev_0 H \sin \theta \text{ or } v_0 = X/H \sin \theta.$$

Thus for every direction θ there will theoretically be only one velocity of ray which can escape. In practice of course there will

[1] Bestelmeyer, *Ann. der Phys.* xxii. p. 429, 1907.
[2] Bucherer, *Ann. der Phys.* xxviii. p. 513, 1909.

be a finite pencil owing to the condenser planes not being infinitely close. After the rays have left the condenser they are acted on by the magnetic field only, and the deflection measured by allowing them to strike the sensitive surface of a photographic film bent into a cylinder, coaxial with the condenser. If z be the deflection and a the distance between the film and the edge of the condenser,

$$\frac{e}{m} = \frac{v_0(a^2 + z^2)}{2H\sin\theta . z} = \frac{X(a^2 + z^2)}{2H^2\sin^2\theta . z}.$$

By reversing X and H a double trace is formed on the film (Fig. 65); each pair of points can be used to find e/m for rays of a known velocity. Bucherer however used chiefly the most deflected rays for which $\theta = \pi/2$ and adjusted the velocity by altering the fields.

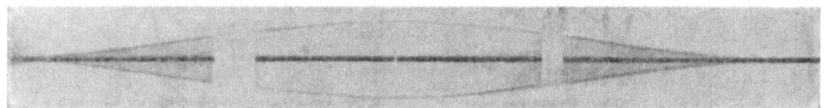

Fig. 65.

In all these and the subsequent experiments the whole apparatus was exhausted to a high vacuum. Bucherer found very definitely better agreement with the Lorentz than with the Abraham formula. In these experiments v/c varied from ·379 to ·678. Wolz[1] modified the method so as to get an accurate absolute value of e/m. For this the correction due to the stray field at the edge of the condenser is appreciable, though it is unimportant in determining the variation of e/m with velocity. He found this experimentally by making experiments with the plate at different distances from the condenser. His results agreed well with the Lorentz formula.

The absolute value, corrected by Neumann for an error in measurement, was $e/m_0 = 1\cdot 7706 \times 10^7$, where m_0 is the mass of the electron for small velocities. Neumann[2] made an extremely careful series of measurements, using the same method, and in part, the same apparatus. He found full agreement with the

[1] Wolz, *Ann. der Phys.* xxx. p. 273, 1909.
[2] Neumann, *Ann. der Phys.* xlv. p. 529, 1914. See also Schaefer, *Ann. der Phys.* xlix. p. 934, 1916.

Lorentz formula except at the highest velocities ($\beta = 0.8$) where the lines on the plate became broad, and the measurements less exact, probably owing to the finite range of velocity in the pencil used which becomes more serious when the mass is varying rapidly with velocity. Hupka[1] has also found agreement with the Lorentz formula, using the method of § 64. The results of the former observers are shown plotted in Fig. 66, taken from Cunningham's *Relativity and the Electron Theory*, p. 67. This shows $\dfrac{e}{m\sqrt{1-\beta^2}}$ plotted against β, so that the points should lie on a line parallel to the axis if the Lorentz formula is true, and it will be seen that they do so within the errors of experiment.

70·5. This result might be supposed to prove that the whole mass of the electron is electrical. If this is so, and the electron is assumed spherical, its radius a can be found from the equation $m_0 = \tfrac{2}{3} e^2/a$ (see *Recent Researches*, p. 21).

Einstein has shown that to conform with the principles of Relativity mass must vary with the velocity according to the law $m_0 \Big/ \sqrt{1 - \dfrac{v^2}{c^2}}$. This is a test imposed by Relativity on any theory of mass. We see that it is satisfied by the conception that the whole of the mass is electrical in origin, and this conception is the only one yet advanced which gives a physical explanation of the dependence of mass on velocity. Using the value of e found by the methods explained in the next chapter, namely

$$4.77 \times 10^{-10} \text{ E.S.U. or } 1.59 \times 10^{-20} \text{ E.M.U.}$$
and
$$e/m_0 = 1.765 \times 10^7,$$

the value of a is 1.87×10^{-13} cm.

An electron might cease to act in this way at a much greater distance if instead of being a spherical mass of negative electricity it had a structure represented by variations in the density of the electricity, varying periodically with the distance from the centre, the total charge being equal to $-e$.

70·6. It may be convenient to summarise in a table the results of the measurements of e/m made by different observers, and with ions produced in different ways. In the upper part of the table

[1] Hupka, *Ann. der Phys.* xxxi. p. 169, 1910.

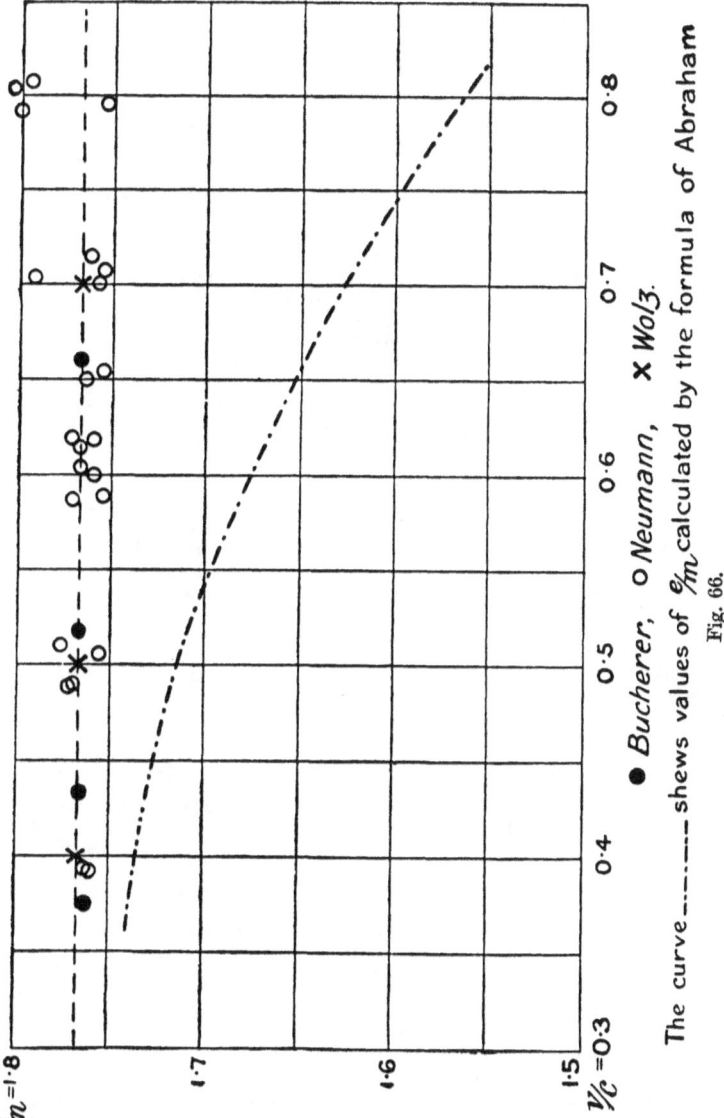

Fig. 66.

● Bucherer, ○ Neumann, × Wolz.

The curve ----- shews values of e/m calculated by the formula of Abraham

are given some of the older measurements of historical interest, and below some of the best of the modern ones, the last two being typical of the optical determinations.

Table of Values of e/m.

Source of Ions	Observer	Date	Method of Determination	Value of e/m	$v \times 10^{-9}$
Cathode rays	J.J.Thomson	1897	Magnetic and electrostatic deflection	7.7×10^6	2.2–3.6
,,	J.J.Thomson	1897	Magnetic deflection and heating effect	1.17×10^7	2.4–3.2
,,	Kaufmann	1897–8	Magnetic deflection and potential difference	1.86×10^7	
,,	Simon	1899	Magnetic deflection and potential difference	1.865×10^7	
,,	Wiechert	1899	Magnetic deflection and velocity of ions	1.01×10^7–1.55×10^7	
,,	Seitz	1901	Magnetic and electrostatic deflection	6.45×10^6	7.03
,,	Seitz	1902	Magnetic and electrostatic deflection, heating effect and potential difference	1.87×10^7	5.7–7.5
,,	Becker	1905	Magnetic deflection and retardation in electric field	1.8×10^7	10
Lenard rays	Lenard	1898	Magnetic and electrostatic deflection	6.39×10^6	
Ultra-violet light	J.J.Thomson	1899	Retardation of discharge by magnetic field	7.6×10^6	
,,	Lenard	1900	Magnetic deflection and potential difference	1.15×10^7	
Incandescent metals	J.J.Thomson	1899	Retardation of discharge by magnetic field	8.7×10^6	
Radium	Becquerel	1900	Magnetic and electrostatic deflection	10^7 approximately	2×10^{10}
Radium	Kaufmann	1901–2	Magnetic and electrostatic deflection	1.77×10^7	
X-rays	Bestelmeyer	1907	Magnetic and electrostatic deflection (crossed fields)	1.72×10^7	6×10^9 9.7×10^9
Incandescent oxide	Bestelmeyer	1911	Magnetic deflection and potential difference	1.767×10^7	1.7×10^9
,,	Classen	1907	Magnetic deflection and potential difference	1.775×10^7	2×10^9 4×10^9
Ultra-violet light	Alberti	1912	Magnetic deflection and potential difference	1.756×10^7 1.766×10^7	7×10^9 8.5×10^9
Radium	Bucherer	1909	Magnetic and electrostatic deflection (crossed fields)	1.763×10^7	1.1×10^{10} 2×10^{10}
,,	Wolz	1909	Magnetic and electrostatic deflection (crossed fields)	¹$1.7706 \times 10^7$	1.2×10^{10} 2.1×10^{10}
,,	Neumann	1914	Magnetic and electrostatic deflection (crossed fields)	1.765×10^7	1.2×10^{10} 2.4×10^{10}
—	Fortrat	1912	Zeeman effect	1.7636×10^7	—
—	Paschen	1916	Bohr's theory	1.7649×10^7	—

The values of e/m in the lower half of the table are corrected to zero velocity.

¹ Corrected by Neumann.

Taking as the most probable value $e/m_0 = 1\cdot 765 \times 10^7$ and as the value of e/m for hydrogen ions in electrolysis 9570·6 we find for the ratio of the mass of an electron to that of a hydrogen atom 1 : 1844.

It is to be noted that these large values of e/m for gases only occur when the pressure of the gas is very low, when in fact there is very little gas for the ion to get entangled with; when the pressure of the gas is high, the ion seems to act as a nucleus round which the molecules of the gas collect; the ion thus gets loaded up, and the ratio of e/m is very small compared with its value at lower pressures.

Value of e/m for the Positive Ions.

71. The first determination of the value of e/m for the positive ions was made by W. Wien[1]. The positive ions he used were those which occur in what are known as 'canal-strahlen.' If an

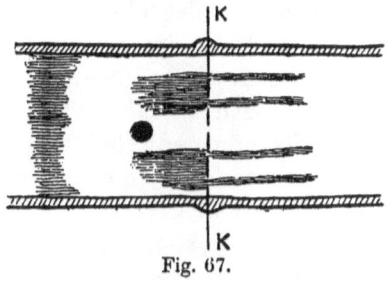

Fig. 67.

electric discharge passes between an anode and a cathode perforated with a number of holes, then behind the cathode, *i.e.* on the side of the cathode opposite to the anode, pencils of light are seen to penetrate through the holes as in Fig. 67[2], producing phosphorescence when they strike the glass. These rays—the canal-strahlen—have been shown by Wien to consist of positively charged ions. He exposed a long pencil of these rays coming through a perforated iron cathode to both an electrostatic and a magnetic field, and measured the corresponding deflections;

[1] W. Wien, *Wied. Ann.* lxv. p. 440, 1898.
[2] Wehnelt, *Wied. Ann.* lxvii. p. 421, 1899.

from these he deduced by the method described in § 60, the values of e/m and v, and found

$v = 3.6 \times 10^7$ cm./sec., while $e/m = 300$.

The 'canal-strahlen' or positive rays are only deflected with great difficulty, and it is necessary to use very strong fields; this increases the difficulty of the investigation; in Wien's experiments the strength of the magnetic field was 3250. It will be seen that the velocity of the positive ions is very much smaller than that of any of the cathode rays hitherto measured, while the value of e/m is of an entirely different order, being only about 1/30000 of the value for the negative ion; moreover the value of e/m for the positive ions in the gas is *of the same order of magnitude* as the value of e/m in the ordinary electrolysis of solutions. Thus if m were the mass of the atom of iron, e the charge carried by an atom of hydrogen, e/m is about 200, or since iron is divalent the value of e/m for the ion in the electrolysis of solutions is about 400.

In some later experiments Wien[1] has measured the value of e/m for the 'canal-strahlen' in tubes filled with different gases; he found that the most deflectible rays gave values for e/m of the order 10^4, *i.e.* the value for the hydrogen atom, not only when the tube was filled with hydrogen but also when it was filled with other gases very carefully prepared: the phosphorescence due to the most deflectible rays was however much greater in hydrogen than in air or oxygen. We shall see that a metallic cathode emits hydrogen with great persistency, so that it is practically impossible to prepare a tube which does not contain hydrogen near the cathode. Along with these deflectible rays Wien found others with a much smaller e/m.

71·1. In Wien's experiments there was apparently a continuous variation of e/m. This is now known to be due to the fact that the positive rays, unlike cathode rays, do not retain their charge unaltered in their passage through the residual gas of the tube, but are often neutralised by collision with a gas molecule while retaining their speed and ability to affect the phosphorescent screen. If the neutralisation occurs while the particle is in the deflecting fields it will experience only a fraction of the proper

[1] Wien, *Ann. der Phys.* viii. p. 241, 1902.

deflection, and so will give a low value of e/m. To obviate this it is necessary to work at very low pressures, a few thousandths of a millimetre at most, and preferably less. A series of experiments on these lines has been made by one of the authors. The form of apparatus finally adopted is shown in Fig. 68. Instead of a fluorescent screen, a photographic plate is used to detect the

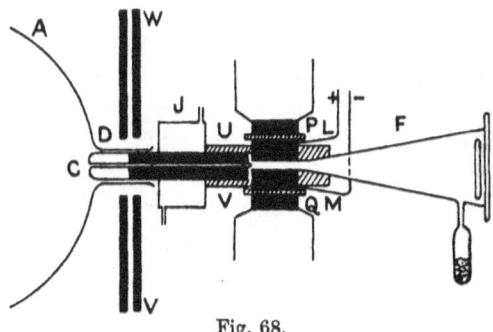

Fig. 68.

rays. As they have slight penetrating power it is advisable to use plates in which the silver bromide is concentrated in a thin layer. Such a plate is the Paget Process plate.

The discharge takes place in a large glass flask A: a volume of from one to two litres is a convenient size for this purpose. The cathode C is placed in the neck of the flask. The position of the front of the cathode has a very considerable influence on the brightness of the positive rays and ought to be carefully

Fig. 69.

attended to. The best position seems to be when the front of the cathode is flush with the prolongation of the wider portion of the flask. The shape of the cathode is represented in section in Fig. 69: the face is made of aluminium, the other portion is soft iron. A hole is bored right through the cathode to admit the fine tube through which the positive rays are to pass. Care should be taken to bore this hole so that its axis is the axis of symmetry of the cathode. The tube through which the positive rays pass is fastened into the cathode in the way shown in Fig. 69.

The bore of this tube will vary with the object of the experiment. If very accurate measurements are required, the diameter of the tube must be reduced to ·1 mm. or less. With these very fine tubes, however, very long exposures ($1\frac{1}{2}$ to 2 hours) are necessary. The length of the tube is about 7 cm. The tubes are prepared by drawing out very fine bore copper tubing until the bore is reduced to the desired size. The tube is straightened by rolling it between two plane surfaces, and great care must be taken to get the tube accurately straight, as the most frequent cause of dimness in the positive rays is the crookedness of the tube. After long use the end of the tube nearest the discharge tube gets pulverised by the impact of the positive rays, and the metallic dust sometimes silts up the tube and prevents the rays getting through. The cathode is fastened in the glass vessel by a little sealing-wax, and a similar joint unites it to the ebonite box, UV. To keep the joints cool and prevent any vapour coming from the wax, the joints are surrounded by a water jacket J through which a stream of cold water circulates.

The electric field is produced between the faces of L and M which are pieces of soft iron with plane faces. These are fitted into the ebonite box UV so that their faces are parallel: the distance between the faces should be small compared with their lengths. In many of the experiments the length of the faces was 3 cm. and their distance apart 1·5 mm. Their faces are connected with the terminals of a battery of small storage cells: in this way any required difference of potential can be maintained between them.

These pieces of soft iron practically form the poles of an electromagnet, for the poles of the electromagnet P and Q are made of soft iron of the same cross-section as L, M; they fit into indentations in the outside of the ebonite box and are only separated from the pieces L, M by the thin flat pieces of ebonite which form the walls of the box. This arrangement makes the magnetic field as nearly coterminous as possible with the electric, which is desirable in several of the experiments. A light-tight vessel F 40 cm. long is fastened by wax to the ebonite box while the other end is fixed to the apparatus which contains the photographic plate. One form of this, designed by Mr Aston, is represented in Fig. 70. The photographic plate is suspended by a silk thread wound round a tap T which fits into a ground glass joint; by turning the tap the thread can be rolled or unrolled and the plate lifted up or let down. The plate slides in a vertical box B made of thin metal; this is light-tight except at the openings A which are placed so that the positive rays can pass through them. The openings are on both sides of the box and about 5 cm. in diameter. When the silk thread is wound up the strip $DEFG$ of photographic plate in the box is above the opening A, so that there is a free way for the rays to pass through A and fall on a willemite screen behind it. This screen is not used for purposes of measurement, but only to see before taking the photograph that the tube is giving an adequate supply of positive rays. The box is sufficiently large to hold a film long enough for two or more photographs; if it is wished to take two photographs, the plate is lowered until the bottom half comes opposite to the opening A, a photograph is taken in this position, the plate is then let down still further until the top half of the plate comes opposite to the opening, then a second photograph is taken. This plan

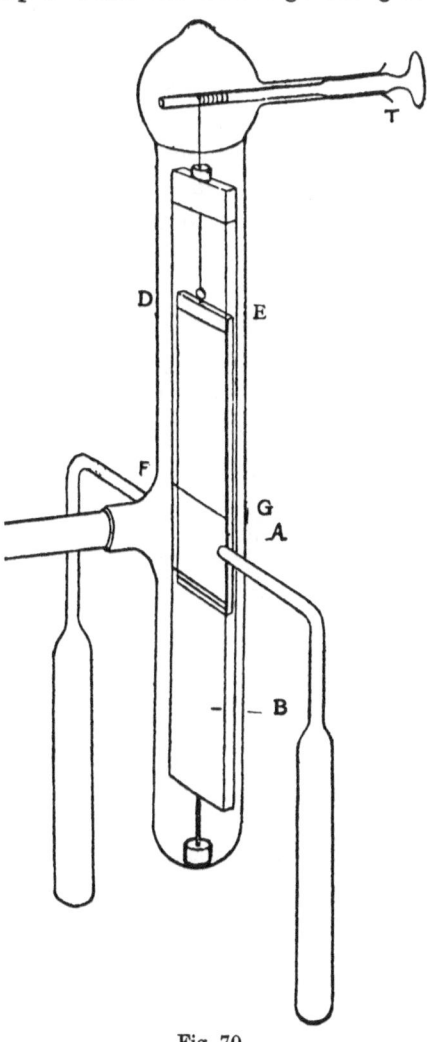

Fig. 70.

is convenient because the deflections of the different kinds of positive rays differ so much that it is difficult to measure them accurately when they are all on one plate. For example the magnetic deflection of the hydrogen atom is about fourteen times that of the mercury one; thus if the deflection of the hydrogen atom is within the limits of the plate, that of the mercury atom would be too small to measure accurately. When we can take two photographs, however, without opening the tube, we may take one with a small magnetic field to get the deflection of the hydrogen atom, and the second with a much larger one to get the deflection of the mercury one.

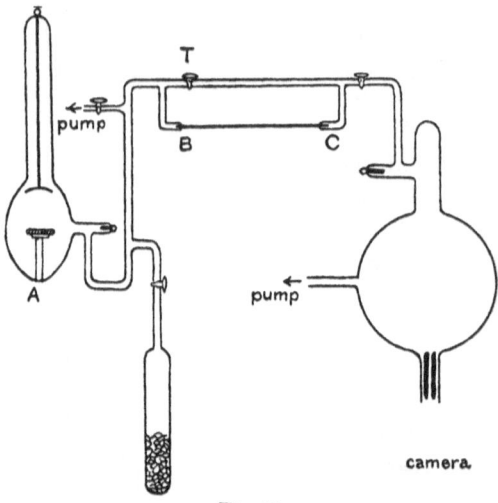

Fig. 71.

Two tubes containing coco-nut charcoal are fused to this part of the apparatus; by immersing these in liquid air the pressure can be made exceedingly small. As the only communication between this part of the apparatus and that through which the discharge passes is through the long and very narrow tube in the cathode, it is possible to have the pressure on the camera side of the apparatus very much less than the pressure on the side through which the discharge is passing.

A Gaede pump worked by a motor is connected with the discharge tube, and keeps the pressure in this part of the apparatus at a suitable value. When the rays in some particular gas are under examination a constant stream of this gas is kept flowing through the discharge tube. The gas is stored in the vessel A, Fig. 71, over a column of mercury (not shown): this vessel is connected with the discharge tube by the system TBC, where BC is an exceedingly fine capillary tube. When the tap T is turned the gas has to pass through this capillary: it does so exceedingly slowly. The rate can be adjusted by raising or lowering a mercury reservoir connected with A; this is held in such a position that

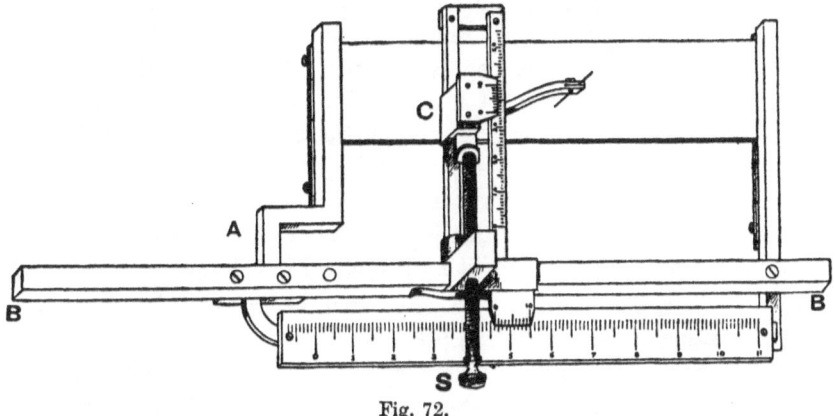

Fig. 72.

when the Gaede pump is in action the pressure in the discharge tube is such as to give well developed positive rays. To screen off the magnetic field due to the electromagnet, thick iron plates V, W, Fig. 68, are placed round the neck of the tube.

The curves on the photographic plates made by the positive particles are measured by the apparatus represented in Fig. 72. The photographic plate is clamped in a holder A, and the position of any point on it is determined by moving the carrier C until the tip of the needle comes just over the point in question. The carrier C has two movements, one parallel to the base BB, and the other, by means of the screw S, at right angles to this direction; the position of the point is read off on the two verniers. The plate is placed in the holder so that the direction of the magnetic deflection is parallel, and that of the electrostatic deflection at right angles, to BB.

71·2. The theory follows at once from equations (1) and (2) of § 59. If z is the magnetic deflection and y the electric, we have

$$z = \frac{e}{vm}\int_0^{op}\left[\int_0^x H dx\right]dx = \frac{Ae}{vm} \text{ and } y = \frac{e}{v^2m}\int_0^{op}\left[\int_0^x Y dx\right]dx = \frac{Be}{v^2m},$$

where A and B can be calculated and are the same for all rays. Hence

$$\frac{z}{y} = \frac{v.A}{B}, \text{ and } \frac{z^2}{y} = \frac{A^2}{B}\cdot\frac{e}{m},$$

from which it follows that all rays with the same velocity lie on a straight line through the spot formed by the uncharged rays which are always present, and that all rays with the same e/m lie on a parabola with this spot as vertex. If a ray is only charged during a fraction of its path in the field one or both of the deflections will be smaller. In the particular case when the fields coincide in position, as is approximately the case with the above arrangement, the deflections are reduced in proportion and the ray lies somewhere on the line joining the origin to the point where the ray would have been if it had retained its charge.

Photographs obtained in this way are shown in Plate I. Those taken at low pressures (·001 mm. or less) show only a series of parabolic arcs, each of which by the above theory corresponds to a definite value of e/m. The outermost parabola (Fig. 1) is found to have a value of e/m corresponding accurately with that found for the hydrogen atom in electrolysis, i.e. 9571 in E.M.U. per gm. In most of the measurements that have been made this parabola is taken as standard and e/m determined for the others by comparison with it. In addition to this there is almost always a parabola for which e/m is $\frac{1}{2}$ the above and others for which it is $\frac{1}{12}$, $\frac{1}{16}$, $\frac{1}{28}$ and $\frac{1}{44}$. These are interpreted as being due respectively to a molecule of hydrogen, an atom of carbon, one of oxygen, and molecules of CO and CO_2 in each case with the same unit charge, the ratios being in agreement with the usual atomic and molecular weights. These parabolas occur whatever gas is in the discharge tube unless very special precautions are taken to remove them, for example by passing a stream of oxygen through the discharge tube for some hours it is possible to obtain a photograph on which the hydrogen parabolas are almost or quite invisible. They return however if the photograph is taken after

PLATE I

Fig. 1.

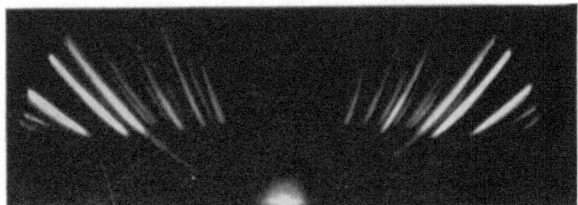

Fig. 2.

Fig. 3.

In Figs. 2 and 3 the symmetry is the result of taking two photographs with opposite magnetic fields.

the stream of oxygen has stopped. These so-called 'residual' gases are largely derived from the metal electrodes which seclude large quantities of gas, especially hydrogen, but no doubt the vapour of tap grease is also responsible for a good deal.

Besides these common constituents, parabolas are found with values of e/m corresponding to almost any gas introduced into the tube. Thus parabolas are known due to all the inert gases, to chlorine and bromine, water, numerous organic and sulphur compounds and mercury. This last is extremely strong if, as is usually the case, a mercury pump is used for the exhaustion of the discharge tube. In most cases any one constituent will give rise to several parabolas. This happens in two ways. If the gas is molecular we find a parabola corresponding to a molecule of the gas with the charge of a hydrogen atom, and in addition others corresponding to similarly charged atoms or portions of the molecule. Thus chlorine gives Cl_{2+} and Cl_{+}, while methane gives CH_{4+}, CH_{3+}, CH_{2+}, CH_{+} and C_{+}, besides of course a strong hydrogen parabola. Numerous other instances might be given, the general rule being that any imaginable decomposition product will occur irrespective of whether it obeys the rules of chemical valency, though there are some exceptions to this. It must be remembered that these decomposition products can only be proved to exist for the very short time which it takes for a ray to travel down the tube, which is usually less than a millionth of a second, and then only in the form of charged ions.

On the other hand if the gas is atomic we find, besides a parabola representing the atom with the unit charge, others with twice, three times or more the value of e/m. These are interpreted as being caused by atoms which have lost more than one electron. Such parabolas are found for all elements which occur in the rays with the exception of hydrogen and, possibly, of helium. The maximum number of charges which an element can receive seems to depend chiefly on its atomic weight, not on its chemical valency. Thus neon, nitrogen, oxygen, carbon, occur with two or, rarely, three. Argon has three, krypton four or five, while mercury can apparently have as many as eight. In all cases the multiple charged parabolas are fainter than the singly charged, and generally the more the charges the fainter the parabola. Molecular rays very rarely are found with more than one charge, but CO with two charges has been observed.

The existence of multiple charges explains a difference which may be noticed between the different parabolas. While most of them are bounded by an imaginary straight line perpendicular to the electric deflection, a few, and notably that due to mercury, extend beyond. Now suppose that the potential difference between the furthest point at which rays are formed and the cathode is V, then if the ray has n positive charges each of amount e, its greatest possible energy is given by $neV = \frac{1}{2}mv^2$. But $y = \dfrac{Bpe}{mv^2}$, where pe is the charge retained in the deflecting fields, therefore the minimum value of y is $\dfrac{Bp}{2nV}$. For the parabolas for which $n = p$ this will be constant, but if $n = 2$, for example, and $p = 1$ the parabola will come twice as near the z axis. Thus if a ray starts with a double charge and loses one charge in the fine tube, it will have energy corresponding to $n = 2$, but the normal e/m and the parabola corresponding to the singly charged atom will show an extension or 'tail' as it is called. This is often a valuable indication that a given parabola is due to an element and not to a compound.

Besides the positively charged particles some of the rays, as first observed by Wien, show deflections in the opposite direction showing that they have acquired a negative charge. These negative parabolas can be seen on Fig. 3, Plate I. They imply that a ray has first lost its positive charge to a molecule of the gas and simultaneously or in a later encounter acquired a negative one. Only some kinds of rays do this, but it should be remembered that the conditions are not very favourable for it to happen. Thus nitrogen, mercury and the inert gases never appear with negative charges; on the other hand, atoms of hydrogen, carbon and oxygen with negative charges appear on most of the photographs. In general, chemically electronegative elements appear with negative charges, chlorine for example being very strong on the negative side. Molecules seldom appear with negative charges, but C_2, O_2, OH and rarely H_2 are known, indications of C_2H have also been found. Beside the parabolas, experiments made at all but the lowest pressures show other lines, Figs. 1 and 3, Plate I. These are called 'secondaries' and are attributed to the loss or gain of charge by the rays during their passage through the fields. By separating the electric and magnetic fields it can be shown that it is possible for a particle both to lose its original positive charge, and after being neutral to regain it. Many of these secondary lines are surprisingly fine, leading up to a single point on the main parabola, showing that the rays which lose or gain their charge have predominantly a single velocity. For a further discussion of these secondaries see vol. II.

Measurements have also been made, using a Faraday cylinder placed behind a parabolic slit, instead of a photographic plate, to detect the rays. By altering the magnetic field the rays of different e/m can be made to enter the Faraday cylinder in succession, and their number found by measuring the rate of charge of an electrometer connected to the Faraday cylinder. Though less convenient to use, this method has the advantage that it gives a quantitative measure of the number of the rays of each kind. It is found that the photographic method greatly exaggerates the numerical importance of the hydrogen rays. Though the hydrogen lines are generally much the strongest on the plates, the electrical method shows that they are not really very numerous unless hydrogen has been deliberately put into the tube.

The most important conclusion from the above experiments is that these charged atoms and molecules always have an exact multiple of the charge on a hydrogen atom, and that there is no evidence of the existence of any subdivision of this. In addition it gives very direct confirmation to the assumption of the chemist that the atom of an element is, in some cases at least, a definite thing and that its weight is not merely a statistical average. That there are numerous cases of exception was the third important discovery resulting from this method. It was found when neon was examined that besides a parabola for which e/m was $\frac{1}{20}$ that for H_+ there was another fainter one for which it was $\frac{1}{22}$. Since the atomic weight found from density measurements is 20·2, this suggested that neon was a mixture of two kinds of atoms of identical chemical properties, but with masses respectively 20 and 22 times that of hydrogen. Such atoms were believed to exist from radioactive considerations and had been named isotopes. The alternatives were that the second line was due to a compound NeH_2 or to CO_2 with two charges, but both these were extremely unlikely. Dr Aston made several only partially successful attempts to separate the two isotopes by distillation or diffusion and so produce a measurable difference in density. To settle the question, he designed an apparatus by which e/m could be measured with sufficient accuracy to decide whether the main line was 20·0 or 20·2.

71·3. This apparatus, which Aston[1] calls the mass spectrograph, is shown in Fig. 73. The rays are produced in the discharge tube B about 20 cm. in diameter and pass through two fine aluminium slits S_1, S_2, each about ·05 mm. × 2 mm., and arranged accurately

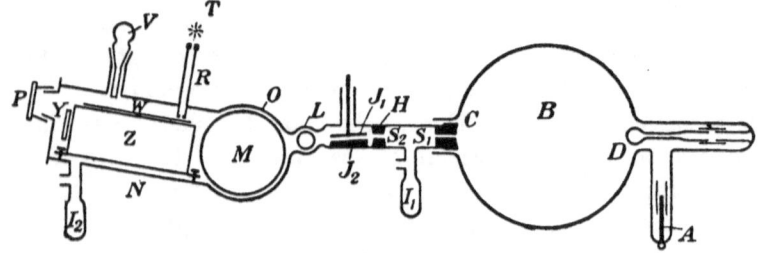

Fig. 73. Mass spectrograph.

parallel. The rays then pass between the plates J_1, J_2 of a condenser by which they are deflected downwards, and a fairly narrow pencil allowed to pass through a diaphragm (not shown) near L. They then pass through a magnetic field perpendicular to the plane of the paper at M arranged so as to give an upward deflection which more than overcomes the downward deflection to the electric field. The rays pass between two earthed plates Z to protect them from stray electric fields and strike the underside of the photographic plate W, which can be moved by means of a rod operated through the ground glass joint V, so that when a photograph is not being taken, it lies above the plane of the paper and out of the track of the rays. Y is a willemite screen on which the rays can be thrown to estimate their brightness. It is looked at through the window P. A fine beam of light can be thrown on to the plate by a light at T so as to produce a fiducial spot

[1] See Aston, *Isotopes* (Arnold).

from which measurements can be made. One great advantage of this method is that it permits of a focusing of the rays. The theory in its simplest form is as follows. In Fig. 74, S_1S_2Z represents the path of the rays before deviation; after deviation by the electric field they can be regarded, to a first approximation, as diverging from the point Z in the middle of the condenser. Call the angle of deviation θ and suppose that a small pencil of angular width $\delta\theta$ is selected by the diaphragm D. These rays are bent

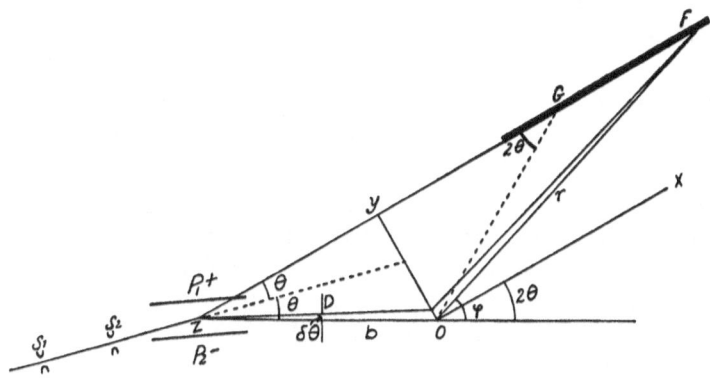

Fig. 74.

again by the magnetic field, which for simplicity we will regard as concentrated at the point O. Let the angle of deviation here be ϕ. Now

$$\theta = \frac{Be}{mv^2}, \quad \phi = \frac{Ae}{mv},$$

for small angles. Hence

$$\frac{\delta\theta}{\theta} + \frac{2\delta v}{v} = 0$$

and

$$\frac{\delta\phi}{\phi} + \frac{\delta v}{v} = 0$$

for a pencil of rays of the same e/m. Thus

$$\frac{\delta\theta}{\theta} = \frac{2\delta\phi}{\phi}.$$

The focusing effect follows from this. Let $ZO = b$, then the width of the pencil at O is $b\delta\theta$, the width at a distance r further on will be

$$b\delta\theta + r(\delta\theta + \delta\phi) = \delta\theta\left[b + r\left(1 + \frac{\phi}{2\theta}\right)\right].$$

Now the angles θ and ϕ are in opposite directions, so $\phi/2\theta$ is negative; write it $-\phi/2\theta'$. Then if $\phi > 2\theta'$ the width will vanish when r is given by
$$r(\phi - 2\theta') = b.2\theta',$$
and at this point the rays will come to a focus to the first order. The position of this focus for rays of different e/m and hence different ϕ (θ being supposed kept constant) can be found by the following construction. Draw rectangular axes OX, OY so that the angle between OX and the direction ZO of the rays is 2θ. Then the coordinates of the focus are
$$r \cos(\phi - 2\theta'), \quad r \sin(\phi - 2\theta');$$
the latter is equal to $r(\phi - 2\theta')$ correct to squares of the angles, and this by the above is $2b\theta'$, a constant. Hence the points of focus will lie on a line parallel to OX and a flat plate placed here will receive all the rays in focus. In the actual apparatus the magnetic field occupies a considerable area, being formed by circular pole pieces of 8 cm. diameter. A detailed investigation shows that this extra complication has little effect on the above result. It can also be shown that the scale of m/e as measured along the plate is approximately linear near the point for which $\phi = 4\theta'$. The use of slits instead of a tube, and the above focusing effect both increase the intensity for the same separating power, or conversely increase the separating power for lines of nearly equal e/m without decreasing the intensity.

71·4. In the use of the instrument no attempt was made to determine e/m absolutely, the process was one of comparison with known reference lines. By putting gases in the tube whose behaviour in the parabola apparatus was known, no difficulty was found in identifying a number of lines, such for instance as C_{++}, O_{++}, C_+, O_+, CO_+, and the group CH_+, CH_{2+}, CH_{3+}, CH_{4+}, formed from methane. In this way a calibration curve could be drawn up showing the connection between mass and distance from the fiducial point. But suppose the fields changed. The magnetic deflection will be unaltered for a ray for which mv/e changes in the same proportion as H; hence the mass which comes to the position formerly occupied by m is $m' = \dfrac{mv.H'}{v'.H}$, where H'/H measures the ratio of the magnetic field at any point of the path, assuming that the distribution of the magnetic field has not changed. Now the electric deflection is inversely proportional to mv^2/e, hence

$$\frac{v^2}{v'^2} = \frac{m'}{m} \cdot \frac{V}{V'},$$

where V, V' are the old and new potentials across the condenser.

Hence
$$\frac{m'}{m} = \frac{V}{V'} \cdot \frac{H'^2}{H^2}.$$

This is accurately true for variation of the *magnetic* field, only if it changes in the same ratio at all points; this is not strictly the case if the permeability of the magnet varies, though a good approximation in practice. Hence by changing V or H, but preferably the former, the same lines appear in new positions, but the ratio of the masses to those which would have occupied the same positions if they had been present in the first photograph, is the same for all points of the scale. This ratio can be found approximately in each case from the rough calibration curve, and a more accurate value found by taking the mean, or if V has been altered, it can be calculated from the ratio $V:V'$. In either case fresh points can be fixed on the calibration curve which can thus be built up, a process facilitated by the fact that the scale is not far from linear in m. While this process of calibration is the most convenient for finding the approximate value of e/m with amply sufficient accuracy to identify lines, a more accurate method when two lines are to be compared which do not differ by more than about $3:1$ is to use the above equation for m'/m directly. Thus to compare carbon (12) with oxygen (16) it is sufficient to take two photographs, one with 320 volts on the condenser, the other with 240. If the ratio of the masses is exactly $12:16$, the carbon line with 320 volts should come to the same place as the oxygen line with 240 volts, as is actually found to be the case. If the coincidence were not exact the difference could readily be found from even a rough knowledge of the calibration curve. This method however assumes that the magnetic field has been kept accurately constant.

A modification of this procedure, known as the method of 'bracketing,' has been used by Aston in most of his later work. Consider, for example, the comparison of two lines nearly in the ratio of 1 : 2, such as, for example, H and H_2 or H_2 and He. Instead of exactly doubling the field, three exposures are taken *on the same plate* with potentials V, $2V + h$ and $2V - h$, where h is a small quantity. If the ratio is exactly 1 : 2, the heavier line will lie symmetrically between two of the lines due to the lighter system, and any deviation from the ratio can be readily detected as an asymmetry, and found with sufficient accuracy from a knowledge of the calibration curve. In his later work Aston has overcome the error due to a possible change in the magnetic field by using a commutator to change the potentials from one to another many times during the exposure, so that all three lines are formed almost simultaneously, and the only effect of a change in the magnetic field is to broaden them all equally.

71·5. The first result obtained by the mass spectrograph was to show conclusively that the two neon lines were 20·0 and 22·0, taking oxygen as 16·00 which, together with the observed density of 20·2, proved that neon was a mixture of two isotopes. Further work showed that many elements were in the same position. Thus chlorine consists of a mixture of atoms of weight 35 and 37, bromine of 79 and 81, argon of 36 and 40, krypton of 78, 80, 82, 83, 84 and 86, while xenon has nine kinds of atoms of weight 124, 126, 128, 129, 130, 131, 132, 134 and 136. In addition the very important result was obtained that all these weights were integral, taking $O = 16·00$, to one or two parts in a thousand. This remarkable fact throws a flood of light on atomic constitution.

The parabola method had shown that hydrogen, alone among the elements, never is found with a double charge, making it extremely probable that its atom contained only one electron, an idea further supported by the measurements of X-ray scattering. Further, the positively charged hydrogen atom, the hydrogen atom deprived of its one electron, is the lightest separate system known, next to the electron. It was natural to suppose that the positive charge required to neutralise the numerous electrons of more complicated atoms was also built of units, and the only possible unit for whose existence there was any independent evidence was this positively charged hydrogen atom, for which Rutherford has suggested the name of 'proton.' Aston's result was a brilliant confirmation of this view, and makes it extremely probable that atoms can be regarded as composed, as far at least as their mass is concerned, merely of the two constituents, electrons and protons, in equal numbers. But there is one serious objection which a few years ago would probably have been sufficient to wreck the theory. The hydrogen atom itself does not conform. The whole-number law, in fact, holds if oxygen is taken as 16·00; on this basis chemical measurements gave hydrogen as about 1·008. This difference might at first sight have been accounted for if hydrogen itself had two isotopes 1 and 2 or 3. Actually there is no evidence for this, and anyhow it would not solve the difficulty, for Aston's measurements showed conclusively, by comparing hydrogen with helium, and helium with doubly charged oxygen, that the ratio between hydrogen and oxygen is not integral. His most recent work gives for hydrogen 1·00778, in good agreement with the results obtained chemically and from density measurements. It would thus appear necessary to suppose that if the other elements are built up of protons and electrons there is a loss of mass in the process, of the same percentage amount in all cases. The conception of electrical mass here comes to the rescue. The mass of a conductor carrying a charge depends on its capacity, and if two conductors with charges of opposite sign are brought close together the electric mass of the whole is less than the sum of the separate masses. We have only to suppose that some of the electrons come very near the protons in the complex atoms for the electrical mass to be reduced below the sum of its components. A similar result follows from the principle of relativity, from which it can be shown that the mass m and total energy E of a system are connected by the relation $E = mc^2$, where c is the velocity of light. Thus if the combination of hydrogen atoms to form a complex atom is associated with sufficient loss of energy, the loss of mass will be explained. That very close packing of some of the constituents of complex atoms occurs is almost certain. Rutherford's experiments on the scattering of α particles, and Moseley's on the wave-lengths of characteristic X-rays, afford very strong evidence that most of the mass of an atom is concentrated in a small region or nucleus, and that this nucleus bears a positive charge Ne, where e is the charge on the electron and N an integer called the atomic number which, with a few exceptions, is the number of the places the element occupies in the list of elements arranged in order of atomic weight. If Z is the atomic weight, then on the above view the atom contains Z protons and Z electrons. If all the massive protons are in the nucleus, then since, except for hydrogen, Z is always greater than N, there must be associated with them in the nucleus a number of electrons $Z - N$, leaving the remaining N electrons for the rest of the atom. Since Rutherford finds that the size of the nucleus is of the order 10^{-12}, and we have seen that the minimum diameter of the electron is $3\cdot 7 \times 10^{-13}$, the packing must be close, and it is easy to understand an appreciable loss of mass.

71·6. Recently Aston[1] has made a very careful investigation of the slight divergencies from the whole-number rule which occur for the elements other than hydrogen and which had been suspected but not proved with the older apparatus. The chief instrumental difference is an increase in the intensity and length of the magnetic field; this is now 15 cm. long and fields up to 15,700 gauss are used, with currents well within the permissible values. This makes it possible to use larger angles of deflection, the electric deflection being now one-sixth of a radian, thus giving greater resolving power. In addition finer slits are used, placed farther apart. The accuracy claimed is 1 in 10,000. The chief difficulty in reaching a high accuracy was found to be in the electric field. At first the 'bracketing' method gave quite contradictory results, and this was eventually traced down to a polarisation of the plates of the condenser, by which the actual potential between them was different from that supplied from the very accurately adjusted battery of small accumulators. By gilding the condenser plates this effect was greatly reduced, but was still appreciable with the very high accuracy aimed at. Accordingly, a modification of the bracketing method was used, two potentials V_1, V_2 were applied alternately by means of the commutator, their ratio being chosen so as to be nearly, but not quite, in the ratio of the masses to be compared, say $x : a$. This gives two lines close together whose distance apart can be measured and translated

[1] Aston, *Proc. Roy. Soc.* cxv. p. 487, 1927.

into a mass-ratio from an approximate knowledge of the calibration. If the potentials were known exactly their ratio multiplied by the above mass-ratio would give the required value of $x:a$. Owing to the polarisation effect the ratio V_1/V_2 is not known with sufficient certainty, and Aston checks it by applying the same process with the same potentials to two other lines, whose exact ratio is known. For standard ratios, atoms which appear with double charges give $2:1$, and it is assumed that the masses of

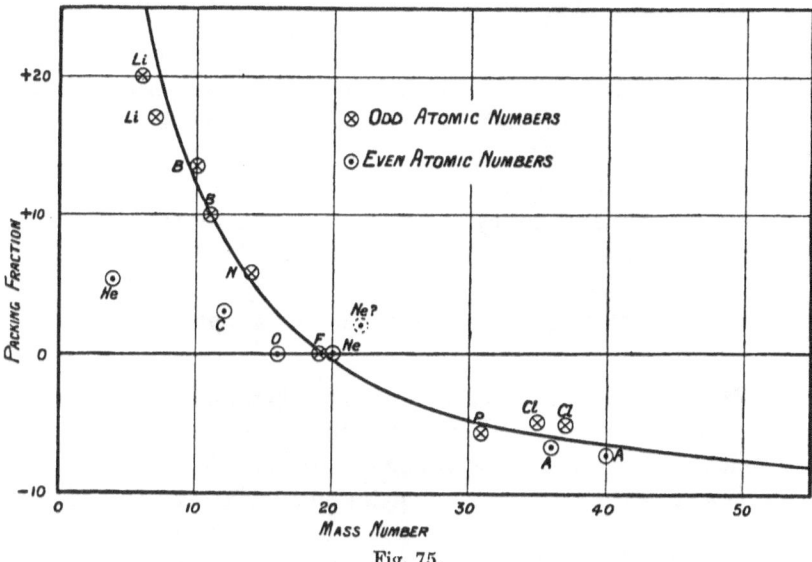

Fig. 75.

chemical compounds are the sums of the masses of constituents. In these ways a series of ratios can be built up gradually, from which one can be found, in most cases, near enough to the value required to act as a suitable standard. In these accurate determinations allowance must be made for the mass of the electron whose loss renders the atom positively charged.

Before giving Aston's results mention should be made of some accurate determinations previously made by Costa[1], using a mass spectrograph of his own design, of the ratios H : He : C, the ratio

[1] Costa, *Ann. der Phys.* iv. p. 425, 1925.

of He to the Li-6 isotope, and that of Li-7 to nitrogen for which the chemical value 14·008 was used. He claims an accuracy of 3 in 10,000, and to this accuracy the first series is identical with Aston's results. He also made a determination of the mass of nitrogen, finding fair agreement with the chemical value. The results for Li are included in the diagram of Aston's results.

Aston expresses the divergence from the whole-number rule by the difference from the nearest integer divided by the mass and multiplied by 10,000. He calls this the packing fraction, reckoned positive in cases such as hydrogen where the mass is greater than the nearest unit. His results are shown in Figs. 75, 76. The isotopes of odd and even atomic numbers behave differently among the light elements, or what is perhaps an easier way of expressing the result, the first three elements whose atomic weights are multiples of 4 (He, C, O) lie off the curve given by the others. It had previously been suggested that the α particle (helium nucleus) is an intermediate unit in the structure of the nucleus of ordinary, as well as of radioactive elements, and the exceptional behaviour of these three elements slightly strengthens this view.

71·7. It may at first seem remarkable that, for example, atoms of masses 35 and 37 should exist together in chlorine and its compounds without being separated by the ordinary chemical processes of crystallisation, distillation and the like. The explanation lies in the minute size assumed for the nucleus, which for the purpose of the rest of the atom is supposed to behave practically as a massive charged point. It is only the *net* charge that matters, and the addition, for example, of two protons and two electrons to the nucleus of chlorine 35 has no appreciable effect on the outer parts of the atom, and hence on its chemical and physical properties. The only exceptions are those properties in which the mass enters directly. Such is, for example, the diffusion of a gas through a narrow aperture, *e.g.* through pipeclay. Here what matters is the velocity of thermal agitation, and this, at a given temperature, is inversely as the square root of the molecular weight. Thus in diffusing a mixture of isotopes the lighter should diffuse the faster, and it is probable that Aston did actually cause a slight separation in neon by repeated use of this process. There are

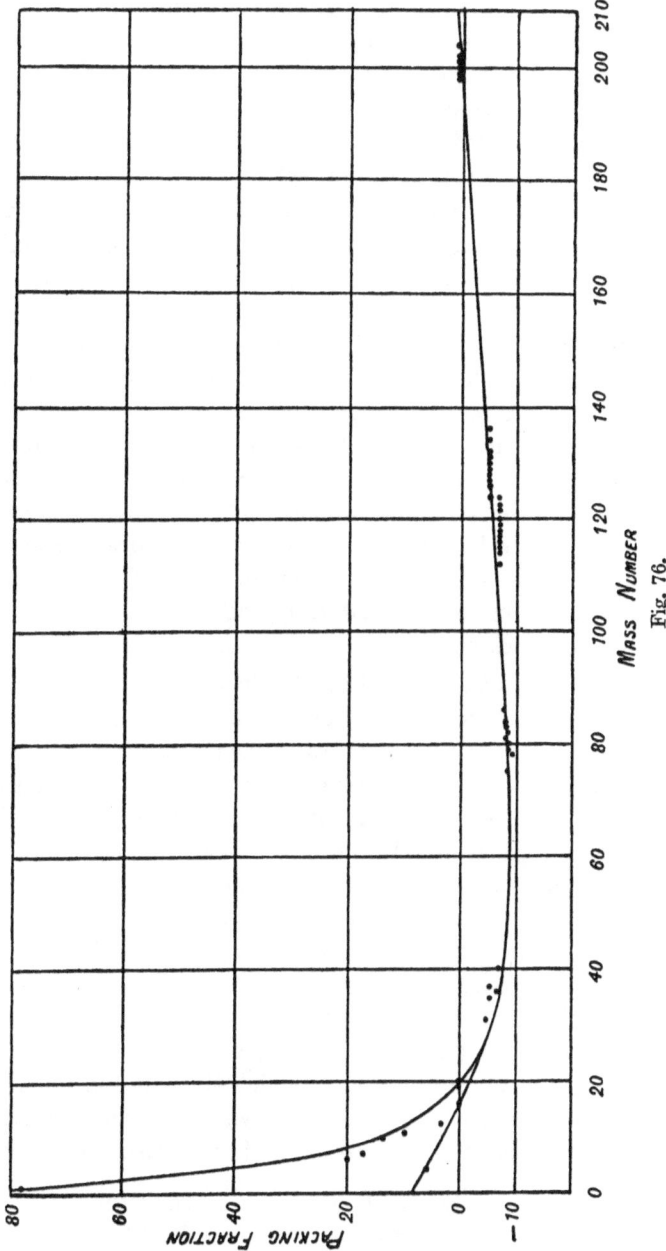
Fig. 76.

serious technical difficulties in ensuring adequate mixing, and the labour in producing any considerable separation is enormous, but Harkins[1] working on HCl succeeded in producing a separation into two gases whose densities indicated a change of ·055 unit in atomic weight. It will be seen that even this is a very long way from complete separation into 35 and 37.

Another property which has been used is the *rate* of evaporation (vapour pressure is the same for both). This comes about as follows. Owing to the greater speed of the lighter isotope atoms, either in liquid or gas, they will reach the bounding surface more often than the heavier. Since their mean energies are the same, the same proportion of liquid molecules which reach the bounding surface will evaporate in both cases. Hence the rates of evaporation of the two isotopes are in the ratio of their velocities, *i.e.* inversely as the square root of the ratio of their masses. By evaporating mercury at a low pressure and condensing the distillate on a surface cooled by liquid air, Bronsted and Hevesy[2] after repeated fractionations succeeded in getting samples of mercury differing by about 5 parts in 10,000 in density, and Hönigschmidt and Birckenbach[3] have found appreciable differences in atomic weight.

The discovery of isotopes has greatly reduced the importance of chemical atomic weight as a fundamental constant. This now appears merely as a weighted mean of the different isotopes of the element, except in the comparatively few cases where it is single. It also explains the anomalous position in the periodic table of a few elements, such as tellurium, whose atomic weight did not accord with the position given by their chemical properties. The chemical properties depend of course on the atomic number, and the periodic table is the table of elements arranged in order of nuclear charge. While the atomic weight generally increases with the nuclear charge, the isotopes of adjoining elements may overlap or two may coincide ('isobars') and then it will depend on what proportion the isotopes occur, as to which element has the highest chemical atomic weight. No case is known in which *all* the isotopes of an element are heavier than all those of the next highest element.

[1] Harkins, *Science*, Oct. 14, 1921; *Nature*, Oct. 3, 1921.
[2] Bronsted and Hevesy, *Phil. Mag.* xliii. p. 31, 1922.
[3] Hönigschmidt and Birckenbach, *Ber.* lvi. B, 1219, 1923.

71·8. The form of apparatus above described is adequate for all elements which have stable volatile compounds, or are themselves gases. It fails however for most of the metals, and in investigating these, other methods of producing a beam of charged atoms must be found. Various methods of producing such ions are known and have been applied to this purpose by Dempster[1], by one of the authors[2], and by Aston. These methods are described in Chap. IX, p. 389. For the measurement of e/m the above de-

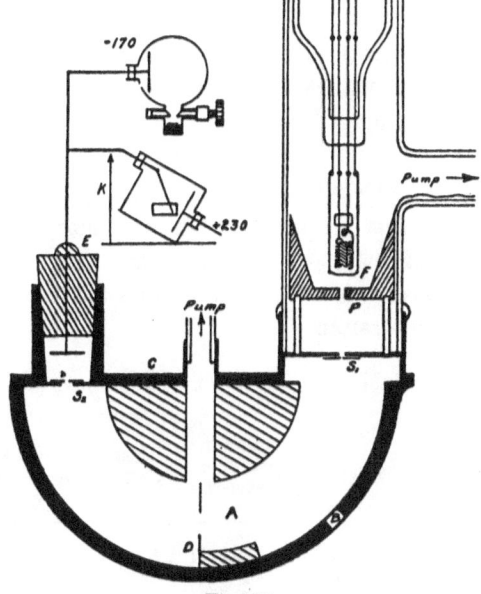

Fig. 77.

scribed parabola and mass spectrograph were used by the two latter authors respectively. The parabola method is lacking in resolving power and only lithium and beryllium could be successfully dealt with. Dempster used a novel method which deserves mention. His apparatus is shown in Fig. 77. The ions are produced between P and F by cathode rays striking the heated metal or its vapour. After diffusing through P they are accelerated by a field of 800–1000 volts and then pass the fine slit S_1 (·37 mm. wide).

[1] Dempster, *Phys. Rev.* xi. p. 316, 1918; xviii. p. 415, 1921; xx. p. 631, 1922.
[2] G. P. Thomson, *Proc. Camb. Phil. Soc.* xx. p. 210, 1920; *Phil. Mag.* vi. 42, p. 857, 1921.

They are then bent into a semicircle by a magnetic field perpendicular to the plane of the paper, and received through the slit S_2 (·42 mm. wide) into a Faraday cylinder connected with an electroscope. By altering the accelerating voltage, keeping the magnetic field constant, rays of various e/m can be made to pass through the second slit. If r is the radius of curvature of the rays,
$$mv = eHr \text{ and } \tfrac{1}{2}mv^2 = Ve,$$
so that
$$e/m = \frac{2V}{H^2r^2}.$$

The apparatus as used has the advantage that all rays which leave S_1 with equal values of e/m and v and make a small angle with the normal to S_1, will describe semicircles of equal radii, and will intersect again, approximately at the other end of a diameter; thus the rays are focussed on S_2.

With this apparatus Dempster showed that magnesium contained isotopes 24, 25, 26, calcium 40 and 44, and zinc 64, 66, 68 and 70, besides confirming values found by the other observers for lithium and potassium.

Table of Elements and Isotopes.

Element	Atomic Number	Atomic Weight	Minimum Number of Isotopes	Mass-numbers of Isotopes in Order of Decreasing Intensity
H	1	1·008	1	1
He	2	4·00	1	4
Li	3	6·94	2	7, 6
Be	4	9·02	1	9
B	5	10·82	2	11, 10
C	6	12·00	1	12
N	7	14·01	1	14
O	8	16·00	1	16
F	9	19·00	1	19
Ne	10	20·20	2	20, 22
Na	11	23·00	1	23
Mg	12	24·32	3	24, 25, 26
Al	13	26·96	1	27
Si	14	28·06	3	28, 29, 30
P	15	31·02	1	31
S	16	32·06	3	32, 34, 33
Cl	17	35·46	2	35, 37
A	18	39·88	2	40, 36
K	19	39·10	2	39, 41
Ca	20	40·07	2	40, 44

Table of Elements and Isotopes (continued).

Element	Atomic Number	Atomic Weight	Minimum Number of Isotopes	Mass-numbers of Isotopes in Order of Decreasing Intensity
Sc	21	45·1	1	45
Ti	22	48·1	1	48
V	23	51·0	1	51
Cr	24	52·0	1	52
Mn	25	54·93	1	55
Fe	26	55·84	2	56, 54
Co	27	58·97	1	59
Ni	28	58·68	2	58, 60
Cu	29	63·57	2	63, 65
Zn	30	65·38	4	64, 66, 68, 70
Ga	31	69·72	2	69, 71
Ge	32	72·38	3	74, 72, 70
As	33	74·96	1	75
Se	34	79·2	6	80, 78, 76, 82, 77, 74
Br	35	79·92	2	79, 81
Kr	36	82·92	6	84, 86, 82, 83, 80, 78
Rb	37	85·44	2	85, 87
Sr	38	87·63	2	88, 86
Y	39	88·9	1	89
Zr	40	91·25	3	90, 94, 92 (96)
Ag	47	107·88	2	107, 109
Cd	48	112·41	6	114, 112, 110, 113, 111, 116
In	49	114·8	1	115
Sn	50	118·70	11	120, 118, 116, 124, 119, 117, 122, 121, 112, 114, 115
Sb	51	121·77	2	121, 123
Te	52	127·5	3	128, 130, 126
I	53	126·92	1	127
X	54	130·2	7	129, 132, 131, 134, 136, 128, 130, (126), (124)
Cs	55	132·81	1	133
Ba	56	137·37	1	138, (136), (137)
La	57	138·91	1	139
Ce	58	140·25	2	140, 142
Pr	59	140·92	1	141
Nd	60	144·27	3	142, 144, 146, (145)
Er	68	167·7	several	164 to 176
Hg	80	200·6	6	202, 200, 199, 198, 201, 204
Pb	82	207·2	3	208, 206, 207, (209), (203), (204), (205)
Bi	83	209·00	1	209

(Numbers in brackets are provisional only.)
Copied (with additions) from p. 120 *The Structure of the Atom*, Andrade.

A list of the isotopes so far known is given above. There appear to be no very simple rules to determine the number or weight of the isotopes of an element, and though many theories of nuclear structure have been brought forward they have not so far (1927) had much success in predicting isotopes. Aston has pointed out that elements of odd atomic number never have more than two isotopes, and that when they have two the weights of these usually differ by two units. Exceptions are lithium and boron. There is a general tendency for the number of isotopes of elements of even atomic number to increase with the atomic number, but there are plenty of exceptions.

72. The value of e/m for the positive ions emitted by a hot iron wire was determined by one of the authors in 1903, using the method of § 67. The apparatus was not adapted to great accuracy, but the result $e/m = 400$ showed that the ions were of molecular dimensions. A number of more recent determinations of e/m for these ions are given in Chap. IX.

The α particles from radioactive substances are doubly charged atoms of helium shot out with a velocity of the order 2×10^9. Measurements have been made of e/m for these rays by Rutherford, Des Coudres, Mackenzie and others, which have been of great importance in settling the nature of the rays. For an account of the methods see Rutherford's *Radio-active Substances and their Radiations*. The results obtained are in satisfactory agreement with the value 4830 calculated for an atom of helium which has lost two electrons.

CHAPTER VII

DETERMINATION OF THE CHARGE CARRIED BY THE NEGATIVE ION

73. WE have seen that the value of e/m for the negative ions in gases at a low pressure is more than a thousand times the greatest value of the ratio of the same quantities for ordinary electrolytes. The question at once arises, is this due to a difference in the masses of the ions, or to a difference in their electrical charges, or to both these causes? to decide these points we must determine the value of m or e. The writer made in 1898[1] and 1899[2] determinations of the value of e for the ions produced in one case by X-rays and in the other by ultra-violet light. The method was based on the discovery made by C. T. R. Wilson[3] (see Chap. VIII) that gaseous ions, whether positive or negative, act as nuclei for the condensation of clouds even in the absence of dust; and that if we have a mass of dust-free gas containing ions in a closed vessel, and cool the gas by a sudden expansion, then a cloud will be produced if the ratio of the volume of the gas after expansion to the volume before is greater than 1·25. An expansion of this amount is quite incapable of producing more than very slight condensation in the gas if it does not contain ions. The water condenses round the ions, and if these are not too numerous each ion becomes the nucleus of a drop of water. Thus by producing a sudden expansion in a gas containing ions we can get a little drop of water round each ion; these drops are visible, and we can measure the rate at which they fall. Sir George Stokes has shown that if v is the velocity with which a drop of water falls through a gas, a the radius of the drop, μ the coefficient of viscosity of the gas, and g the acceleration due to gravity, then

$$v = \tfrac{2}{9} \frac{ga^2}{\mu};$$

thus if we measure v we can determine a, and hence the volume

[1] J. J. Thomson, *Phil. Mag.* v. 46, p. 528, 1898.
[2] J. J. Thomson, *Phil. Mag.* v. 48, p. 547, 1899.
[3] C. T. R. Wilson, *Phil. Trans.* A, p. 265, 1897.

of each drop. If q is the volume of water deposited from each cubic centimetre of the gas, n the number of the drops, we have
$$q = n\tfrac{4}{3}\pi a^3.$$
To find q we may proceed as follows: the gas after being cooled by the very rapid expansion is supersaturated and moisture is deposited on the ions; during the condensation of the water, heat is given out which warms the gas, so that the temperature of the gas rises above the lowest temperature reached during the expansion before condensation has taken place. Let t_2 be the lowest temperature reached during the expansion, t the temperature when the drops are fully formed, then if L is the latent heat of evaporation of water, C the specific heat of the gas at constant volume, M the mass of unit volume of the gas after expansion, we have
$$Lq = CM(t - t_2) \quad \ldots\ldots\ldots\ldots\ldots (1);$$
we neglect the heat required to raise the temperature of the water in the gas in comparison with that required to raise the temperature of the gas itself. We have further
$$q = \rho_1 - \rho,$$
where ρ_1 is the density of the water vapour before condensation begins, and ρ the density at the temperature t. Substituting this value for q in equation (1), we get
$$\rho = \rho_1 - \frac{CM}{L}(t - t_2) \quad \ldots\ldots\ldots\ldots (2).$$
Since ρ is a known function of t this equation enables us to find t when t_2 is known, ρ_1 and M being found as follows.

If x is the ratio of the final to the initial volume of the gas and T the temperature in degrees centigrade of the gas before expansion, then since the mass of 1 cubic centimetre of air at the temperature $0°$ C. and under a pressure of 760 millimetres of mercury is $\cdot 00129$ grm., we have
$$M = \frac{\cdot 00129}{x} \times \frac{273}{273 + T} \frac{P}{760},$$
where P is the initial pressure of the gas expressed in millimetres of mercury.

Again,
$$\rho_1 = \frac{\rho'}{x},$$

where ρ' is the density of water vapour at the temperature T before expansion; as the air was saturated with water vapour at this temperature ρ' can be obtained directly from the Tables of the vapour pressure of water vapour.

The cooling caused by the adiabatic expansion is determined by the equation
$$\log \frac{273 + T}{273 + t_2} = \cdot 41 \log x \quad \ldots\ldots\ldots\ldots (3).$$
For in such an expansion pv^γ is constant, where p is the pressure, v the volume and γ the ratio of the specific heat at constant pressure to that at constant volume: but $pv = R\theta$, where θ is the absolute temperature and R a constant, hence we have during an adiabatic expansion
$$v^{\gamma-1}\theta = \text{a constant};$$
hence if $v_1\theta_1$, $v_2\theta_2$ are the initial and final values of v and θ, we have
$$v_1^{\gamma-1}\theta_1 = v_2^{\gamma-1}\theta_2,$$
or
$$\log \frac{\theta_1}{\theta_2} = (\gamma - 1) \log \frac{v_2}{v_1}.$$
Since $\gamma = 1\cdot 41$ this is equivalent to equation (3). From (3) we determine t_2, and then since
$$C = \cdot 167, \quad L = 606,$$
equation (2) becomes
$$\rho = \frac{\rho'}{x} - \frac{\cdot 167}{606} \times \frac{\cdot 00129}{x} \cdot \frac{273}{273 + T} \frac{P}{760} (t - t_2) \quad \ldots. (4).$$

As an example of how this equation is applied let us take a case which occurred in one of the experiments. Here
$$T = 16°, \quad P = 760, \quad x = 1\cdot 36.$$
To get t_2 we have
$$\log \frac{273 + 16}{273 + t_2} = \cdot 41 \log 1\cdot 36 = \log 1\cdot 134,$$
hence $\quad\quad 273 + t_2 = 254\cdot 8, \text{ or } t_2 = -18°\cdot 2.$
We find from the Tables that at $16°$
$$\rho' = \cdot 0000135,$$
hence equation (4) becomes
$$\rho = 99\cdot 3 \times 10^{-7} - 2\cdot 48 \times 10^{-7} (t + 18\cdot 2) \quad \ldots.. (5).$$

To solve this equation we keep substituting various values for t until we find one for which the corresponding value of ρ given by (5) is the same as the value of the vapour pressure of water at the temperature t. We find by this process of trial and error that the solution of equation (5) is $t = 1\cdot2$, and the corresponding value of ρ is $51\cdot5 \times 10^{-7}$. Substituting this value for ρ we find $q = 47\cdot7 \times 10^{-7}$ grms.

When we know q and a, n the number of drops is at once determined by the equation

$$n = q/\tfrac{4}{3}\pi a^3.$$

In this way we can determine the number of ions per cubic centimetre of gas. When we know the number of ions and also the velocity of the ions under unit electric force, we can very easily deduce the charge carried by an ion by measuring the current carried by these ions across each unit of area under an electric force E. For if n is the whole number of ions of both signs per c.c., U the mean of the velocities of the positive and negative ions under unit electric force, the current through unit area is equal to

$$ne\,E\,U,$$

where e is the charge on the ion; the electric force E ought to be so small that the current is proportional to the electric force. When this is not the case the number of ions is diminished by the action of the electric field, and depends upon the magnitude of the electric force.

We can easily measure the current through the ionised gas and thus determine $ne\,E\,U$, and as n, E, U are known we can deduce the value of e.

74. This method was first applied by the author to determine the charge on the ions produced by X-rays. The method used for making the cloud and measuring the expansions is the same as that used by C. T. R. Wilson[1]: the apparatus for this and the electrical part of the experiment is represented in Fig. 78. The gas which is exposed to the rays is contained in the vessel A; this vessel is connected by the tube B with the vertical tube C, the lower end of which is carefully ground so as to be in a plane perpendicular to the axis of the tube, and is fastened

---

[1] C. T. R. Wilson, *Proc. Camb. Phil. Soc.* ix. p. 333, 1897.

down to the india-rubber stopper D. Inside this tube there is an inverted thin-walled test tube P with the lip removed and the open end ground so as to be in a plane perpendicular to the axis of the tube. The test tube slides freely up and down the larger tube and acts as a piston. Its lower end is always below the surface of the water which fills the lower part of the outer tube; a tube passing through the india-rubber stopper puts the inside of

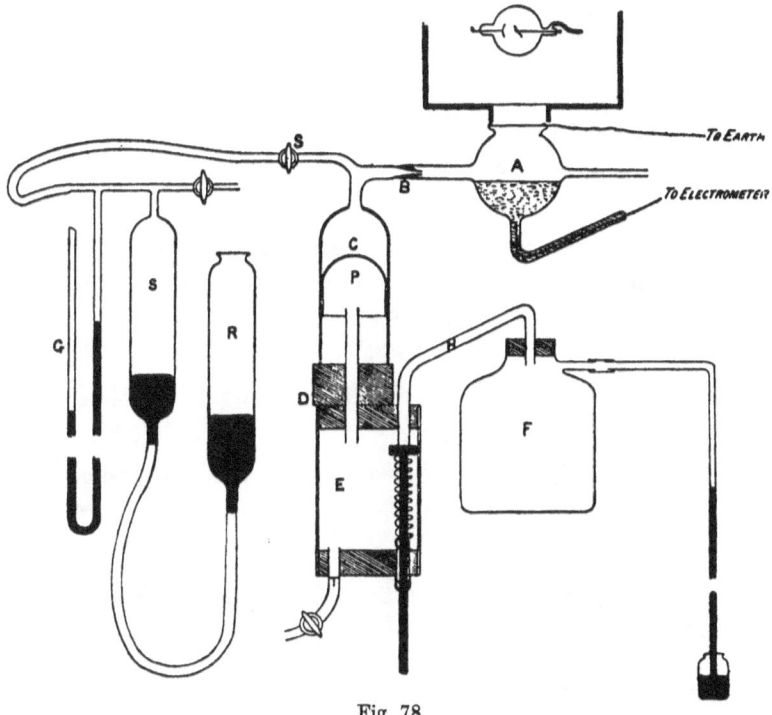

Fig. 78.

the test tube in communication with the space E. This space is in connection by the tube H with a large vessel F in which the pressure is kept low by a water-pump. The end of the tube H is ground flat and is closed by an india-rubber stopper which presses against it; the stopper is fixed to a rod, and by pulling this rod down smartly the pressure inside the test tube is lowered and the test tube falls rapidly until it strikes against the india-rubber stopper. The tube T, which can be closed by a stop-cock, admits air into E and allows us to force the test tube back into its place for another expansion. The tubes R and S are for the purposes of regulating the amount of expansion. To do this the mercury vessel R is raised or lowered when the test tube is in its lowest position until the gauge G indicates that the pressure in A is the desired amount below the atmospheric pressure. The stop-cock S is then closed and air is admitted into the interior of the piston by opening the stop-cock T. The piston then rises until the pressure in A differs from atmospheric pressure only by the amount required to support the weight of the piston; this pressure is only that due to a fraction of a millimetre of mercury.

If Π is the barometric pressure, then P_1, the pressure of the air before expansion, is given by the equation

$$P_1 = \Pi - \pi,$$

where π is the maximum vapour pressure of water at the temperature of the experiment. The pressure of the air P_2 when the piston is at the bottom is given by

$$P_2 = P_1 - p,$$

where p is the pressure due to the difference of level of the mercury in the two arms of the gauge G.

Thus if v_2 is the final and v_1 the initial volume of the gas,

$$\frac{v_2}{v_1} = \frac{P_1}{P_2} = \frac{\Pi - \pi}{\Pi - \pi - p}.$$

The vessel in which the rate of fall of the fog and the conductivity of the gas are tested is at A. It is a glass tube 36 millimetres in diameter covered with an aluminium plate; to avoid the abnormal ionisation which occurs when X-rays strike against a metal surface, the lower part of the aluminium plate is coated with wet blotting-paper, and the electric current passes from the blotting-paper to the horizontal surface of the water beneath. The induction coil and the focus bulb for the production of the X-rays are placed in a large iron tank, in the bottom of which a hole is cut and closed by an aluminium window. The vessel A is placed underneath this window and the bulb giving out the rays some distance above it so that the beam of rays escaping from the tank is not very divergent. The intensity of the rays can be reduced to any required degree by inserting leaves of tinfoil or sheets of aluminium between the bulb and the vessel.

In these experiments it is necessary to work with very weak rays, so that the number of ions is comparatively small; when the number of ions is large some of them seem to escape from being caught by the cloud produced by the expansion, and when this is the case the number of ions deduced from the time of fall of the cloud will be too small; it is therefore advisable to work with such weak ionisation of the gas that the first cloud clears away all the ions.

To find the current passing through the gas, the tank and the aluminium plate on the top of the vessel A are connected with one pair of quadrants of the electrometer, the other pair of quadrants is connected with the water surface in the vessel A; this surface is charged up to a known potential by connecting it with one of the terminals of a battery, the other terminal of which is connected with the earth. After the surface has been charged it is disconnected from the battery and the insulation of the system tested by observing whether there is any leak when the X-rays are shut off; the insulation having been found satisfactory, the rays are turned on and the charge begins to leak from the electrometer; by measuring the rate of leak the quantity of electricity which in one second passes through the gas exposed to the rays can be determined. For suppose that in a second the electrometer reading is altered by p scale divisions, and that one scale division of the electrometer corresponds to a potential difference V between the quadrants, and that C is the capacity of the system consisting of the electrometer, the water surface and the connecting wires, then the quantity of electricity which passes in one second through the gas exposed to the rays is pVC. If n is the total number of ions positive as well as negative per cubic centimetre of the gas, u_0 the mean of the velocities of the positive and negative ions under a potential gradient of a volt per centimetre, E the potential gradient in volts per centimetre acting on the ionised gas, A the area of the water surface, the current through the gas is equal to $Aneu_0E$; but as this current is equal to pVC, we have

$$pVC = Aneu_0E,$$

an equation by means of which we can determine ne, and as from the experiments on clouds we know the value of n we can at once deduce the value of e. Proceeding in this way the author found in 1898 that for the ions produced by X-rays passing through air, using electrostatic units,

$$e = 6.5 \times 10^{-10} \text{ (gr.)}^{\frac{1}{2}} \text{ (cm.)}^{\frac{3}{2}} \text{ (sec.)}^{-1}.$$

A similar series of experiments on the ions produced by X-rays passing through hydrogen gave for e the charge on the hydrogen ion the value

$$6.7 \times 10^{-10} \text{ (gr.)}^{\frac{1}{2}} \text{ (cm.)}^{\frac{3}{2}} \text{ (sec.)}^{-1}.$$

The difference between this and the value of the charge on the ion in air is much less than the error of experiment, so that the charges on the ions are the same in these gases. This was shortly afterwards confirmed by the experiments made by Townsend on the rates of diffusion of the ions; an account of these experiments has already been given in Chap. II. For his determination of the charge on the heavy ions in electrolytic gases see later (vol. II).

75. The author in 1901–2 repeated these experiments on the charges carried by the ions, making some modifications in the method. In the first place, the ionisation was produced by the radiation from radium instead of by the X-rays; this was done to get a more uniform rate of ionisation than is possible with X-ray tubes, the irregularity of which gave a great deal of trouble in the earlier investigation. Secondly, the electrometer used in the new experiments was much more sensitive than the old one, the new electrometer was of the Dolezalek type and gave a deflection of 20,000 scale divisions for a potential difference of one volt.

The measurements made by C. T. R. Wilson[1] (see Chap. VIII) show that with expansions between 1·25 and 1·3 negative, and only negative, ions act as nuclei for cloudy condensation, while with expansions greater than 1·3 both negative and positive ions are brought down by the cloud. It was feared that when the expansions were sufficiently large to bring both sets of ions into play the more active negative ions might have a tendency to monopolise the aqueous vapour, and that therefore the whole of the positive ions might not be brought down with the cloud. This fear was found to be justified, for with the expansion apparatus

[1] C. T. R. Wilson, *Phil. Trans.* cxciii. p. 289.

used in the earlier experiments it was found that with expansions greater than 1·3 the number of particles in the cloud formed in the ionised gas was not, as it should have been if all the ions had been caught by the cloud, twice as great as when the expansion was less than this value. The apparatus was modified so as to make the rate of expansion very much more rapid than in the earlier experiments; with the new apparatus the number of particles in the cloud when the expansion was greater than 1·3 was twice as great as when the expansion was less than this value; this confirms the view that with this apparatus all the ions are caught by the cloud. The result of a number of determinations of e with the new apparatus, using different samples of radium and different intensities of radiation, was that

$$e = 3 \cdot 4 \times 10^{-10} \, (\text{gr.})^{\frac{1}{2}} \, (\text{cm.})^{\frac{3}{2}} \, (\text{sec.})^{-1}.$$

76. Another method of finding e has been used by H. A. Wilson[1]. C. T. R. Wilson found that clouds could be deposited round negative ions by an expansion which was insufficient to produce condensation on positive ions. It is thus possible to adjust the expansion so as to get a cloud in which all the drops are negatively charged. H. A. Wilson arranged his experiment so that such a cloud was formed between two horizontal plates; these plates could be maintained at different electrical potentials so that a uniform field acted between the plates, this field acting on the charged drops produced a vertical force in addition to that due to the weight of the drop and so affected the rate of fall.

Let X be the vertical force, e the charge on the drop, v_1 the rate of fall of the drop under this force, and v the rate of fall when there is no electric field; then since the rate of fall is proportional to the force on the drop, if a is the radius of the drop, ρ its density,

$$\frac{Xe + \tfrac{4}{3}\pi\rho g a^3}{\tfrac{4}{3}\pi\rho g a^3} = \frac{v_1}{v},$$

or

$$Xe = \tfrac{4}{3}\pi\rho g a^3 \frac{(v_1 - v)}{v}.$$

But

$$v = \frac{2}{9} \frac{g a^2 \rho}{\mu},$$

so that

$$Xe = \sqrt{2 \cdot 9\pi} \sqrt{\frac{\mu^3 v^{\frac{3}{2}}}{g\rho}} \frac{(v_1 - v)}{v}.$$

[1] H. A. Wilson, *Phil. Mag.* vi. 5, p. 429, 1903.

Thus if X, v, and v_1 are known, e can be determined.

By this method Wilson found $e = 3 \cdot 1 \times 10^{-10}$ electrostatic units.

Wilson found that some of the drops in the cloud carried a charge $2e$ and others a charge $3e$.

76·1. Great improvements have been made in Wilson's method by Millikan[1], who has carried out the most accurate series of researches so far made on the value of e. Millikan has modified Wilson's method so that a single charged particle can be observed,

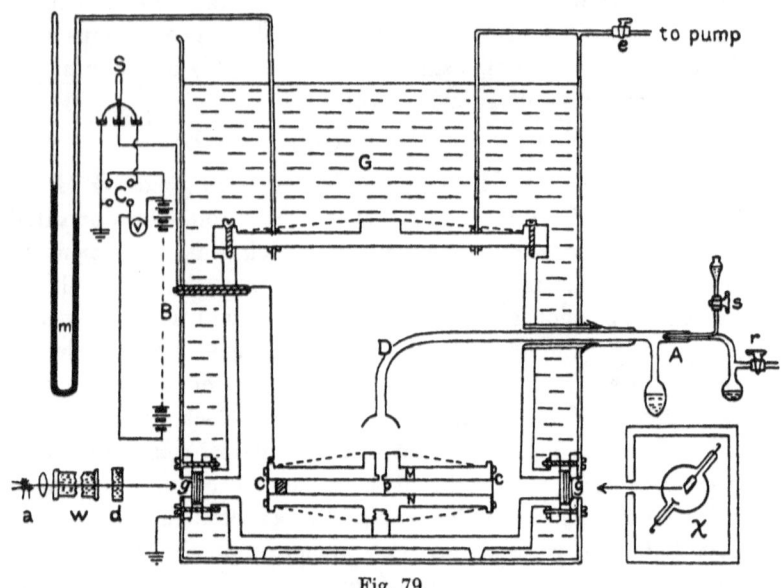

Fig. 79.

and in this way, besides greatly increasing the accuracy, has been able to give an extraordinarily direct proof of the atomic nature of electric charges. Millikan's apparatus in its final form is shown in Fig. 79. The central feature of the apparatus is the parallel plate condenser M, N, consisting of two optically flat metal plates 22 cm. in diameter and separated by three pieces of echelon plates 14·9174 mm. thick. In the top of M are five small holes P through which fall the drops (usually of oil, sometimes

[1] Millikan, *Phil. Mag.* xix. 1909; xxxiv. p. 1, 1917; *Phys. Rev.* xxxii. p. 349, 1911; i. p. 219, 1913; ii. p. 109, 1913.

mercury) formed in the 'atomizer' A. The oil bath G maintains the whole at a constant temperature. The fall of the drops was made visible by light from the arc a filtered through a trough of water w, and one of cupric chloride d, for the removal of heat rays, and their motion was observed in a specially designed telescope (not shown in the fig.), and timed over a distance of 1·0220 cm. with a chronograph. The air in the condenser could be ionised when required by means of X-rays through the window g. The procedure was to time the fall of a particle in the absence of an electric field, and then to time its rise under a field strong enough to overcome gravity. Often the drops carried a charge produced by friction in the atomizer, if not they were allowed to collect a charge from ions in the air of the condenser. The process of rise and fall could be repeated many times, the same particle being kept under observation, sometimes for hours at a time.

The first result of this method was to give a very direct test of the atomic nature of electric charges. Thus, assuming only that the force on the particle due to viscosity is proportional to the velocity, we have

$$mg = kv_g, \quad nXe - mg = kv_F,$$

where mg is the weight in air of the particle, v_g, v_F the velocities of fall and rise, X the field strength and n the number of units of charge. Thus

$$ne = \frac{k}{X}(v_g + v_F).$$

If therefore for any one particle the quantity $\frac{1}{t_g} + \frac{1}{t_F}$ is calculated, where t_g and t_F are the times of fall and rise respectively, it should be an integral multiple of a quantity constant for a given particle. Again, if t_F and $t_{F'}$ are the times of rise with different charges n and $n \pm n'$, then $\frac{1}{n'}\left(\frac{1}{t_F} - \frac{1}{t_{F'}}\right)$ should be equal to the same constant. Some thousands of changes of charge were observed, but in no case did the times fail to obey this rule. One of the shorter series of observations is shown below.

Millikan claims that the evidence obtained in this way for the atomic nature of electric charges is as strong as any of the chemical evidence for the law of multiple proportions and the atomic theory of matter.

Drop No. 8.

t_g	t_F	n'	$\frac{1}{n'}\left(\frac{1}{t_{F'}}-\frac{1}{t_F}\right)$	n	$\frac{1}{n}\left(\frac{1}{t_g}+\frac{1}{t_F}\right)$	
55·69	29·97			3	·017094	$V_i = 6669$ volts
55·90	29·83					$V_f = 6657$ volts
56·18	29·69					$t = 23°·01$ C.
		1	·017251			
55·91	60·89			2	·017131	$p = 75·40$ cm.
56·12	29·76	1	·017179			$v_1 = ·018230$ cm./sec.
55·90	29·49					$a = ·0001250$ cm.
56·29	29·51			3	·017211	$\frac{1}{pa} = 106·3$
		1	·017072			
56·26	59·49					$\frac{l}{a} = ·07608$
56·33	59·44			2	·017327	$e_1^{2/3} = 65·13$
Mean 56·06			·017167		·017191	$e^{2/3} = 61·11$

In practically all Millikan's experiments the change of charge was a single unit, and he is inclined to doubt Franck and Westphal's result that multiply charged ions are present in appreciable numbers. In order that a drop should experience a change of charge it was usually necessary not only to ionise the air but to switch off the electric field, otherwise the relative velocity of ion and drop was so great that there was little chance of their coming together. Millikan found that a drop was about as likely to get an extra charge of the same sign as it already possessed as to get one of opposite sign, in spite of the electrostatic repulsion in the former case. He shows that for particles of the size used, mostly from 10^{-4} to 5×10^{-4}, the electrostatic energy is small compared with the kinetic energy of thermal agitation of the ions, even when the drop has a considerable number of charges.

In order to deduce the value of e it is necessary to know the value of k. According to Stokes' law, $k = 6\pi\mu a$ for a sphere and, since $m = \frac{4}{3}\pi a^3 \rho$, where a is the radius,

$$ne = \tfrac{4}{3}\pi \left(\frac{9\mu}{2}\right)^{\frac{3}{2}} \left(\frac{1}{g\rho}\right)^{\frac{1}{2}} \frac{(v_g + v_F) v_g^{\frac{1}{2}}}{X}.$$

Millikan used this equation and found that the value of e was greatest for the smallest drops, *i.e.* for those which fell slowest.

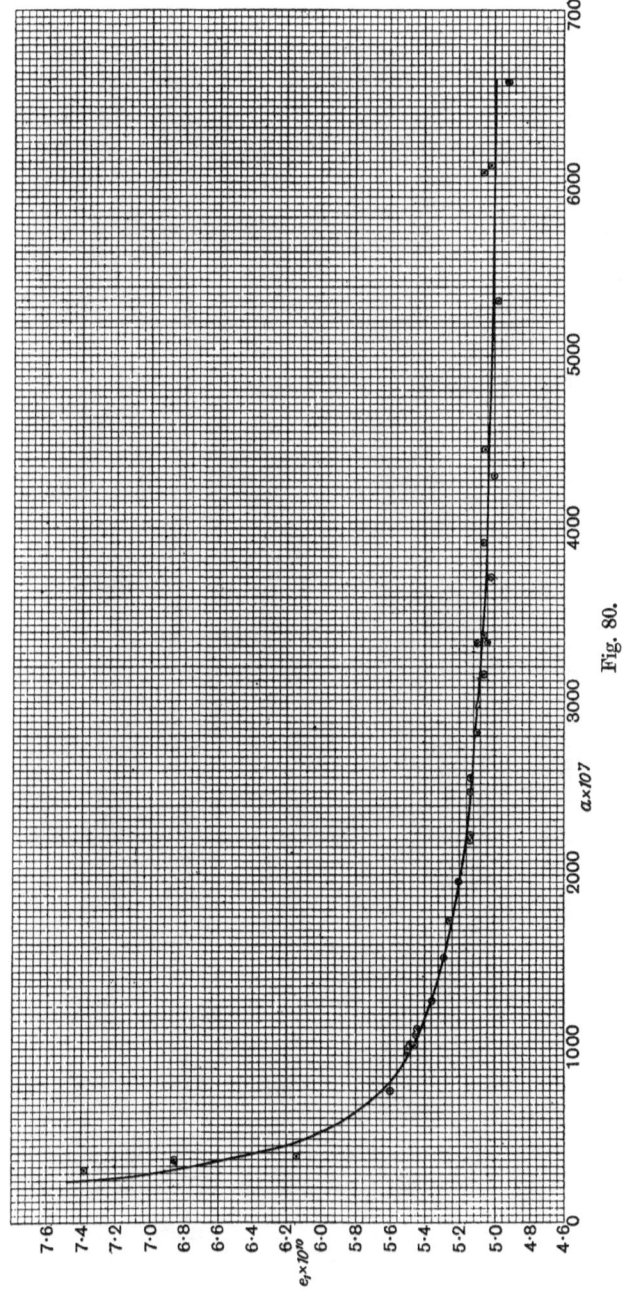

Fig. 80.

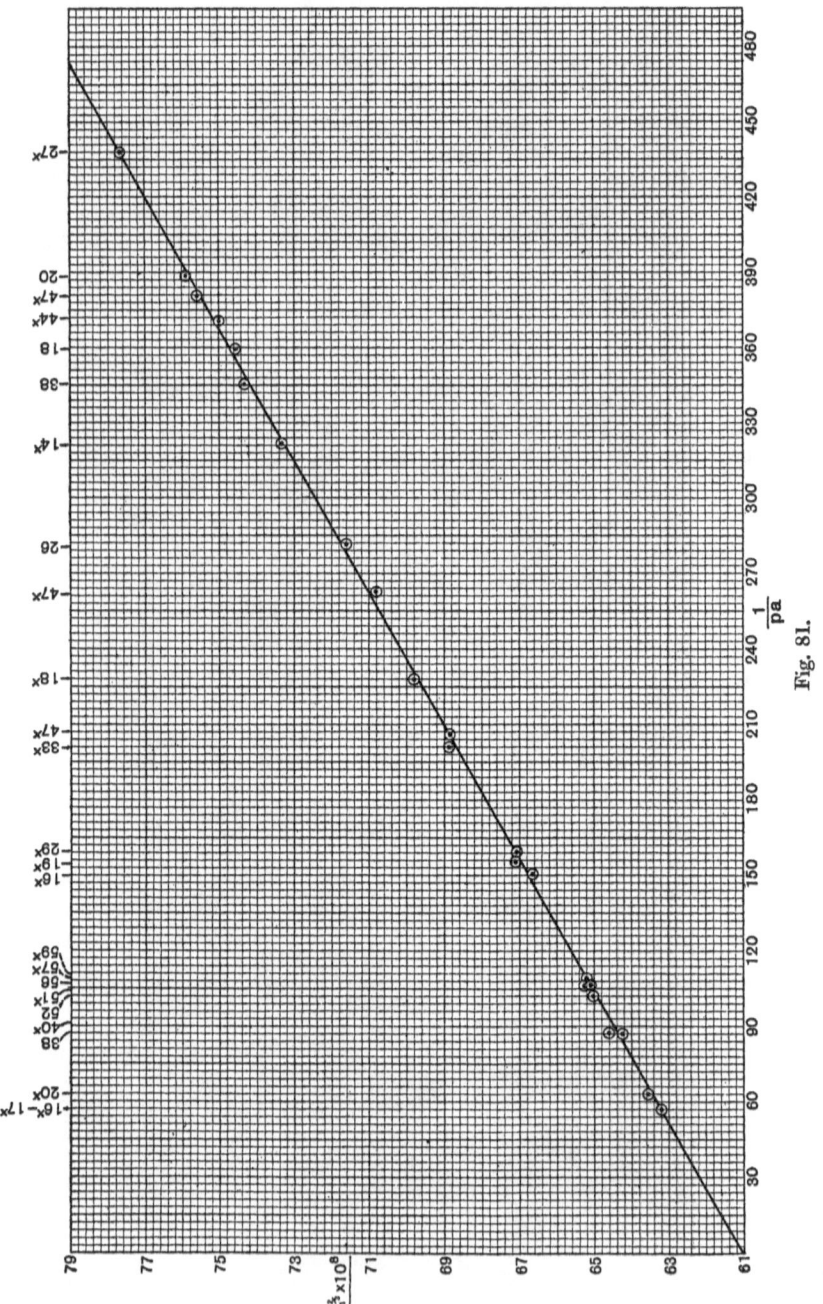

Fig. 81.

Fig. 80 shows the relation between the apparent value of e and the radius a computed from Stokes' law. This result is to be explained by the fact that Stokes' law is deduced from hydrodynamical principles of continuous flow and absence of slip, and ceases to be exact for spheres so small that their size is comparable with the mean free path of the molecules of the gas. For a first order correction Millikan takes

$$k = 6\pi\mu a \left\{1 + A\frac{l}{a}\right\}^{-1}, \text{ see § 41·3,}$$

where l is the mean free path, or since l varies inversely as p the pressure,

$$k = 6\pi\mu a \left\{1 + \frac{b}{pa}\right\}^{-1}.$$

If e_1 is the value found for e from the uncorrected Stokes' law, and e that as corrected, it is easy to show that

$$e^{\frac{2}{3}}\left(1 + \frac{b}{pa}\right) = e_1^{\frac{2}{3}}.$$

Thus if $e_1^{\frac{2}{3}}$ is plotted against $1/pa$ the result should be a straight line whose intercept on the $e_1^{\frac{2}{3}}$ axis gives the true value of $e^{\frac{2}{3}}$. The line in Fig. 81 shows that this expectation is justified. Millikan's final value for e is $4·774 \times 10^{-10}$ E.S.U., taking μ at 23° C. as ·00018227.

Regener[1] working on a similar method has found $e = 4·86 \times 10^{-10}$ as the mean of experiments on oil drops and on drops formed round the gaseous ions obtained in the rapid electrolysis of caustic potash. Mattauch[2], also working on oil drops, has found $4·758 \times 10^{-10}$.

76·2. It should be mentioned that Ehrenhaft considers that he has obtained evidence of charges, which, if multiples of a unit at all, are multiples of a much smaller one than that found by other experimenters. The resulting controversy has fully confirmed Millikan's result but has led to some interesting conclusions regarding the behaviour of small particles and in especial the corrections necessary in Stokes' law. It may be mentioned that most of Ehrenhaft's particles (oil and mercury drops and

[1] Regener, *Phys. Zeits.* xii. p. 135, 1911.
[2] Mattauch, *Zeitschrift f. Phys.* xxxii. p. 439, 1925.

particles spluttered from an arc between silver electrodes) were markedly smaller than Millikan's. This makes the correction to Stokes' law more important, and also introduces difficulties in measurement owing to the Brownian movements. A remarkable piece of research was carried out by Meyer and Gerlach[1] who were able, not only, as Millikan had done, to follow the motion of a particle during many changes of charge, but actually to keep it in view while the pressure in the observation chamber was changed and so to examine the same particle at different pressures. Using small particles from a platinum arc they found it necessary to suppose that the platinum had about half the normal density, but of course there is no proof that the particles were spherical. They charged the particles photoelectrically, and found the usual atomic relation for changes of charge, but e_1 was a function of the pressure. These experiments have been followed up by work by Bär[2] and by Mattauch[3], the latter of whom was even able to

	Mattauch	Millikan
A	0·898	0·864
B	0·312	0·290
C	2·37	1·25

keep the same particle in view during a change of the gas from nitrogen to carbon dioxide. Bär found that the correcting factor $1 + Al/a$, which had been found on theoretical grounds by Cunningham, was not adequate for the larger values of l/a, and adopted the form $1 + \dfrac{l}{a}(A + Be^{-c\frac{a}{l}})$ which had previously been used by Millikan[4]. Mattauch, working with drops of oil and of mercury, was able to find constants which held for oil in both nitrogen and carbon dioxide over a range of l/a between 0·1 and 5. Mercury

[1] Meyer and Gerlach, *Elster-Geitel Festschrift*, 1915, p. 196; also *Ann. der Phys.* xlv. p. 177, 1914.
[2] R. Bär, *Ann. der Phys.* lxvii. p. 157, 1922.
[3] Mattauch, *Zeitschrift f. Phys.* xxxii. p. 439, 1925.
[4] Millikan calculates l from the formula $\mu = \cdot 3502 \rho \bar{c} l$, where $\bar{c}$ is the *average* molecular velocity. On this basis Cunningham's A is equal to ·788. See Millikan, *Phys. Rev.* xxxii. p. 380, 1911.

drops showed a curious separation into two classes, those which evaporated, which gave normal results, and those which did not, which behaved as though they had a density much below that of normal mercury. If these latter are not really dust particles, the effect must apparently be due to an adsorbed layer of gas altering the density, though it is not clear why this should form in some cases and not in others. For a further account of these and other experiments on this subject see the article by W. Gerlach on 'electrons,' vol. XXII of Geiger and Scheel's *Handbuch der Physik*, 1926.

76·3. Other methods of finding e have been used, based on measurements of radioactivity. Thus Rutherford and Geiger[1] counted the number n of α particles emitted per sec. from a known small quantity of radioactive material by detecting the current produced as each particle entered a specially designed ionisation chamber. They then determined nE the total charge carried by the particles. This gave $E = 9\cdot3 \times 10^{-10}$, which on the assumption $E = 2e$ gives $e = 4\cdot65 \times 10^{-10}$. Regener[2] used a similar principle but found n by counting the scintillations produced by α particles from polonium on a fluorescent crystal. He found $E = 9\cdot58 \times 10^{-10}$ or $e = 4\cdot79 \times 10^{-10}$ E.S.U. For a fuller account of these methods and for other less exact methods which have been used, see Rutherford, *Radio-active Substances and their Radiations*.

76·4. Having found the value of e, let us compare it with E the charge carried by the hydrogen ion in the electrolysis of solutions. If N is the number of molecules in a cubic centimetre of a gas at a pressure of 760 mm. of mercury and at 0° C., then we know as the result of experiments on the liberation of hydrogen in electrolysis that
$$NE = 1\cdot22 \times 10^{10}.$$
In treatises on the Kinetic Theory of Gases (for example, O. E. Meyer, *Die kinetische Theorie der Gase*) it is shown how by the aid of certain assumptions as to the nature and shape of the molecules it is possible to find N. The values found in this way vary considerably, the best determinations of N lying between

[1] Rutherford and Geiger, *Proc. Roy. Soc.* lxxxi. p. 162, 1908; *Phys. Zeits.* x. p. 42, 1909.
[2] Regener, *Berl. Ber.* ii. p. 948, 1909.

$2 \cdot 1 \times 10^{19}$ and 10^{20}; this would make E lie between $6 \cdot 1 \times 10^{-10}$ and $1 \cdot 29 \times 10^{-10}$; the value of e is well between these limits. Hence we conclude that the charge carried by any gaseous ion is equal to the charge carried by the hydrogen ion in the electrolysis of solutions.

This conclusion is also confirmed by the experiments of Townsend already referred to. In these experiments the charges on the ions in air, hydrogen and carbonic acid gas were directly compared with E, and proved to be equal to it (see p. 80). Starting with this result we can by direct experiment on gases determine the value of E, and then by the aid of the equation

$$NE = 1 \cdot 22 \times 10^{10},$$

the number of molecules in a cubic centimetre of the gas, and hence the mass of a molecule of the gas; proceeding in this way we avoid all those assumptions as to the shape and size of the molecules of the gas, and the nature of the action which occurs when two molecules come into collision, which have to be made when the same quantities are determined by means of the Kinetic Theory of Gases.

Using Millikan's value for e we find

$$N = 2 \cdot 705 \times 10^{19},$$

and the number in a gr. molecule is $6 \cdot 06 \times 10^{23}$.

It is now generally accepted that these particles or electrons are at once the fundamental unit of negative electricity and a universal constituent of matter. The neutrality of ordinary matter and results of Millikan's experiments show that positive electricity must exist in equal units, a result we have already had occasion to refer to in connection with the experiments on positive rays.

77. The determinations of e described above have been made on ions produced by X-rays or radium rays. The properties of the ions in gases are the same, however, whether the ions are produced by X-rays, radium, Lenard, or cathode rays, or by the agency of ultra-violet light. Evidence in support of this is afforded by the fact that, as we have seen, the velocity of the ions in the electric field is the same in whichever of the above-mentioned ways they are produced. We shall see too (Chap. VIII) that they behave in exactly the same way with respect to their power of producing condensation of clouds. We have thus strong reasons for thinking that the charge on the ion does not depend upon the kind of radiation used to liberate the ion. I have made some direct experiments on this point, and have made measurements of the charge on the negative ions produced by the incidence of ultra-violet light on metals; the method used was the same as in the case of the ions produced by X-rays, and the result was that within the limits of experimental error the charge on the negative ion produced by the action of ultra-violet light was the same as that on the ion produced by X-rays[1]. See also Meyer and Gerlach's experiments above.

The case of the ions produced by ultra-violet light is interesting, as it is the one in which both the values of e and of e/m (when the pressure is low) have been measured when the ions are the same in the two experiments.

78. As e is the same as E the charge on the hydrogen ion, while e/m is about eighteen hundred and forty times E/M, where M is the mass of the atom of hydrogen, it follows that m is only about $1/1840$ of M, so that the mass of the carrier of the negative charge is only $1/1840$ of that of the atom of hydrogen.

79. Let us now sum up the results of the determinations of e and of e/m which have been made for the ions produced in gases by radiations of different kinds. We have seen that in all the cases in which e has been determined it has been found equal to E, the charge on a hydrogen ion in liquid electrolysis. The charge on the gaseous ion does not, like that on the ions in liquids, depend on the substance from which the ions are produced; thus in the case of the ions produced by X-rays or analogous radiation, the charge on an ion produced from oxygen is the same as that on one produced from hydrogen, though in liquids the charge on an oxygen ion is twice that on a hydrogen one.

Again, at very low pressures, when the negative ion can escape getting entangled with the molecules of the gas by which it is surrounded and at all pressures in certain gases, the *mass* as well as the charge of the negative ion is invariable and much smaller than the mass of the smallest portion of ordinary matter, *i.e.* that of an atom of hydrogen, recognised in the Kinetic Theory of Gases.

[1] J. J. Thomson, *Phil. Mag.* v. 48, p. 547, 1899.

CHAPTER VIII

ON SOME PHYSICAL PROPERTIES OF GASEOUS IONS

80. ONE of the most striking effects produced by ions is the influence they exert on the condensation of clouds. One instance of this is the discovery by R. von Helmholtz[1] of the effect of an electric discharge on a high pressure steam jet. When steam rushes out from a jet placed near a pointed electrode connected with an electric machine or an induction coil, a remarkable change in the appearance of the jet takes place when electricity is escaping from the electrode. This can conveniently be shown by throwing the shadow of the jet on a screen; when there is no escape of electricity the jet is nearly transparent and the shadow is very slight; as soon however as electricity begins to escape, the opacity of the jet increases to a remarkable extent, the shadow becomes quite dark and distinct, and colours arising from the diffraction of the light by the small drops of water make their appearance, the jet sometimes presenting a very beautiful appearance. For an account of the ways of arranging the experiments so as to observe these colours to the best advantage and of a method by which the size of the drops of water can be deduced from the colour phenomena, we must refer to a paper by Barus[2]. This effect evidently shows that the electrification makes the steam condense into water drops.

In a later paper by R. von Helmholtz and Richarz[3], published after the death of the former, the authors show that a steam jet is affected by making or breaking the current through the primary of an induction coil, even when the terminals of the secondary placed in the neighbourhood of the jet are separated by much more than the sparking distance, and that the effects persist even when the terminals are wrapped in moist filter-paper so as to catch any metallic particles that might be given off from them.

[1] R. v. Helmholtz, *Wied. Ann.* xxxii. p. 1, 1887; see also Bidwell, *Phil. Mag.* v. 29, p. 158, 1890.
[2] Barus, *American Journal of Meteorology*, ix. p. 488, 1893.
[3] R. v. Helmholtz and Richarz, *Wied. Ann.* xl. p. 161, 1890.

R. von Helmholtz and Richarz (*loc. cit.*) showed that the steam jet was affected by gases from the neighbourhood of flames whether these were luminous or not; the very cool flames of burning ether and alcohol are exceptions to this statement.

A platinum wire raised to a dull red heat affected the jet when electrified, and if raised to a bright yellow heat affected the jet even when unelectrified, except when the wire was surrounded by hydrogen, in which case the unelectrified wire had no effect. Coal gas passed through platinum gauze raised to a dull red heat also influenced the jet.

The jet is also affected by the presence in its neighbourhood of certain substances such as sulphuric acid, also by gases which are dissociating or undergoing chemical changes in the air, such as N_2O_4 or NO_2; it is not affected by ozone or hydrogen peroxide. If however ozone is destroyed by bubbling through such substances as solutions of potassium iodide or potassium permanganate, the gas which emerges has the power of affecting the jet; this gas has also the power of forming clouds when it comes into contact with moist air, as was first shown by Meissner[1]; experiments on this point have also been made by R. von Helmholtz and Richarz and by J. S. Townsend[2]. The action in this case and in other cases of the effect of chemicals is, as we shall see, probably due to the formation of some substance which dissolves in the drops of water and lowers their vapour pressure; thus the drops in this case are not formed of pure water, but of more or less dilute solutions.

Moist air drawn over phosphorus, sodium or potassium also affects the jet.

Lenard and Wolf[3] also showed that the incidence of ultra-violet light on a zinc plate or on some fluorescent solutions in the neighbourhood of a steam jet produced condensation in the jet; a similar effect was produced by ultra-violet light passing through quartz. Richarz[4] showed that the incidence of X-rays produced condensation in the jet. There was for some time con-

[1] Meissner, *Jahresber. f. Chemie*, 1863, p. 126.
[2] J. S. Townsend, *Proc. Camb. Phil. Soc.* x. p. 52, 1898.
[3] Lenard and Wolf, *Wied. Ann.* xxxvii. p. 443, 1889.
[4] Richarz, *Wied. Ann.* lix. p. 592, 1896.

siderable difference of opinion as to the cause of this behaviour of the steam jet; the earliest researches on this subject came at a time when the experiments of Aitken[1], of Coulier[2] and of Kiessling[3] had drawn attention to the great effect produced by dust on cloudy condensation. These physicists had shown that the clouds produced by the lowering of temperature resulting from a small adiabatic expansion of the damp dusty air of an ordinary room entirely disappeared if the dust were filtered out of the air: the drops in the cloud were shown to collect round the particles of dust, the water drops were thus able to start with a finite radius—that of the dust particle—and so had not to pass through the stage when their radius was of molecular dimensions, when, as Lord Kelvin has shown, the effect of surface tension would lead to such intense evaporation as soon to cause the disappearance of the drops.

The discovery of the effect of dust on the condensation of water vapour produced a tendency to ascribe the formation of clouds in all cases to dust and to dust alone; in fact, to use the indication of the steam jet as a measure of the dustiness of the air; thus, for example, Lenard and Wolf ascribed the effect which they found was produced by the incidence of ultra-violet light on metals to metallic dust given off by the metal under the influence of the light. On the other hand, R. von Helmholtz, and later Richarz, strongly maintained the view that many of the effects they observed were not due to dust, but to ions, and they gave strong arguments and made some striking experiments in support of this view: as however this evidence is somewhat indirect, and as the truth of their view has been indisputably proved by the direct experiments made later by C. T. R. Wilson[4], we shall proceed at once to a description of his researches.

81. The method used by Wilson was to cool the moist gas suddenly by an adiabatic expansion, so that the gas which was saturated with water vapour before cooling became supersaturated

---

[1] Aitken, *Nature*, xxiii. pp. 195, 384, 1880; *Trans. Roy. Soc. Edin.* xxxiii. p. 337, 1881.
[2] Coulier, *Journal de Pharm. et de Chimie*, xxii. p. 165, 1875.
[3] Kiessling, *Naturw. Verein d. Hamburg-Altona*, viii. 1, 1884.
[4] C. T. R. Wilson, *Phil. Trans.* clxxxix. p. 265, 1897.

afterwards. One of the arrangements used by Wilson to produce the expansion is shown in Fig. 82: the way in which the apparatus works has already been explained (see p. 295). It is very important in these experiments that the expansions which produce the cloud should be as rapid as possible, for with slow expansions

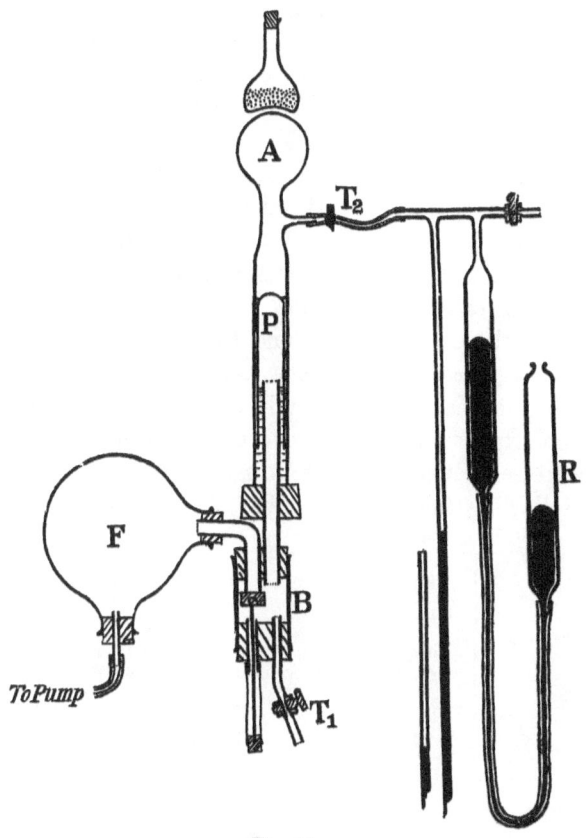

Fig. 82.

as soon as the supersaturation is sufficient for the first drops to be formed, if these have time to grow before the expansion is completed, they will rob the air of its moisture, and the supersaturation will not rise much above the value required for the formation of the first drops. To ensure this rapid expansion, the piston P, Fig. 82, should be light and able to move freely up and down, and the arrangement by which the difference of pressure between the inside and outside of the cylinder is produced should work very rapidly.

82. Using an arrangement of this nature, Wilson found that when dusty air filled the expansion chamber a very slight expansion was sufficient to produce a dense fog; if this was allowed to settle and the process repeated, the air by degrees got deprived of the dust which was carried down by the fog; when the air became dust-free no fogs were produced by small expansions. If we take as the measure of the expansion the ratio of the final to the initial volume of the gas, no cloud was produced in the dust-free air until the expansion was equal to 1·25. When the expansion was between 1·25 and 1·38, a few drops made their appearance; these drops were very much fewer in hydrogen than in air. On increasing the expansion beyond 1·38 a much denser cloud was produced in the dust-free gas, and the density of the cloud now increased very rapidly with the expansion. Thus we see that even when there is no dust, cloudy condensation can be produced by sudden expansions if these exceed a certain limit. This limit appears to be independent of the nature of the gas, as is shown by the following table, which gives the ratio of the volumes re-

Gas	Rain-like condensation		Cloud-like condensation	
	Final/initial volume	Super-saturation	Final/initial volume	Super-saturation
Air.........	1·252	4·2	1·375	7·9
Oxygen.....	1·257	4·3	1·375	7·9
Nitrogen....	1·262	4·4	1·375	7·9
Hydrogen...	—	—	1·375	7·9
Carbonic acid	1·365	4·2	1·53	7·3
Chlorine	1·30	3·4	1·44	5·9

quired to produce the first or rain-like stage of condensation and the supersaturation, *i.e.* the ratio of the pressure of the aqueous vapour actually present when the condensation begins to the saturation vapour pressure at that temperature: the third and fourth columns give the corresponding quantities for the second stage of the condensation, *i.e.* when the expansion produces a dense cloud.

The rain-like condensation is absent in hydrogen.

83. The description given above relates to the behaviour of gas in the normal state; on exposing the gas to X-rays, Wilson found that, as in the normal gas, there were no drops until the expansion was equal to 1·25; on passing this limit however the density of the cloud was very greatly increased by the rays, and if these were strong the few drops which were all that were formed when the rays were absent were replaced by a dense and almost opaque cloud. The strength of the rays does not affect the expansion required to produce the cloud; no matter how strong the rays may be there is no cloud produced unless the expansion exceeds 1·25; the strength of the rays increases the number of drops in the cloud, but does not affect the stage at which the cloud begins. The effect of the rays in producing a cloud lasts some few seconds after the rays have been cut off. Wilson[1] has shown that the radiation from uranium and other radioactive substances produces the same effect as X-rays, as does also ultra-violet light when incident upon such a metal as zinc: the effects produced by ultra-violet light are however somewhat complicated and we shall have to return to them again.

84. That the effect produced by X-rays and uranium rays is due to the production of charged ions formed in the gas can be shown directly by the following experiment. If the ions produced by the X-rays act as nuclei for the water drops, then since these ions can be withdrawn from the gas by applying to it a strong electric field, it follows that a cloud ought not to be formed by the rays when the air which is expanded is exposed to a strong electric field while the rays are passing through it. This was found to be the case, and the experiment is a very striking one. Two parallel plates were placed in the vessel containing the dust-free air; these plates were about 5 cm. apart, and were large enough to include the greater part of the air between them[2]. The

[1] C. T. R. Wilson, *Phil. Trans.* cxcii. p. 403, 1899.
[2] J. J. Thomson, *Phil. Mag.* v. 46, p. 528, 1898.

plates could be connected with the terminals of a battery of small storage cells giving a potential difference of about 400 volts. X-rays passed through the gas between the plates; the gas had previously been freed from dust. When the plates were disconnected from the battery a suitable expansion produced a dense cloud; when however the plates were connected with the battery only a very light cloud was produced by the expansion, and this cloud was almost as dense when the X-rays did not pass through the air as when they did.

85. When a dense cloud has been produced by X-rays by an expansion between 1·25 and 1·38, or by an expansion without X-rays greater than 1·38, then for some little time after drops can be produced by expansions less than 1·25, and these are not eliminated by the action of an electric field. A dense fog apparently leaves behind it little drops of water, which, though too minute to be visible, act in the same way as particles of dust, producing cloudy condensation with very slight expansions. Wilson[1] has also shown that when electricity is discharged from a pointed electrode in the expansion chamber, cloudy condensation is, as in the case of exposure to X-rays, much increased for expansions between 1·25 and 1·38. When the discharge was stopped before the expansion took place, it was found that fogs could be produced for 1 or 2 minutes after the cessation of the discharge; the expansion required to produce the fog diminished as the interval after the cessation of the discharge increased, showing that some of the nuclei produced had grown during this interval. This effect is probably due to the formation of some chemical compound during the discharge, perhaps nitric acid, which by dissolving in the drops lowers their vapour pressure.

86. Wilson (*loc. cit.*) showed that the passage of ultra-violet light through a gas (as distinct from the effects produced when it is incident on a metallic surface) produces very interesting effects on the condensation of clouds. If the intensity of the light is small, then no clouds are produced unless the expansion equals that (1·25) required to produce clouds in gases exposed to X-rays. If however the ultra-violet light is very intense, clouds are produced in air or in pure oxygen, but not in hydrogen, by very much smaller expansions, and the expansion required decreases as the time of exposure to the light increases; thus the nuclei producing the clouds grow under the influence of the light. If the light is exceedingly strong, clouds are produced in air or oxygen without any expansion at all; these clouds are exceedingly fine and may last for hours after the light is cut off. Wilson was even able to produce these clouds in air standing over a 17 per cent. solution of caustic potash, and which therefore was not saturated with water vapour; in this case the drops lasted for three hours after the light was cut off, so that there could be very little evaporation from the drops; this, as Wilson points out, shows that the drops cannot be pure water. These clouds are probably analogous to those observed many years ago by Tyndall[1], when ultra-violet light passed through air containing the vapours of certain substances of which amyl-nitrite was the one which gave the most striking effects. The effects can be explained by the formation under the influence of the ultra-violet light of some substance—Wilson suggests that in his experiments it was H_2O_2—which by dissolving in the drops as they form lowers the equilibrium vapour pressure, and thus enables the drops to grow under circumstances which would make drops of pure water evaporate. This explanation is supported by the fact that ultra-violet light does not produce these clouds in water vapour by itself or in hydrogen: and also by the fact that, unlike the clouds due to X-rays, these clouds formed by ultra-violet light do not diminish in density when a strong electric field is applied to the gas, showing that the nuclei are either not charged or that if they are charged they are so loaded with foreign molecules that they do not move perceptibly in the electric field. Vincent[2] has observed movements of these drops in a strong electric field; he found that some drops moved in one, others in the opposite direction, while there were some which did not move at all. Thus some drops are uncharged, others positively or negatively charged. It would thus seem that the charges have nothing to do with the formation of these drops, the drops merely forming a home for the ions produced by the ultra-violet light.

[1] J. Tyndall, *Phil. Trans.* cvi. p. 333, 1870.
[2] Vincent, *Proc. Camb. Phil. Soc.* xii. p. 305, 1904.

87. Buisson[1], who examined this question with great care, could not detect any conductivity in the air through which the ultra-violet light passed. Lenard[2] has however shown that a certain kind of ultra-violet light which is absorbed so quickly by the air as to be extinguished, when the air is at atmospheric pressure, within a space of a few centimetres, does produce electrical conductivity in the gas through which it passes, and that a charged conductor placed in the neighbourhood of air traversed by these rays loses its charge, and does so much more rapidly when the charge is positive than when it is negative. Lenard determined the velocity of the negative ions by a method analogous to that described on p. 87 and found this velocity through air at atmospheric pressure to be 3·13 cm./sec. under a potential gradient of a volt per cm.: this is about twice the velocity of the ions produced by X-rays; on the other hand the velocity of the positive ions under the same potential gradient was not more than ·0015 cm./sec., which is only about one-thousandth part of the velocity of the positive ion produced by X-rays. The greater mobility of the negative ions explains why the leak from a positively charged body in the neighbourhood of the ionised gas is so much more rapid than that from a negatively charged one. We shall return to this point in the chapter on the effect of ultra-violet light on gases.

88. The results obtained by Wilson and Lenard seem to point to the conclusion that when gas is exposed to the action of ordinary ultra-violet light, we have some chemical action taking place, resulting in the formation of a product which by dissolving in water lowers the vapour pressure over the drops and thus facilitates their formation. When these drops are exposed to the influence of ultra-violet light of the kind investigated by Lenard, they lose, as so many other bodies do when illuminated by light of this kind, negative electricity, and it is the negative ions liberated in this way which produce the electrical conductivity investigated by Lenard. The difference between the action of ultra-violet light and X-rays is that the former when very intense can produce

[1] Buisson quoted by Perrin, *Thèses présentées à la Faculté des Sciences de Paris*, p. 31, 1897.
[2] Lenard, *Ann. der Phys.* i. p. 486; iii. p. 298, 1900.

clouds with little or no expansion, while the latter cannot; this on the theory given above is due to ultra-violet light being more efficient than X-rays in promoting chemical action; there are many examples of this, *e.g.* the combination of hydrogen and chlorine.

The influence of minute traces of soluble substances in promoting the formation of clouds has been shown in a very straightforward way in some experiments made by H. A. Wilson[1]. The writer[2] has shown how drops, even if their existence is very transient, would facilitate the progress of chemical combination between the gases surrounding them, and how this action would afford an explanation of the remarkable fact investigated by Baker[3] and Pringsheim[4], that the occurrence of some of the best known cases of chemical combination between gases depends upon the presence of moisture and does not take place in gases dried with extreme care.

89. *Nuclei from metals.* C. T. R. Wilson[5] has shown that certain metals produce nuclei which cause cloudy condensation when the expansion exceeds 1·25, although the effects are much more marked when the expansion is increased to 1·30. The amount of this effect depends greatly upon the kind of metal used: amalgamated zinc gives comparatively dense clouds, polished zinc and lead also show the effect well; on the other hand polished copper and tin produce no appreciable effect. The order of the metals in respect to their power of producing nuclei for cloudy condensation is the same as their order in respect to their power of affecting a photographic plate placed at a small distance from their surface, a subject which has been studied by Russell[6] and Colson[7]. The effect produced by the presence of a metal on clouds in hydrogen is very slight.

Although the expansion required to produce cloudy condensation when metals are present may be the same as when charged

[1] H. A. Wilson, *Phil. Mag.* v. 45, p. 454, 1898.
[2] J. J. Thomson, *Phil. Mag.* v. 36, p. 313, 1893; *B.A. Report*, 1894.
[3] Baker, *Phil. Trans.* clxxix. p. 571, 1888.
[4] Pringsheim, *Wied. Ann.* xxxii. p. 384, 1887.
[5] C. T. R. Wilson, *Phil. Trans.* cxcii. p. 403, 1899.
[6] Russell, *Proc. Roy. Soc.* lxi. p. 424, 1897; lxiii. p. 102, 1898.
[7] Colson, *Comptes Rendus*, cxxiii. p. 49, 1896.

ions are produced by X-rays, the metal effect differs from the X-ray effect inasmuch as it is not diminished by the application of an intense electric field. These clouds are apparently the result of a chemical action between the metal and the gas. If a compound is formed which can exist in the air in small drops, normally invisible, these may grow to visible size in the presence of sufficient moisture, the small size of the original drops being compensated by the smaller vapour pressure of a solution as compared with pure water. In some cases these clouds are formed even when the water vapour is not supersaturated.

90. The few nuclei that produce rain-like condensation with expansions between 1·25 and 1·38 in gases not exposed to any external ionising agent arise from the ionisation which as we have seen is present in all gases shut up in closed vessels. In his earlier experiments Wilson was not able to detect any diminution in the number of drops when the expansion took place in a strong electric field. In some later experiments (*Phil. Mag.* June, 1904) in which he used very much larger vessels he was able to show that an electric field produced a great diminution in the number of nuclei. The absence of this diminution in the small vessels is due to the great diminution in the number of free ions produced by their diffusion to the walls of the vessel, the diffusion producing much greater effect in small than in large vessels. This reduction in the number of the ions not only makes variation in their number more difficult to detect on account of the rapidity with which the big drops formed round them fall, but it also enables very weak electric fields to remove them from the vessel, so that in small vessels small accidental differences of potential might have produced saturation before the external field was applied.

Comparative Efficiency of Positive and Negative Ions in producing Condensation of Clouds.

91. The writer[1] in 1893 made an experiment with a steam jet which showed that negative electrification had a decidedly greater effect in promoting condensation than positive. The following arrangement was used. A vertical glass tube dipped into the steam chamber, and to the top of this tube was fused a horizontal

[1] J. J. Thomson, *Phil. Mag.* v. 36, p. 313, 1893.

cross-piece, the steam issued from nozzles at the ends of the cross-piece; into these nozzles pointed platinum wires were fused, and these wires were connected with the terminals of a small induction coil. When the coil was in action there was great condensation in the two jets, but the jet at the nozzle connected with the negative terminal of the coil was always denser than that connected with the positive; this was not due to any want of symmetry in the tubes or differences in the nozzles, for on reversing the coil the denser cloud passed from one nozzle to the other. No sparks passed between the platinum electrodes, the strength of the coil being only sufficient to give a non-luminous discharge from their points.

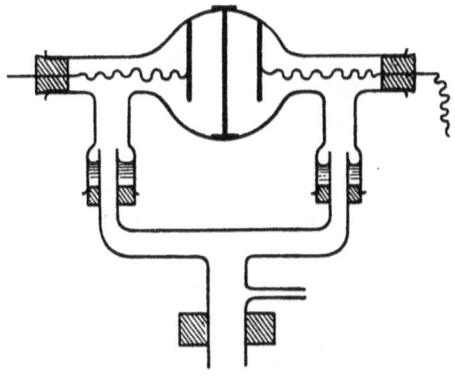

Fig. 83.

Later in 1898[1] I observed indications of a similar effect when clouds were produced by expansion, but the subject was first systematically investigated by C. T. R. Wilson[2] in 1899. Wilson investigated the amount of expansion required to make positive and negative ions act as nuclei for the condensation of water drops; he used several methods, the arrangement of the apparatus in one of these being shown in Fig. 83. The vessel in which the clouds were observed was nearly spherical and about 5·8 cm. in diameter. It was divided into two equal chambers by a brass partition (about 1 mm. thick) in the equatorial plane; the vessel

[1] J. J. Thomson, *Phil. Mag.* v. 46, p. 528, 1898.
[2] C. T. R. Wilson, *Phil. Trans.* cxciii. p. 289, 1899.

was cut in two and the edges of the two halves ground smooth, to allow them to be easily cemented against the face of the partition. The latter was circular and had a narrow strip of brass soldered to each face extending all round the circumference except for a gap at the top. When the halves of the glass vessel were cemented against these strips, a slit was left at the gap about 4·5 cm. long and 2·5 mm. wide on each side of the partition. This slit was covered with a thin piece of aluminium cemented to the outer surface of the glass and to the edge of the brass partition. A thin layer of air in contact with each surface of the partition could thus be exposed to X-rays from a source vertically over the dividing plate. Each half of the apparatus contained a second brass plate parallel to the central plate and 1·8 cm. from it. There was room between the sides of these plates and the walls of the vessel for the air to escape when the expansion was made. To keep the beam of X-rays parallel to the surface of the partition a lead screen with a slit 4 mm. wide was placed about 2 cm. above the aluminium window of the glass vessel: this screen was moved until when both plates were kept at the same potential exactly equal fogs were obtained on the two sides. The metal plates were covered with wet filter-paper to get rid of any ions due to the metal. Suppose now that the middle plate is earthed while the left-hand plate is at a lower and the right-hand plate at a higher potential. Then it is evident, since the ionisation is confined to a layer close to the middle plate, that under these circumstances the left half of the vessel will contain positive ions and the right half negative ones. Wilson found that with an expansion of 1·28 there was a dense fog in the half containing the negative ions, and only a few drops in the half containing the positive ones, and that this excess of condensation in the negative half continued until the expansion was equal to 1·31, when little or no difference was to be seen in the clouds in the two halves. Care was taken that the potential of the positive plate should exceed that of the middle one by the same amount as this exceeded the potential of the negative plate.

The difference between the effects produced by positive and negative ions is shown in the following table, where the time of fall of the drops is used to measure the number of nuclei which produce condensation; if this number is small, then the water drops formed round them will be large and will therefore fall rapidly, while if the number of nuclei be large, since there is only the same quantity of water to be distributed among them, the drops will be small and will fall slowly. In the experiments referred to in the table there was a potential difference equal to

Expansion	Time of fall Left side	Time of fall Right side	Ratio of times negative/positive
1·28	positive 5 negative 15	negative 16 positive 3	$\left.\begin{matrix}3\cdot2\\5\cdot0\end{matrix}\right\}$ 4·1
1·30	negative 15 positive 5 negative 10 positive 2	positive 2 negative 15 positive 2 negative 10	$\left.\begin{matrix}7\cdot5\\3\cdot0\\5\cdot0\\5\cdot0\end{matrix}\right\}$ 5·1
1·31	positive 7 negative 14	negative 12 positive 7	$\left.\begin{matrix}1\cdot7\\2\cdot0\end{matrix}\right\}$ 1·8
1·32	negative 8 positive 8 negative 14 positive 12	positive 5 negative 10 positive 8 negative 17	$\left.\begin{matrix}1\cdot6\\1\cdot2\\1\cdot7\\1\cdot4\end{matrix}\right\}$ 1·5
1·33	negative 12 positive 12	positive 10 negative 13	$\left.\begin{matrix}1\cdot2\\1\cdot1\end{matrix}\right\}$ 1·15
1·35	negative 10 positive 10	positive 10 negative 10	$\left.\begin{matrix}1\cdot0\\1\cdot0\end{matrix}\right\}$ 1·0

that due to two Leclanché cells between the middle plate and either of the outer ones. The words 'positive' and 'negative' in the table indicate that the positive or negative ions respectively were in excess in the region referred to.

The difference in the rates of fall of the drops with the same expansions is due to irregularities in the action of the bulb used to produce the X-rays. The negative ions begin to act as nuclei for foggy condensation when the expansion is about 1·25, corresponding to about a fourfold supersaturation; while we see from the table that the positive ions do not begin to act as nuclei until the expansion is equal to 1·31, corresponding to about a sixfold supersaturation. Wilson has shown that all the negative ions are caught when the expansion is equal to 1·28, but that it is not until the expansion reaches 1·35 that all the positive ions are caught. This is not due to the negative ions having a larger electrical charge than the positive; to show this take an expansion vessel such as that shown in Fig. 82 and ionise the gas in it by X-rays; first produce a fog with an expansion of 1·28 (which only brings down the negative ions), and determine the number of ions from the time of fall in the way already explained; then with the same intensity of radiation produce a cloud by an expansion of 1·35, which brings down both the positive and negative ions, and again calculate the number of ions; we shall find it twice as great as in the first case, thus showing that the numbers of positive and negative ions are equal. As the gas as a whole has no charge, the total charge on the positive ions must be equal to that on the negative; hence as there are as many positive ions as negative, the charge on a positive ion must be the same as that on a negative one. We shall return to the origin of the greater efficiency of the negative than of the positive ions when we discuss the theory of the action of ions in promoting condensation. In the meantime we may point out that this difference between the ions may have very important bearings on the question of atmospheric electricity; for if the ions were to differ in their power of condensing water around them, then we might get a cloud formed round one set of ions and not round the other. The ions in the cloud would fall under gravity, and thus we might have separation of the positive and the negative ions and the production of an electric field, the work required for the production of the field being done by gravity[1]. An action of this kind would tend to make the charge in the air positive, as more negative ions than positive would be carried down by water drops: for a further consideration of this effect we may refer the reader to the papers by Elster and Geitel[2] on the ionic theory of atmospheric electricity.

[1] J. J. Thomson, *Phil. Mag.* v. 46, p. 528, 1898.
[2] Elster and Geitel, *Phys. Zeits.* i. p. 245, 1900.

Theory of the Effect of Ions on Condensation.

92. The effect of electrification on the evaporation of drops of water was investigated by the writer in *Applications of Dynamics to Physics and Chemistry,* p. 165. The general tendency of this effect can easily be seen from elementary principles: for if we have a drop of water of radius a, carrying a charge e of electricity, its potential energy is equal to $\frac{1}{2}e^2/Ka$, where K is the specific inductive capacity of the dielectric surrounding the drop. Now as the drop evaporates the electricity remains behind, so that e does not change while a diminishes, hence the potential energy due to the electrification of the drop increases as the drop evaporates; thus to make the charged drop evaporate more work has to be available than when it is uncharged, so that electrification will diminish the tendency of the drop to evaporate, and the drop will be in equilibrium when the vapour pressure of the water vapour around it would not be sufficient to prevent the evaporation of an uncharged drop. The surface tension of the water will, as was shown by Lord Kelvin, produce the opposite effect; for the potential energy due to the surface tension is equal to $4\pi a^2 T$, where T is the surface tension; thus as the drop evaporates the energy due to surface tension diminishes, so that the work required to vaporise a given quantity of water in a spherical drop is less than if surface tension were absent or inoperative, as it would be if the surface were flat. Thus a curved drop will evaporate when a flat one would be in equilibrium.

It is shown in *Applications of Dynamics to Physics and Chemistry,* p. 165, that when δp, the change in the vapour pressure due to the electrification and surface tension, is only a small fraction of the original vapour pressure p,

$$\frac{\delta p}{p} = \frac{1}{R\theta}\frac{1}{4\pi a^2}\frac{d}{da}\left(4\pi T a^2 + \frac{e^2}{2Ka}\right)\frac{1}{\sigma - \rho} \quad \ldots\ldots(1);$$

or when T does not vary with a,

$$\frac{\delta p}{p} = \frac{1}{R\theta}\left\{\frac{2T}{a} - \frac{e^2}{8\pi K a^4}\right\}\frac{1}{\sigma - \rho};$$

σ is the density of water, ρ that of the vapour, θ the absolute temperature, R the constant which occurs in the equation for a 'perfect' gas, $p = R\theta\rho$; in the investigation this equation is assumed to hold for the water vapour. When the change in the pressure is not a small fraction of the equilibrium vapour pressure for an infinitely large drop, then the investigation already alluded to shows that the preceding equation has to be replaced by

$$R\theta \log_e \frac{p}{P} + \frac{R\theta(\rho' - \rho)}{\sigma} = \left(\frac{2T}{a} - \frac{e^2}{8\pi Ka^4}\right)\frac{1}{\sigma},$$

where p and ρ are the equilibrium vapour pressure and density for a drop of radius a, P and ρ' the corresponding quantities for a drop of infinite radius. Since $\rho' - \rho$ is exceedingly small compared with σ, this equation becomes approximately

$$R\theta \log_e \frac{p}{P} = \left(\frac{2T}{a} - \frac{e^2}{8\pi Ka^4}\right)\frac{1}{\sigma} \quad \ldots\ldots\ldots(2).$$

We see from this equation that if e is zero the equilibrium vapour pressure p for a drop of finite size is always greater than P, so that such a drop would evaporate unless the vapour around it were supersaturated; when however the drop is electrified this is no longer the case, for we see from equation (1) that in this case if the vapour is saturated, i.e. if the vapour pressure is P, the drop will grow until its radius a is given by the equation

$$\frac{2T}{a} - \frac{e^2}{8\pi Ka^4} = 0.$$

Thus if the drop were charged with the quantity of electricity carried by a gaseous ion, i.e. $4 \cdot 77 \times 10^{-10}$ electrostatic units, and if the surface tension of the small drop were equal to 76, which is the value for thick water films, then a would be equal to $1/2 \cdot 6 \times 10^7$, and thus each gaseous ion would be surrounded by a drop of water of this radius; if we call this radius c, then equation (2) may be written

$$R\theta c \log_e \frac{p}{P} = 2Tx(1 - x^3) \quad \ldots\ldots\ldots(3),$$

where $x = c/a$. This equation enables us to find the size of a drop corresponding to any vapour pressure.

For water vapour at $10°$ C., $R\theta$ is equal to $1 \cdot 3 \times 10^9$. Putting for c the value previously found and $T = 76$, equation (3) becomes approximately

$$\cdot 33 \log_e \frac{p}{P} = x(1 - x^3) \quad \ldots\ldots\ldots(4).$$

From this equation we see that even in a space far from saturated with water vapour, *i.e.* when p is only a fraction of P, drops will be formed, and that the size of these drops diminishes only very slowly as the quantity of water vapour in the surrounding air diminishes; thus if we diminish the quantity of water vapour in the air to $1/e$, *i.e.* $1/2 \cdot 7$ of that required to saturate it, we see from equation (4) that the radius of the drops formed round the ions would only be rather more than 10/11 of the radius of the drop formed in saturated air: and that to reduce the drop to half the radius corresponding to saturation, we should have to dry the air so completely that p/P was only about $1/3 \times 10^{18}$. We have seen that there are always some ions in the air, hence if there is any water vapour in the air some of it will be condensed into fine drops. It has been suggested that these drops play a part in certain cases of chemical combination; the preceding numerical example will show the difficulty of getting the gas dry enough to produce a substantial reduction in the volume of these charged drops.

Supersaturation required to make one of the Charged Drops grow to a Large Size.

93. As the radius of the drop increases from c to an infinite size, x diminishes from unity to zero. Now the right-hand side of equation (4) vanishes at each of these limits, but between them it reaches a maximum value which occurs when $4x^3 = 1$ or $x = \dfrac{1}{1 \cdot 58}$, when $x(1 - x^3)$ reaches the value $\cdot 471$; hence we see from equation (4) that for the drops to increase to a large size $\log_e p/P$ must reach the value $1 \cdot 4$ approximately. Hence for the drops to grow p/P must be about $4 \cdot 1$: this, on the theory we have given, is the amount of supersaturation required to make large drops grow round the ions. We have seen from Wilson's experiment that it actually requires a fourfold supersaturation, a very good agreement with theory.

93·1. Some justification is perhaps required for the use of formulae involving the electric energy for a charged sphere in a case like this, where the charge usually consists of a single indivisible unit, while the ordinary electric theory assumes a continuous distribution of charge. But if we take the case of a small conducting sphere it will usually contain a large number of free electrons, and the addition of another electron will have the effect of crowding the electrons on the surface closer together. This will give rise to a repulsion which can do work if the sphere is allowed to expand, and this is all that is required to make the argument valid. In the actual case of a water drop of the size under consideration it may perhaps be doubted if there are sufficient free ions to produce a sensibly uniform distribution of charge over the surface, but it must be remembered that the large dielectric constant of water makes it act in electrostatic problems very like a conductor quite apart from the existence of ions. The dielectric constant is due to the polar molecules and these are sufficiently numerous to act sensibly as a continuous medium, the charges at their ends taking the place of the free electrons in the conducting case.

94. Wilson showed that even when there is no external ionisation, a dense cloud the nuclei of which are not charged is produced by an eightfold supersaturation: we can by the aid of equation (2) determine the radii of these nuclei, supposed spherical; putting in that equation $e = 0$, $T = 76$, $R\theta = 1.3 \times 10^9$, and $p/P = 8$, we find that a, the radius of the nucleus which produces this kind of condensation, is equal to $1/1.9 \times 10^7$. This nucleus is thus slightly *larger* than the drop which collects round an ion, as we found that the radius of this drop is $1/2.6 \times 10^7$. With regard to the nature of the nuclei which produce the cloud corresponding to the eightfold supersaturation, Wilson has proved that the amount of supersaturation required to produce the cloud is the same in air, oxygen, hydrogen and carbonic acid; the size of the nuclei is therefore the same in all these gases; it is thus very improbable that they consist of aggregations of the molecules of the gas; it would seem most likely that they are minute drops of water which are continually being formed from the saturated vapour and then evaporating, but lasting sufficiently long to enable them to be caught during the sudden expansion, and to act as the nuclei round which the drops in the cloud condense. These minute drops of water are not however all of the same size, for after passing the expansion 1·38 the density of the cloud increases very rapidly as the expansion increases, showing that many more nuclei become efficient when the expansion increases. This behaviour of the cloud indicates that there are little drops of water of different sizes, the small ones being more numerous than the larger ones, and that there is a fairly definite limit to the size of the drop, the number of drops whose size exceeds this limit being too small to produce an appreciable cloud. This collection of drops of different sizes is what we might expect if we regard the little drops as arising from coalescence of molecules of water vapour, and the larger drops from the coalescence of the smaller ones.

95. The fact that the drops are of different sizes indicates that they are not in a state of equilibrium with regard to evaporation and condensation, and the drops have probably a very ephemeral existence. The following considerations show that on the view of the relation between surface tension and the thickness of water films, to which Reinold and Rücker were led by their experiments on very thin films, drops of pure water of a definite radius might be in equilibrium with saturated water vapour even if they were not charged.

Effect on the Condensation of Variation of Surface Tension with the Radius of the Drop.

96. When a liquid film gets very thin its surface tension no longer remains constant but depends upon its thickness. For very thin films theory indicates (see Lord Rayleigh, *Phil.*

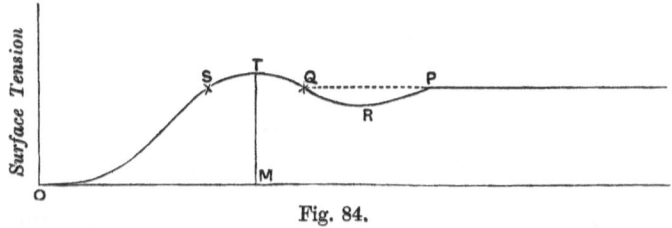

Fig. 84.

Mag. v. 33) that the surface tension would be proportional to the square of the thickness, while the experiments of Reinold and Rücker show that the surface tension has a maximum value for a thickness comparable with that of the dark spot in soap films; more recent experiments by Johannot have made it probable that, considering the surface tension as a function of the thickness, there is more than one maximum. Taking for simplicity the case when there is only one maximum the relation between the surface tension and the thickness is represented by a curve of the character of Fig. 84, the ordinates representing the surface tension, and the abscissæ the thickness.

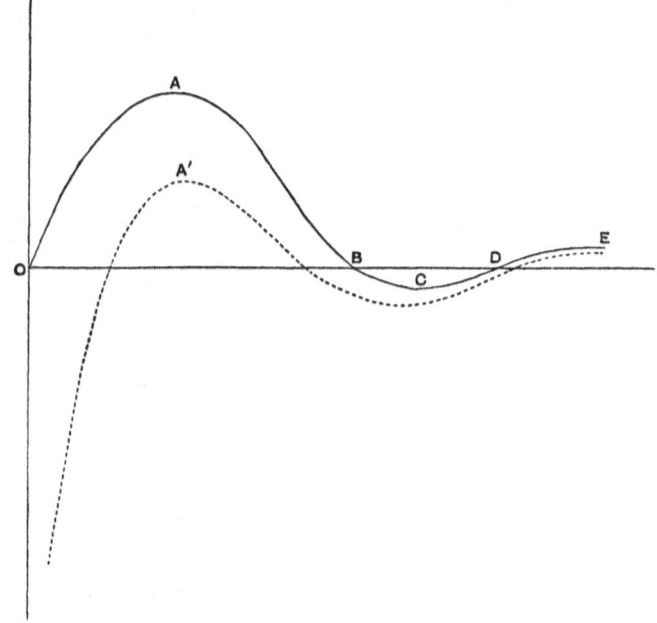

Fig. 85.

When the surface tension varies with the radius of the drop equation (1) becomes

$$\sigma R\theta \log_e \frac{P}{p} = \frac{2T}{r} + \frac{dT}{dr} - \frac{e^2}{8\pi K r^4} \quad \ldots\ldots\ldots(5).$$

From Fig. 84, taking r to represent the thickness of the film, we see that when $r = 0$, $\frac{2T}{r} + \frac{dT}{dr}$ vanishes, and that this quantity will attain a maximum and then diminish as r increases; we shall take the case when it goes on diminishing until it vanishes and changes sign. Then $\frac{2T}{r} + \frac{dT}{dr}$ will be represented by a curve of the type of Fig. 85.

Let us first take the case when the drop is unelectrified. Equation (5) shows that the radius of a drop when the vapour pressure is p can be got by drawing a horizontal line at a distance $\sigma R\theta \log_e \frac{p}{P}$ above the horizontal axis of coordinates and finding where it cuts the curve.

We see now, when we take the variation of surface tension into account, that even the slightest supersaturation will produce some condensation, but that these drops will be exceedingly small[1], and will evaporate as soon as the supersaturation ceases. Let us suppose however that we produce enough supersaturation to carry the drop represented by the point A on Fig. 85. When once the drop has got past this point the equilibrium vapour pressure gets smaller as the drop gets bigger, so that water will condense on the drop and the drop will increase in size, but the bigger it gets the smaller the equilibrium vapour pressure and the faster it grows; in fact the region between A and C is unstable, and when once the drop has got past A it will become large enough to be visible. This I think is the explanation of the dense cloud which C. T. R. Wilson found is produced even in the absence of ions by an eightfold supersaturation—this supersaturation is required to carry the drop past A.

Let us now consider the reverse process, the evaporation of a drop already formed. To fix our ideas let us suppose that the drop is originally in an atmosphere saturated with water vapour and that the temperature is gradually raised so that the surrounding atmosphere is no longer saturated. The drop will evaporate until it gets to the state represented by the point E; now evaporation becomes more difficult because as the drop gets smaller the equilibrium vapour pressure gets smaller and to get the drop past the state represented by the point C will require a finite rise in temperature (*i.e.* will require the temperature to be raised until the amount of water vapour in the air is only a definite fraction of the saturation amount at that temperature), just as in the reverse process it required the temperature to be lowered a finite amount to carry the drop past A. If the rise in temperature is not

[1] Of course drops would not be formed if the radius indicated by theory were smaller than the radius of a water molecule.

sufficient to carry the drop past C the drop will not wholly evaporate, but will be in equilibrium when its radius is between OD and OC, thus all these residual drops will be within comparatively narrow limits of size; when the temperature of the drop is the same as that of the air all these drops will have a radius OD. These small and invisible drops of water will greatly facilitate the formation of a cloud when next supersaturation takes place, since these drops to grow to visibility will only require the supersaturation corresponding to E, while if they were not present to begin with, the supersaturation required to produce a cloud would be that corresponding to A. This explains the well-known fact that when once a cloud has been formed, it only requires very slight supersaturation to produce another after a short interval.

So far we have only been dealing with uncharged drops; when the drops are charged we must use

$$\frac{2T}{r} + \frac{dT}{dr} - \frac{e^2}{8\pi K r^2}$$

instead of $\quad\dfrac{2T}{r} + \dfrac{dT}{dr}$;

the dotted curve represents the graph of this quantity, the supersaturation required to produce visibility corresponds to A' and is less than that for an uncharged drop.

The principles of the preceding theory ought to apply to the phenomena attending the supersaturation of salt solutions; we see that only a finite and definite amount of supersaturation could occur without deposition of salts, and that amount would be diminished by the presence of ions.

Difference between the Actions of Positive and Negative Ions in producing Condensation.

97. The production of electrification by the splashing of drops and bubbling through water suggests that at the surface of a drop of water there is a double layer of electrification, *i.e.* a layer of one sign at the surface of the drop and a layer of the opposite sign in the gas, the distance between the two layers being very small. If a layer of this kind existed it would produce a difference between the condensing powers of positive and negative ions. Perhaps the easiest way to see this is to notice that by equation (2) $\sigma R\theta \log_e \frac{P}{p}$ is equal to the excess of pressure at the surface of the water over the atmospheric pressure; for $2T/r$ is the pressure due to the surface tension, while $\frac{KR^2}{8\pi}$ or $\frac{e^2}{8\pi Kr^4}$ is the *tension* due to the electric field. If there is a double layer at the surface of the drop the expression for the tension must be modified; if V is the difference of potential between these layers due to their charges, and d the distance between the layers, then even when the drop is electrically neutral there is a tension equal to

$$\frac{K}{8\pi} \frac{V^2}{d^2}$$

on the surface of the water, so that

$$\sigma R\theta \log_e \frac{P}{p} = \frac{2T}{r} - \frac{1}{8\pi} \frac{KV^2}{d^2} \quad \ldots\ldots\ldots\ldots (6).$$

Now suppose the drop of water has a charge e, the electric force at the surface of the water will be

$$\frac{V}{d} + \frac{e}{Kr^2},$$

and the tension on the surface will be

$$\frac{K}{8\pi} \left\{ \frac{V}{d} + \frac{e}{Kr^2} \right\}^2.$$

Thus
$$\sigma R\theta \log_e \frac{P}{p} = \frac{2T}{r} - \frac{K}{8\pi} \left(\frac{V}{d} + \frac{e}{Kr^2} \right)^2$$
$$= \frac{2T}{r} - \frac{1}{8\pi} \frac{e^2}{Kr^4} - \frac{V}{4\pi d} \frac{e}{r^2} - \frac{K}{8\pi} \frac{V^2}{d^2}.$$

Comparing this expression with (6) we see that the effect of the charge is to diminish the right-hand side by

$$\frac{1}{8\pi} \frac{e^2}{Kr^4} + \frac{Ve}{4\pi dr^2};$$

now if $\frac{Ve}{4\pi dr^2}$ is positive the effect of the double layer will be to promote condensation, while if this term is negative the layer will retard condensation. Thus the layer will make an ion of the sign which produces at the surface of the drop an electric field in the same direction as that due to the double layer, more effective as

334   ON SOME PHYSICAL PROPERTIES OF GASEOUS IONS   [97

a condenser than an ion of the opposite sign. Thus if the double layer in the case of water had the negative coating inside, a negative ion would be more efficient in producing condensation than a positive one. When a fresh surface of water is exposed to air we have seen that the air gets negatively electrified; we may regard this as indicating that an equal quantity of positive goes to the water surface to form the outer coating of the double layer, so that this double layer has the negative side next the water, the positive side next the air.

We know that in some liquids the electrification produced by bubbling air through them is positive instead of negative; in such liquids the outer coating of the double layer should be negative, and for these the positive ion should be a better condenser than the negative. Przibram[1] has shown that clouds formed in the vapour of alcohol and of some other organic substances condense more easily on positive ions than on negative; he considers however that the difference is due to differences in the relative sizes of the ions.

Laby[2] and Andrén[3] have measured the expansions required to give clouds in the vapours of a number of organic substances.

97·1. A most important development of the cloud method is due to C. T. R. Wilson[4], who has used the property of condensation on ions to make visible the ionisation produced by various agencies such as α-rays, fast electrons and X-rays. His method is to make an expansion immediately after the gas in the chamber has been ionised, and to take an instantaneous photograph of the drops formed on the ions before the air currents in the chamber have had time to move them appreciably from the positions in which they were formed. In this way the distribution of the ionisation due to the rays can be studied in a peculiarly direct manner. The expansion chamber used was generally larger than that in the experiments described above; thus in some recent work it was a flat cylinder 3 cm. high and 16·5 cm. in diameter. The plunger, whose movement produces the expansion, was of brass and flat,

1 Przibram, *Wien. Bericht.* Feb. 1906.
2 Laby, *Phil. Trans.* A, ccviii. p. 445, 1908.
3 Andrén, *Ann. der Phys.* (4), lii. p. 1, 1917.
4 *Proc. Roy. Soc.* A, lxxxv. p. 285; lxxxvii. p. 277; civ. pp. 1, 192.

forming the whole base of the chamber. In order to determine the distribution in space of the drops, photographs were taken simultaneously from two directions and the result viewed stereoscopically. An electric field of about 3 volts per cm. was maintained between the top and bottom of the chamber. This served to remove old ions which would otherwise cause confusion.

With this apparatus the track of each α-ray or fast electron appears as a line of drops. In the case of the α-ray the line is straight for most of its length, and the drops are densely crowded along it so that it appears as a continuous streak. In some cases clearly defined bends occur showing that the ray has been appreciably deflected by a single collision. Frequently a short spur occurs at the bend attributed to the recoil of the atom causing the deflection. Occasionally the track forks, usually near the end, which is attributed to the ray having given sufficient energy to an atom of the gas in the chamber for it also to ionise (Fig. 1, Plate II).

With very fast electrons the tracks are also straight, but the drops are much more sparsely spaced along it, as would be expected from the fact that such electrons produce far fewer ions per cm. of path than an α-ray. The slower electrons show irregularly curved tracks with the drops spaced at irregular intervals, there being usually a concentration at the end of the track (Fig. 2, Plate II).

The passage of a beam of X-rays through the chamber shows up as a network of tracks identical with those of electrons whose speed is the greater the harder the X-rays. This shows that the great bulk of the ionisation by X-rays is indirect, the effect of the X-rays being to eject from the molecules of the gas a few fast electrons which then produce the bulk of the ionisation by collision with other molecules. In addition to these long tracks Wilson finds a number of isolated short ones, some visible merely as spheres, others, which he calls 'fish tracks,' consisting of a fine trail of drops leading up to a collection of drops as a head. These last two classes merge into each other. They are attributed to much slower electrons produced either by secondary X-rays excited in the gas or as a kind of 'recoil' by the scattering of the primary X-rays (Fig. 5, Plate II).

In order to increase the distance between successive ions Wilson has worked at pressures much below atmospheric, down to a final pressure of about 10 cm. of mercury. In this way it has been possible to detect each individual pair of ions, the positives and negatives being slightly separated by the action of the field during the short interval between ionisation and taking the photograph. Such a track is shown in Fig. 4, Plate II.

The importance and beauty of this method of investigation can hardly be exaggerated, as it gives such a direct insight into the actual formation of the ions which can be studied individually,

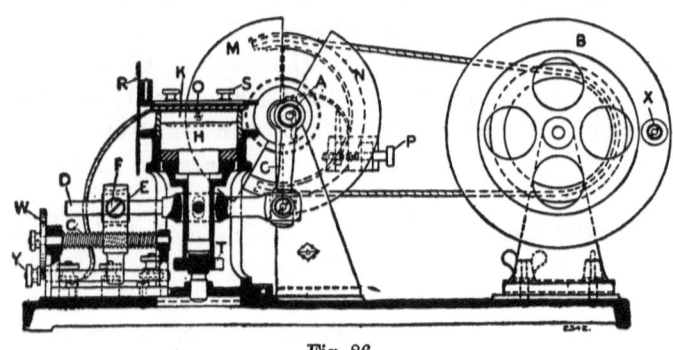

Fig. 86.

and not merely statistically as in almost all other methods. The consequences deduced from it and the interpretation of the different kinds of track will be discussed in detail when we come to consider the mechanism of the different types of ionisation concerned.

As some of the most interesting phenomena are of very rare occurrence it is important to be able to make a large number of experiments in a reasonable time. Shimidzu[1] has devised a modification of the apparatus in which the expansion is made by moving the piston mechanically with a reciprocating motion, which obviates the necessity of resetting it by hand each time. This was especially intended for the investigation of α-ray tracks and in this case the source of ionisation is kept permanently in position, the old ions being removed by switching on a strong electric field between each expansion. Two or three expansions

[1] Shimidzu, *Proc. Roy. Soc.* xcix. p. 425, 1921.

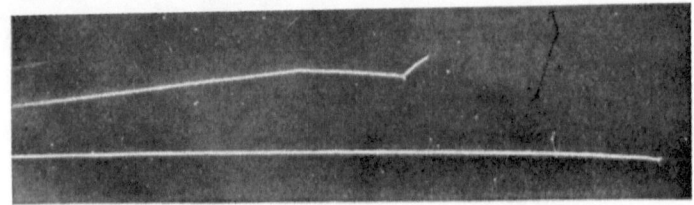

Fig. 1. Tracks of α-rays.

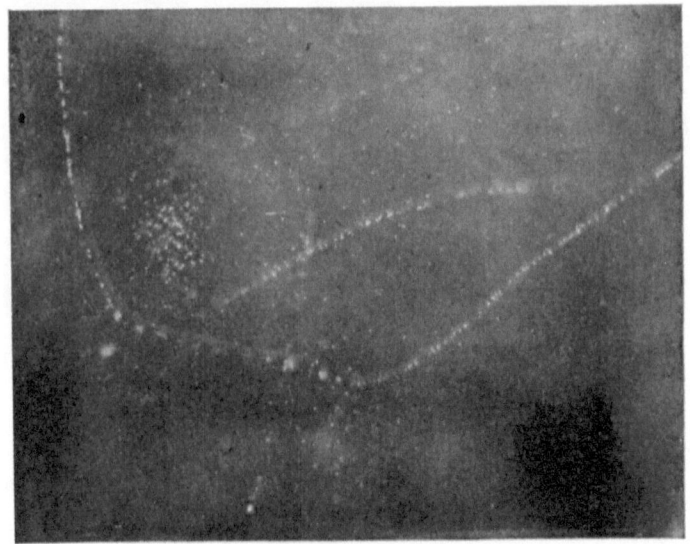

Fig. 2. Fast β-ray tracks.

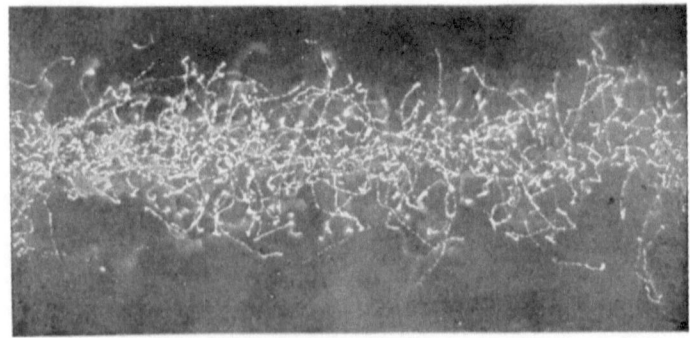

Fig. 3. Beam of X-rays.

PLATE II

Fig. 4. Pairs of ions (magnified).

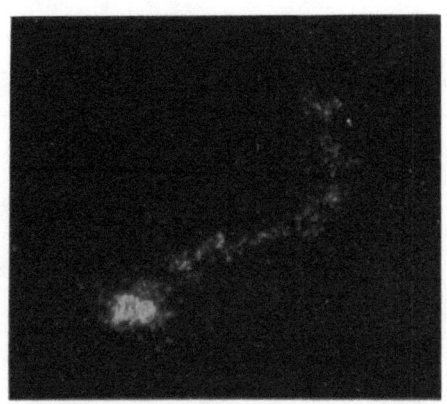

Fig. 5. 'Fish' track (magnified).

can be made per second, and the tracks are photographed by two cinematograph cameras viewing the chamber from directions at right angles. The apparatus is shown in Fig. 86. Here H is the piston, and K the space in which the expansion occurs, while the amount of expansion can be altered by moving the pivot F about which the arm D oscillates. This is done by means of the screw C.

CHAPTER IX

IONISATION BY INCANDESCENT SOLIDS

98. WE shall now proceed to the study of some special cases of ionisation, beginning with that due to incandescent metals. That the air in the neighbourhood of red-hot metals is a conductor of electricity has been known for about two centuries; the earliest observations seem to have been made by Du Fay[1] in 1725, by Du Tour[2] in 1745, by Watson[3] in 1746, by Priestley[4] in 1767, and by Cavallo[5] in 1785. Becquerel[6] in 1853 showed that air at a white heat would allow electricity to pass through it even when the potential difference was only a few volts. Blondlot[7] confirmed and extended this result, and proved that air at a bright red heat was unable to insulate under a difference of potential as low as 1/1000 of a volt; he showed, too, that the conduction through the hot gas was not in accordance with Ohm's law. Recent researches have thrown so much light on the causes at work in the ionisation of gases in contact with glowing solids, that it is unnecessary to enter into these earlier investigations in greater detail. Guthrie[8] seems to have been the first to call attention to one very characteristic feature of ionisation by incandescent metals, *i.e.* the want of symmetry between the effects of positive and negative electrification. He showed that a red-hot iron ball in air could retain a charge of negative but not of positive electrification, while a white-hot ball could not retain a charge of either positive or negative electrification.

[1] Du Fay, *Mém. de l'Acad.* 1733.
[2] Du Tour, *Mém. de Mathématique et de Physique,* xi. p. 246, 1755.
[3] Watson, *Phil. Trans.* abridged, x. p. 296.
[4] Priestley, *History of Electricity,* p. 579.
[5] Cavallo, *Treatise on Electricity,* i. p. 324.
[6] Becquerel, *Annales de Chimie et de Physique,* iii. 39, p. 355, 1853.
[7] Blondlot, *Comptes Rendus,* xcii. p. 870, 1881; civ. p. 283, 1887.
[8] Guthrie, *Phil. Mag.* iv. 46, p. 257, 1873.

99. The ionisation produced by incandescent metals was investigated systematically in great detail by Elster and Geitel[1], who used for this purpose the apparatus represented in Fig. 87. This is a glass vessel containing an insulated metal plate A, which is connected with one pair of quadrants of an electrometer. Underneath this plate there is a fine metallic wire, which can be raised to incandescence by an electric current passing through the leads C, D; to prevent any disturbing effects arising from the

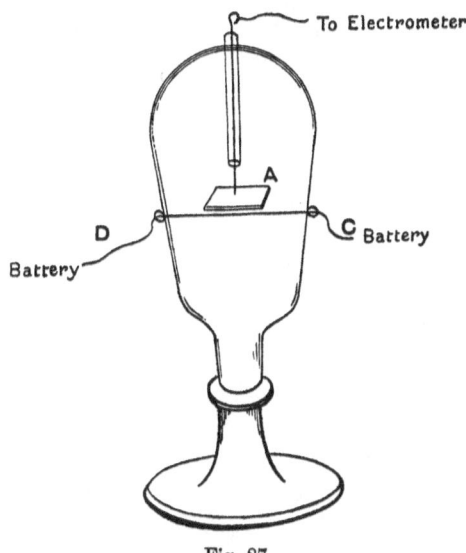

Fig. 87.

change produced by the current in the electric potential of the wire, the middle point of the wire was connected with the earth. Let us first take the case when the gas in the vessel is air or oxygen at atmospheric pressure, then, as soon as the glow of the hot wire begins to be visible, the metal plate receives a *positive* charge; this charge increases until the potential of the plate reaches a value which varies very much with the dimensions of the apparatus used; in Elster and Geitel's experiments it was of the order of a few volts. This potential increases as the

[1] Elster and Geitel, *Wied. Ann.* xvi. p. 193, 1882; xix. p. 588, 1883; xxii. p. 123, 1884; xxvi. p. 1, 1885; xxxi. p. 109, 1887; xxxvii. p. 315, 1889; *Wien. Bericht.* xcvii. p. 1175, 1889.

temperature of the wire increases, until the wire is at a yellow heat; at this stage the potential of the plate is a maximum. After passing this stage the potential diminishes as the wire gets hotter and hotter, until at a bright white heat the charge received by the plate is very small.

The electrification on the plate is very much influenced by the pressure of the gas. Starting at atmospheric pressure and gradually exhausting the vessel, we find that at first the change of pressure does not produce any great effect upon the potential of the plate A, but when we approach very high exhaustion, such as that in Crookes' tubes, the potential of the plate begins to diminish, until at very low pressures it changes sign and may as the exhaustion proceeds reach a very large negative value. The pressure at which the change in sign of the electrification of the plate takes place depends upon the temperature of the wire, the higher the temperature the higher the pressure at which the reversal of the electrification occurs. Again, long-continued incandescence of the wire favours the negative electrification of the plate; the physical condition of the platinum wire is changed by long-continued heating, and the wire becomes brittle. The following experiment, due to Elster and Geitel[1], seems to indicate that the gases absorbed in the platinum wire and which are gradually, but only very gradually, expelled by long-continued heating, play a considerable part in the electrical phenomena connected with the incandescence of metals. They found that if the platinum wire was kept glowing in a fairly good vacuum long enough for the metal plate to receive a negative charge, the introduction of a very small quantity of fresh gas reversed the sign of electrification on the metal plate, and the pressure had to be reduced far below the original value for the negative electrification to be recovered.

100. The effects are also complicated by the dust and vapour given off by the glowing platinum, which form a deposit on the walls of the vessel. The production of this dust can very easily be shown by the study of clouds formed by the method described in Chap. VIII. If a fine platinum wire is fused into the expansion apparatus, and the air rendered dust-free in the usual way, so that no clouds are produced by an expansion less than 1·25, dense

[1] Elster and Geitel, *Wien. Bericht.* xcvii. p. 1175, 1889.

clouds will be formed by comparatively small expansions after a current has been sent through the wire strong enough to raise it to incandescence[1]; indeed it is not necessary to make the wire so hot as to be luminous, an increase in the temperature of the wire to 200 or 300° C. is sufficient to produce the cloud.

The sign of the electrification produced by glowing substances is influenced by the nature of the substances and of the gas surrounding them; thus in hydrogen Elster and Geitel[2] showed that the plate above the incandescent wire became negatively electrified even when the hydrogen was at atmospheric pressure. This electrification continually increased with the temperature. To get the negative electrification, however, the wire must be at least at a bright yellow heat; at lower temperatures the electrification is positive; a clean copper wire, on the other hand, gives a positive electrification in hydrogen, unless the pressure is very low.

Elster and Geitel showed that the sign of the electrification in water vapour and the vapours of sulphur and phosphorus was the same as in air; they could detect no electrification in mercury vapour.

101. The influence of the nature of the incandescent substance is shown by the fact that with incandescent carbon filaments the electrification on the metal plate is always negative. It is also shown clearly by some experiments made by Branly[3]. Branly's method was as follows: he hung up a charged insulated conductor in the neighbourhood of the incandescent body; he found that when the latter was a piece of platinum at a dull red heat the insulated conductor lost a negative but not a positive charge; when the platinum was white hot the conductor was discharged whether electrified positively or negatively. If the incandescent body was an oxide and not a pure metal, at any rate if it was an oxide of one of the metals tried by Branly, viz. lead, aluminium or bismuth, then it would discharge a positively electrified body but not a negatively electrified one, which is exactly opposite to the effect produced by a pure metal at a dull red heat.

[1] R. v. Helmholtz, *Wied. Ann.* xxxii. p. 1, 1887. Lodge, *Nature*, xxxi. p. 267, 1884.
[2] Elster and Geitel, *Wied. Ann.* xxxi. p. 109, 1887.
[3] Branly, *Comptes Rendus*, cxiv. p. 1531, 1892.

102. McClelland[1] sucked the gases from the neighbourhood of the incandescent wire and then investigated their properties. He found that as soon as the wire began to glow the gas would discharge a negatively but not a positively electrified body; when the temperature of the electrified body was increased by about 400° C. the gas began to discharge a positively electrified body, though not so freely as it did a negatively electrified one; when the wire got to a bright yellow heat the gas discharged both positive and negative electricity with equal facility. McClelland investigated the laws of conduction of electricity through the gas which had been in contact with the glowing wire; he found that it showed all the characteristics of conduction through a gas containing ions; thus the relation between the current and the electromotive force is represented by a curve like Fig. 5, the current soon reaching saturation. McClelland also determined the velocity in an electric field of the ions produced by the incandescent metal. He found that their velocity was small compared with that of the ions produced by X-rays, and that the hotter the wire the *smaller* was the velocity of the ions.

102·1. When the experiments are made *in vacuo* the nature of the ions is very different. The ratio of the charge e to the mass m of the carriers of negative electricity from an incandescent wire has been determined by the method of Art. 67. The results, given in Art. 69, are conclusive in showing that the value of e/m for these ions is the same as its value for cathode rays and for the carriers of negative electricity from metals placed in a good vacuum and illuminated by ultra-violet light. Thus the negative electric discharge from a hot wire is due to the emission of electrons.

The small mobilities found by McClelland for the negative ions must be attributed to the loading up of the electrons originally formed by molecules of air and probably also by particles sputtered from the hot metal. The mobilities of the positive ions, the existence of which depends on the presence of traces of salts on the metal, are considered further on p. 397.

103. The account we have already given of the effects observed in the neighbourhood of an incandescent wire shows that the

[1] McClelland, *Proc. Camb. Phil. Soc.* x. p. 241, 1900.

electrification produced in this way is a very complicated phenomenon, and depends:

(1) On the temperature of the wire.
(2) On the pressure of the gas around the wire.
(3) On the nature of the gas.
(4) On the nature of the incandescent wire.

We shall simplify the investigation of the cause of this electrification if we study a case in which as many as possible of these effects are eliminated. Now (2) and (3) are eliminated if we work with the highest attainable vacuum; in this case the phenomena are greatly simplified and exhibit points of remarkable interest. To investigate them we may use a piece of apparatus like that shown in Fig. 88. It consists of a straight piece of fine wire AB,

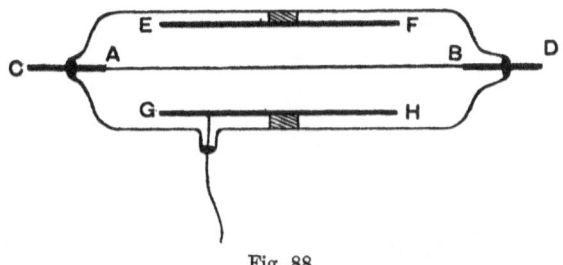

Fig. 88.

which can be heated to any desired temperature by an electric current led in through the leads CA, DB. Around this wire and insulated from it is a metallic cylinder, shown in section in EF and GH; this cylinder should be longer than, and coaxial with, the wire. This system is sealed into a glass vessel connected with an air-pump and the pressure reduced as low as possible. It is desirable to keep the wire red hot for a very considerable time (I have found a week not too long), in order to expel gases absorbed in the wire; until these are got rid of the behaviour of the wire is very irregular. The vessel should be pumped from time to time while the wire is hot, to get rid of the gases coming out of the wire; it will be necessary to exhaust the vessel from time to time, even after these have been expelled, as the heat coming from the wire seems to liberate gas from the walls of the glass vessel and the metal cylinder. Connect the hot wire to one terminal of a battery and the cylinder to the other, and place in the circuit a sensitive galvanometer. If now the wire be made red hot and connected with the negative pole of the battery, an appreciable current will go through the galvanometer; if, however, the terminals are reversed so that the hot wire is connected with the positive pole of the battery, the current which passes is too small to be detected by the galvanometer; thus there can be a current through the exhausted vessel when the negative electricity goes from the hot wire to the cold cylinder, but not an appreciable one when the positive electricity would have to go from the wire to the cylinder; the system can thus transmit a current in only one direction. The current does not obey Ohm's law: at first it increases with the electromotive force, but it soon reaches a saturation value beyond which it does not increase, even though the electromotive force is increased, provided the increase in the electromotive force is not sufficient to enable the electric field itself to ionise the gas.

103·1. The value of the field required to produce saturation in a vacuum increases with the temperature of the filament. With a given difference of potential a stage is reached when further increase in the temperature causes no increase in the current flowing. This is due to the effect of the volume charge of the electrons in opposing the current, so that this is determined by the power of the electric field to get the electrons away from the neighbourhood of the hot wire and not by the number which the wire is capable of emitting.

104. The saturation current increases very rapidly with the temperature. This is well shown by the curve in Fig. 89, which represents the results of the experiments made by O. W. Richardson[1], in the Cavendish Laboratory, on the saturation current between a hot platinum wire and a metal cylinder surrounding it in a high vacuum. The temperatures were obtained by measuring the resistance of the wire. Richardson found that the relation

---

[1] O. W. Richardson, *Proc. Camb. Phil. Soc.* xi. p. 286, 1902; *Phil. Trans.* cci. p. 516, 1903.

between the saturation current I and the absolute temperature θ could be expressed by an equation of the form

$$I = a\theta^{\frac{1}{2}} \epsilon^{-\frac{b}{\theta}};$$

for the curve in Fig. 89,

$$a = 1\cdot 51 \times 10^{26}, \quad b = 4\cdot 93 \times 10^4.$$

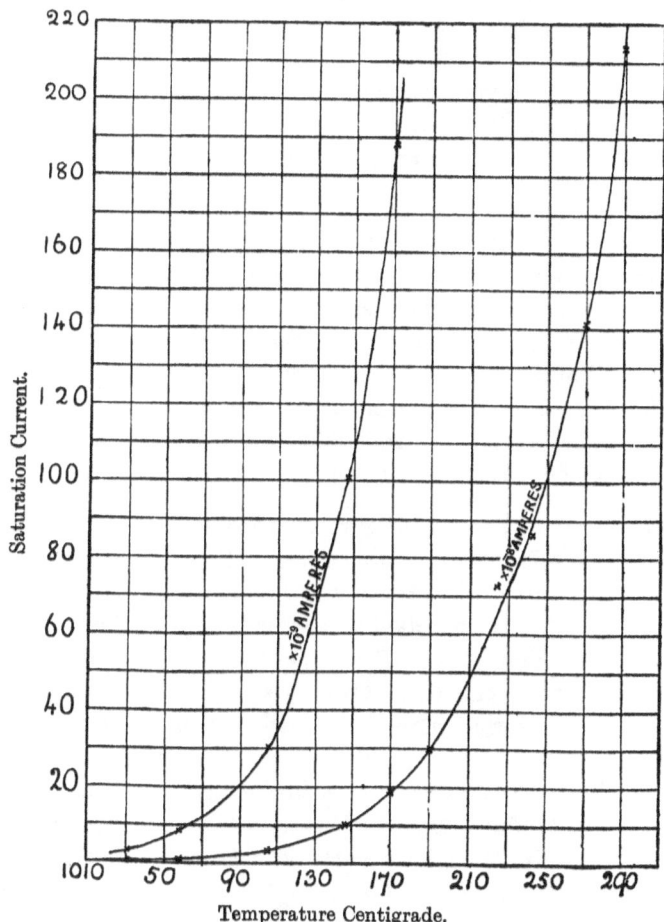

Fig. 89.

In the case of this wire the current amounted to about 4×10^{-4} amperes at the temperature $1500°$ C., which represents a rate of emission of negative electricity from the hot wire of above one milliampere per square centimetre of surface. If the same formula held up to the melting point of platinum, which we shall take to be 2000° C., the rate of emission of negative electricity from the glowing wire would be about 1/10 of an ampere per square centimetre.

The rate of escape of negative electricity from glowing carbon in some cases greatly exceeds that from glowing platinum. This is no doubt chiefly owing to the fact that the carbon can be raised to a much higher temperature than the platinum. Richardson has obtained from carbon filaments in a good vacuum currents of the order of an ampere per square centimetre of surface.

104·1. Tungsten, which melts at 3270° C. and is the most refractory material known, can give enormous currents. Thus Richardson records a current of 0·4 amp. from a wire which took 0·8 amp. heating current. In this case the thermionic current density was 4 amp. per sq. cm. Richardson[1] has investigated the leak from hot sodium and found that it is very much greater than that from platinum at the same temperature and could be detected at temperatures as low as 200°–300° C. With most substances no current can be detected with a galvanometer below about 1000°.

104·2. It was early discovered that some compounds emit negative electricity even more copiously than most metals at the same temperature. Thus Wehnelt[2] found very large emissions from the oxides of the alkaline earths. Horton showed that the emission from lime between 700° C. and 1400° C. was much greater than that from calcium. Owen[3] and Horton examined the emission from Nernst filaments, which is smaller than that from the alkaline earths. In all these cases the emission varies with temperature in the same way as for metals. Wehnelt used his discovery to produce large discharges in gases with small potential differences. By using as a cathode a strip of platinum heated by an auxiliary current and with a little lime or barium oxide on it, the cathode fall of potential is greatly reduced owing to the electrons emitted neutralising the positive space charge.

[1] Richardson, *Phil. Trans.* cci. p. 516, 1903.
[2] Wehnelt, *Ann. der Phys.* xiv. p. 425, 1904.
[3] Owen, *Phil. Mag.* viii. p. 230, 1904.

There was some difficulty in the early experiments with coated metals owing to the oxide splitting off from the metal. Filaments are prepared by the Western Electric Co. of America which have overcome this difficulty. Deininger[1] found that the emission from wires of platinum, carbon tantalum and nickel covered with lime was the same in all cases. Owing to the commercial importance in wireless telegraphy of valves using a hot wire as a source of electrons, many experiments have been made with the object of obtaining wires with a high emission at a low temperature. It is found that tungsten can be covered with a thin film of caesium an atom thick, which shows the very high emission to be expected of caesium. Similar films can be obtained from thorium originally present in the metal in the form of oxide and made to diffuse to the surface, and be reduced by suitable heat treatment[2]. It is even possible to obtain wires in which a thin, probably monomolecular layer of oxygen is covered by the monatomic layer of caesium.

Richardson[3] has found that other compounds beside oxides have the power of emitting electrons when heated, and considers that it is probably a common property of all forms of matter stable enough to exist at a high temperature. He found that the following salts emitted electrons at a comparatively low temperature: the iodides of calcium, strontium, barium and cadmium, calcium fluoride, calcium bromide, manganous chloride and ferric chloride. In these cases there was, in addition to the electronic emission, an emission of negative ions of atomic dimensions. The proportion varied with the temperature and other conditions, the tendency being for the proportion of electrons to increase with rising temperature.

105. The escape of negative electricity from glowing carbon in high vacua is the cause of an effect observed in incandescent electric lamps, known as the Edison effect, and which has been studied by Preece[4] and in great detail by Fleming[5]. The 'Edison

[1] Deininger, *Ann. der Phys.* xxv. p. 285, 1908.
[2] Langmuir, *Phys. Rev.* xxii. p. 357, 1923.
[3] Richardson, *Phil. Mag.* xxvi. p. 458, 1913.
[4] Preece, *Proc. Roy. Soc.* xxxviii. p. 219, 1885.
[5] Fleming, *Proc. Roy. Soc.* xlvii. p. 118, 1890; *Phil. Mag.* xlii. p. 52, 1896.

effect' is as follows: Suppose that ABC represents the carbon filament of an incandescent lamp, and that an insulated metal plate is inserted between the filaments; then if the positive end A of the filament is connected with a wire D leading from the metallic plate and a galvanometer inserted between A and D, a considerable current, amounting in some of Fleming's experiments to three or four milliamperes, passes through the galvanometer, the direction of the current being from A to D through the galvanometer. If, however, the metal plate is connected with the negative electrode of the lamp and a galvanometer inserted in this circuit, the current through the galvanometer is exceedingly small compared with that observed in the preceding case. We see that this is what would occur if there was a vigorous discharge of negative electricity from the negative leg of the carbon filament, and no discharge or a much smaller one from the positive leg; this would tend to make the potential of the metal plate differ but little from that of the negative leg of the carbon loop, while the difference of potential between the positive leg and the plate would be nearly that between the electrodes of the lamp, and consequently the current through a circuit connecting the positive electrode to the metallic plate would be much greater than through one connecting the negative electrode to the plate.

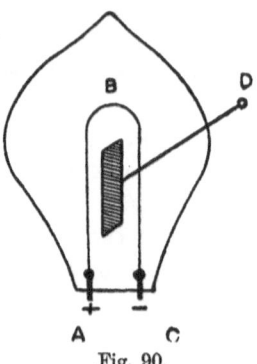

Fig. 90.

Fleming showed that when the negative leg of the carbon loop was surrounded by a cylinder made either of metal or of an insulating substance, the Edison effect disappeared almost entirely. Fleming too found, as Elster and Geitel had previously shown by a somewhat different method, that a current of electricity could pass between an incandescent carbon filament and a cold electrode, if the direction of the current was such as to cause the negative electricity to pass from the hot filament to the cold plate, and that a current would not pass in the opposite direction. Elster and Geitel showed, too, that a plate placed near an incandescent filament received, even in very high vacua, a charge of negative electricity. The behaviour of the hot filament shows that it, like the incandescent platinum wire, emits negative electrification. That the emission from the carbon filament is much greater than that from the platinum wire—great as we have seen the latter to be—is shown by the fact that although, as Fleming (*loc. cit.*) has shown, the 'Edison effect' can be observed with an incandescent platinum wire in place of the carbon filament, the effect from platinum is exceedingly small compared with that from carbon, and is only appreciable when the platinum is so hot that it is on the point of melting.

Selenyi[1] has shown the existence of an appreciable electron current in an ordinary electric lamp by observing a decrease in the current in a strong magnetic field.

106. We have been led to the conclusion that from an incandescent metal or glowing piece of carbon electrons are projected, and the rate of emission amounts in the case of a carbon filament at its highest point of incandescence to a current equal to several amperes per square centimetre of surface. This fact may have an important application to some cosmical phenomena, since, according to the generally received opinion, the photosphere of the sun contains large quantities of glowing carbon; this carbon will emit electrons unless the sun by the loss of its electrons at an earlier stage has acquired such a large charge of positive electricity that the attraction of this is sufficient to prevent the negatively electrified particles from getting right away from the sun; yet even in this case, if the temperature were from any cause to rise above its average value, electrons would stream away from the sun into the surrounding space. We may thus regard the sun, and probably any luminous star, as a source of negatively electrified particles which stream through the solar and stellar systems. Now when electrons moving at a high speed pass through a gas they make it luminous; thus when the electrons from the sun meet the upper regions of the earth's atmosphere they will produce luminous effects. Arrhenius[2] has shown that we can explain in a satisfactory manner many of the periodic variations in the Aurora Borealis if we assume that it is caused

[1] Selenyi, *Zeitschrift f. techn. Phys.* v. 9, p. 412, 1924.
[2] Arrhenius, *Physikalische Zeitschrift*, ii. pp. 81, 97, 1901.

by electrons from the sun passing through the upper regions of the earth's atmosphere.

The emission of electrons from incandescent metals and carbon is readily explained by the view—for which we find confirmation in many other phenomena—that electrons are disseminated through metals and carbon, not merely when these are incandescent, but at all temperatures; the electrons being so small are able to move freely through the metal, and they may thus be supposed to behave like a perfect gas contained in a volume equal to that of the metal. The electrons are attracted by the metal, so that to enable them to escape into the space surrounding it they must have sufficient kinetic energy to carry them through the layer at its surface, where its attraction of the electrons is appreciable. If the average kinetic energy of an electron like that of the molecule of a gas is proportional to the absolute temperature, then as the temperature increases, more and more of the electrons will be able to escape from the metal into the air outside.

Rate at which the Electrons escape from the Metal.

107. We can without much difficulty find an expression for this quantity if we assume that the electrons in the metal behave like a perfect gas. Let AB, CD represent two planes parallel to the surface of the metal including between them the region in which the metal exerts an appreciable force upon the electron. Let us take the axis of x at right angles to these planes, the positive direction of x being from the air to the metal; then if p is the pressure due to the electrons, n the number of electrons in unit volume, X the force acting on an electron, we have when there is equilibrium

$$\frac{dp}{dx} = Xn \quad \ldots\ldots\ldots\ldots\ldots\ldots(1);$$

but if the electrons behave like a perfect gas $p = \beta\theta n$, where θ is the absolute temperature and β a constant which is the same for all gases. Substituting this value for p in equation (1), we get

$$\beta\theta \frac{dn}{dx} = Xn \quad \ldots\ldots\ldots\ldots\ldots\ldots(2);$$

integrating this equation from CD to AB, we get

$$\log \frac{n'}{N} = -\frac{w}{\beta\theta},$$

or
$$n' = N\epsilon^{-\frac{w}{\beta\theta}} \quad \ldots\ldots\ldots\ldots\ldots\ldots(3),$$

where n' and N are respectively the numbers of electrons in unit volume of the air and metal, and

$$w = \int X dx;$$

thus w is the work required to drag an electron out of the metal.

Equation (3) gives the number of electrons in the air when things have attained a steady state. To find the number of electrons coming from the metal in unit time let us proceed as follows: regard the steady state as the result of a dynamical equilibrium between the electrons going from the metal to the air and those going from the air to the metal. If n' is the number of electrons in unit volume of the air, the number which in one second strike against unit area of the metal is by the Kinetic Theory of Gases equal to

$$\sum_{0}^{\infty} u\,dn,$$

dn being the number of electrons which have velocities between u and $u + du$, and the summation is to be taken for all positive values of u. Now if n' is the total number of electrons in unit volume

$$dn = n'\sqrt{\frac{hm}{\pi}}\,\epsilon^{-hmu^2}\,du,$$

where m is the mass of an electron: hence

$$\sum_{0}^{\infty} u\,dn = n'\sqrt{\frac{hm}{\pi}} \int_{0}^{\infty} \epsilon^{-hmu^2} u\,du$$

$$= \frac{1}{2}\frac{n'}{\sqrt{\pi hm}}$$

$$= \frac{n'c}{\sqrt{6\pi}},$$

where c is the velocity of mean square and is equal to $\alpha\,(\theta/m)^{\frac{1}{2}}$,

a being a constant which is the same for all gases. Substituting the value of n' from equation (3) we find that the number of electrons coming from the air and striking against unit area of the metal in unit time is equal to

$$\frac{a}{\sqrt{6\pi}}\left(\frac{\theta}{m}\right)^{\frac{1}{2}} N\epsilon^{-\frac{w}{\beta\theta}}.$$

If we suppose that all the electrons which strike against the metal enter it, this will be the number of electrons entering the metal, and therefore in the steady state the number leaving it; the number may be written in the form

$$a\theta^{\frac{1}{2}}\epsilon^{-\frac{b}{\theta}}:$$

this number multiplied by e will be the quantity of negative electricity leaving unit area of the metal in unit time, and therefore will be the saturation current from a hot wire at the temperature θ. Richardson's measurements of the saturation current at different temperatures agree well, as we have seen, with a formula of this form. From the values of a and b determined by experiments on the escape of electricity from a hot wire we can deduce the values of N and w. Richardson finds that for platinum

$$a = 7\cdot 5 \times 10^{25}, \quad b = 4\cdot 93 \times 10^4;$$

this gives $N = 6\cdot 5 \times 10^{20}$ and $w = 8 \times 10^{-12}$ ergs.

In this experiment no special precautions were taken to rid the wire of gas beyond keeping the pressure below ·01 mm.

107·1. The above investigation was given in the second edition of this book and leads to the result obtained by Richardson theoretically and experimentally in his early work.

Modern views of the electron theory of metals are not in accordance with the view that the electrons inside a metal have the energy of a perfect gas at the temperature of the metal. It is therefore important to see how far the result can be deduced by thermodynamical arguments.

The following argument is given by Richardson[1]:

He considers an evacuated enclosure at constant temperature containing a body emitting electrons. The emission will continue till there is dynamic equilibrium between the electrons leaving the

[1] Richardson, *Emission of Electricity from Hot Bodies*, p. 30.

body and those returning to it. Let p be the pressure of electrons at which this occurs and suppose it so small that the mutual repulsion of the electrons, which is proportional to the square of their density, is negligible compared with the pressure due to the electrons regarded as a gas. Then $p = n'k\theta$. Suppose the enclosure provided with a cylindrical extension in which an insulating piston can move backwards and forwards, so that p can be made to do work. Consider the change in entropy dS due to a motion of the piston. If ϕ is the change of energy of the system which accompanies the transference of each electron from the hot body to the surrounding enclosure of volume v, then

$$dS = \frac{1}{\theta}\{d(n'v\phi) + p\,dv\}$$

$$= \frac{1}{\theta}\left\{\left(p + n'\phi + v\frac{\partial(n'\phi)}{\partial v}\right)dv + v\frac{\partial(n'\phi)}{\partial \theta}d\theta\right\}.$$

Thus
$$\left(\frac{\partial S}{\partial v}\right)_\theta = \left(p + n'\phi + v\frac{\partial(n'\phi)}{\partial v}\right)\Big/\theta,$$

$$\left(\frac{\partial S}{\partial \theta}\right)_v = \frac{v}{\theta}\frac{\partial(n'\phi)}{\partial \theta}.$$

Now dS is a total differential; thus equating the values of $\frac{\partial^2 S}{\partial v \partial \theta}$, we find $\theta\frac{\partial p}{\partial \theta} = p + n'\phi$, neglecting $\frac{\partial(n'\phi)}{\partial v}$ which is zero unless the piston is quite close to the emitting surface.

Substituting $p = n'k\theta$, we have

$$\frac{dn'}{n'} = \frac{\phi}{k\theta^2}d\theta,$$

or
$$n' = A_1 \epsilon^{\int^\theta \frac{\phi}{k\theta^2}d\theta},$$

where A_1 is independent of θ.

Applying the kinetic theory as before we find

$$i = eA_1\sqrt{\frac{k}{2\pi m}}\,\theta^{\frac{1}{2}}\,\epsilon^{\int^\theta \frac{\phi}{k\theta^2}d\theta},$$

where $\sqrt{3k}$ has been substituted for α in the previous expression.

This result assumes that all the electrons which return to the metal are absorbed by it. Richardson and v. Baeyer[1] have shown that it is not actually the case, a considerable proportion being

[1] Richardson and v. Baeyer, *Verh. der Deutsch. Phys. Gesell.* x. p. 96, 1908.

'reflected.' Richardson[1] states that the proportion is about 50 per cent. in many cases, so that in these cases the saturation thermionic current would be halved.

In this, as in the previous work, the mutual repulsion of the electrons has been neglected. While this is actually small at low temperatures, it is by no means necessarily so always. Laue has investigated the equilibrium of electron atmospheres in these cases and finds that the result given above for the density is generally applicable.

The above expression for i requires us to know the way in which ϕ varies with the temperature. The simplest assumption, that it is constant, leads to the formula $i = a\theta^{\frac{1}{2}}\epsilon^{-\frac{b}{\theta}}$ as before. But ϕ depends on both the initial and final kinetic energies of the electrons, and there is reason to suppose, from considerations of specific heat, that the electrons in a metal have very much less kinetic energy than corresponds to the temperature of the metal. If they have none at all, ϕ will contain a term $\frac{3}{2}k\theta$ representing the difference between their kinetic energies in metal and gas. If $\phi - \frac{3}{2}k\theta = \phi_0$, supposed constant,

$$i = eA_1 \sqrt{\frac{k}{2\pi m}} \theta^2 \epsilon^{-\phi_0/k\theta} = A\theta^2 \epsilon^{-\phi_0/k\theta},$$

where A is a constant.

This law has been deduced as an approximation by Richardson as the result of thermodynamic considerations. The approximations involve the neglect of terms of the nature of the Thomson effect. These are small for pure metals and hence the equation would be expected to hold. The small value of the specific heat of electricity in fact, is evidence of the kinetic energy of electrons in a metal being small, and the small Peltier effect shows that if ϕ changes with the temperature to any marked extent it at least does so approximately equally for all metals[2].

It might be supposed that it would be easy to distinguish experimentally between the two types of equation, as they have different mathematical forms. Actually it is very difficult, because the temperature variation due to the exponential is so rapid that

[1] Richardson, *Emission of Electricity*, p. 56.

[2] It seems unlikely that the conclusions of this and the following paragraphs would be affected by adopting Fermi-Dirac statistics, owing to the small density of the electrons in free space.

it overpowers the slower change due to $\theta^{\frac{1}{2}}$ or θ^2. In fact if we consider the way in which we arrived at the second equation it will be seen that to distinguish between the two is equivalent to determining ϕ for ranges near two temperatures θ_1, θ_2 with such accuracy that the error in the difference of the ϕ's is less than $\frac{3}{2}k(\theta_1 - \theta_2)$. As ϕ is generally of the order of 5 volts, and $\frac{3}{2}k\theta$ is only about ·25 volt at 2000° K. this is not easy. Until quite recently the experimental results fitted either equation within the limits of experimental error, though of course the values of the constants a, b and A, ϕ_0 are different according as to which formula is assumed. Some recent results of Davisson and Germer[1] however indicate rather better agreement with the second formula and this is generally now used in expressing the results of experiments, but it must be admitted that the evidence is chiefly theoretical.

The quantity ϕ_0 can be determined experimentally much more accurately than A. Thus for tungsten ϕ_0/k is about 53,000; at 1200° K. the exponential is ϵ^{-44}, thus a 10 per cent. error in ϕ_0 would give a factor $\epsilon^{4\cdot4}$ and alter A by a *factor* of 81. The enormous variation with temperature is shown by the fact that K. K. Smith[2] in experiments on tungsten obtained currents varying over a range of 10^{11} to 1. Even over this range it was impossible to distinguish between the two formulae.

The probability that the equation $i = a\theta^{\frac{1}{2}}\epsilon^{-b/\theta}$ is incorrect deprives calculations of N, such as that on p. 352, of weight. Even in cases, such as the emission from metallic oxides, where the kinetic theory may apply for the electrons inside the solid, Richardson has shown that the probable rapid change of N with the temperature makes it mathematically impossible to determine N at any one temperature from purely thermionic data.

108·1. There is an intimate relation between ϕ and the contact potential difference between the metal and a standard one. Thus consider two metals in an equal temperature enclosure in equilibrium with their electrons, and suppose them joined at some point. In general the pressure of electrons with which the metals are in equilibrium will be different, and electrons will distil over from one to the other till there is a difference of potential between

[1] Davisson and Germer, *Phys. Rev.* xx. p. 300, 1922.
[2] Smith, *Phil. Mag.* xxix. p. 802, 1915.

two points, each just outside the surface of one of the metals, sufficient to balance this tendency. This is what is spoken of as the contact potential difference V. By the kinetic theory

$$n_1/n_2 = \epsilon^{-eV/k\theta},$$

where n_1, n_2 are the values of n' outside the two metals, supposed small enough that mutual repulsion can be neglected. Thus

$$V = \frac{k\theta}{e} \log n_2/n_1 = \frac{k\theta}{e} \log i_2/i_1,$$

where i_2, i_1 are the saturation currents per sq. cm.

Taking logarithms of the expression for the currents and subtracting, we have

$$\int \frac{\phi_1 - \phi_2}{k\theta^2} d\theta = \log i_1/i_2 - \log A_1/A_2 = -\frac{eV}{k\theta} - \log A_1/A_2.$$

Differentiating, $\quad \phi_1 - \phi_2 = eV - e\theta \dfrac{dV}{d\theta}.$

There are two views of the origin of contact potential difference: one regards it as inherent in the metal and the other as due mainly to charged double layers at the surface. On the former view V is the actual difference of potential between two points in the metals, and $e\theta \dfrac{dV}{d\theta}$ differs from the Peltier coefficient only by terms depending on the Thomson effect. If all thermoelectric effects are small compared with eV, as they are in metals, $\phi_1 - \phi_2 = eV$. On the other view there must still be small true differences of potential at the junction of metals to account for the thermoelectric effects, but most of V is caused by the surface layers. If it can be assumed that these do not change with temperature the last result for ϕ still holds, apart from the thermoelectric terms. If $\phi_1 - \phi_2$ and V are independent of the temperature it is easily seen that $\log A_1/A_2 = 0$, or A is a universal constant. Using Sackur and Tetrode's theory for the chemical constant of the electron gas Dushman[1] has found $A = 2\pi m e k^2/h^3 = 60 \cdot 2$ amp. per sq. cm. per $(°K.)^2$. Experiments with metals, especially tungsten, seem to agree with this, but the uncertainty of determination is very great[2].

[1] Dushman, *Phys. Rev.* xxi. p. 623, 1923.

[2] The expression takes no account of 'reflected' electrons and so requires to be reduced by a factor to allow for this effect.

For metallic oxides there appear to be greater discrepancies. Here however thermoelectric effects may be large. Thus S. L. Brown has found that a copper-copper oxide couple with junctions at 20° C. and 530° C. gives an E.M.F. of more than half a volt. Any linear variation of ϕ with θ other than that assumed above will lead to a power of θ in the final formula other than the square, and it is quite possible that the ordinary equation should be modified for non-metallic substances. Also with these complex substances it is quite possible that definite reversible temperature changes occur in the surface layers. Thus Davisson and Germer[1] found such changes with a wire of platinum coated with BaO and SrO; in this case they were slow enough to be detected and allowed for, but a more rapid change would simply appear as an abnormal variation of ϕ with θ, leading to a false value of A if the equation were assumed of the usual type.

The exact conditions which the thermoelectric and other constants of a material must obey in order that A may be constant, have been investigated by H. A. Wilson[2] and by Bridgman[3].

109·1. Certain peculiarities in the emission of the alkali metals have been investigated by Richardson and Young[4]. These substances show appreciable deviations from the usual law, and it appears that their surfaces are 'patchy' with two different values of the work function. The thermionic emission of potassium satisfies an expression of the form

$$i = A_1 \theta^2 \epsilon^{-\omega_1/\theta} + A_2 \theta^2 \epsilon^{-\omega_2/\theta},$$

and a similar relation probably holds for sodium. The values of ω for the thermionic emission are usually less than those found for the photoelectric effect, but by special treatment a photoelectric threshold has been obtained at the commonest value of ω. The probable explanation is that the small values of ω occur only over a very small portion of the surface, so that they have little chance of being affected by the incident light, while the enormous increase of thermionic emission as ω decreases makes them the most important for this effect. Young finds that the constants

[1] Davisson and Germer, *Phys. Rev.* xxiv. p. 666, 1924.
[2] Wilson, *Phys. Rev.* xxiv. p. 38, 1924.
[3] Bridgman, *Phys. Rev.* xxvii. p. 180, 1926.
[4] Richardson and Young, *Proc. Roy. Soc.* A, cvii. p. 377, 1925.

for potassium, taking an approximate one term formula, are unchanged on melting or solidifying. Goetz[1] has also made experiments on the changes of thermionic emission at melting and transition points and finds appreciable changes in the work function for copper, iron and manganese.

Kingdon[2] working on monatomic films of Cs and Th on tungsten finds a more complicated formula than that of Richardson and Young, and believes that the maximum emission in the case of Cs occurs when the film is not quite complete. In any case it is certain[3] that as the film increases a point of maximum emission is reached.

It should be remembered that, as in the corresponding photoelectric phenomenon, a large part of the work required to remove an electron from a conducting surface is due to the attraction between the electron and its electric image in the surface (see p. 445).

Energy of Emitted Electrons.

110·1. The first experiments on the energy of the emitted electrons were made by Richardson, working partly in collaboration with F. C. Brown. The theory on which they were based is as follows. At pressures low enough for their mutual repulsion to be negligible the electrons behave like a perfect gas, and accordingly their velocities should be distributed according to Maxwell's Law. The distribution of velocities among the electrons crossing any surface in the space containing them can then be calculated, and if the electron gas is in equilibrium with an emitting surface this will also be the distribution for the electrons entering the space. Since the emission is not affected by the presence of the electrons, when these are too few to exert an appreciable electric force, this will be the distribution for the emitted electrons whether the external conditions are those of equilibrium or not. The only assumption is that a state of equilibrium between free electrons and the emitting surface is possible, in which the free electrons obey dynamical principles as used in the Kinetic Theory of Gases. It is not necessary that the

[1] Goetz, *Phys. Zeits.* xxiv. p. 377, 1923.
[2] Kingdon, *Phys. Rev.* xxiv. p. 510, 1924.
[3] Becker, *Phys. Rev.* xxvii. p. 112, 1926.

electrons in the solid should do so. On this view the average energy of the emitted electrons is $2k\theta$. This is greater than the average equilibrium kinetic energy $\tfrac{3}{2}k\theta$, because the more rapidly moving ones occur more frequently in the emitted stream than in a random volume. If the emitting surface is taken perpendicular to the axis of x, and u, v, w are the velocity components of an electron parallel to x, y and z respectively, then the number emitted in unit time with velocity components between u and $u + du$ is

$$Nu\,du = N2hmu\epsilon^{-hmu^2}du.$$

The number with velocity components between v and $v+dv$ is

$$Nv\,dv = N\sqrt{\frac{hm}{\pi}}\,\epsilon^{-hmv^2}dv,$$

where $h = (2k\theta)^{-1}$, and N is the total number emitted per unit time, the distribution for w being given by a similar expression.

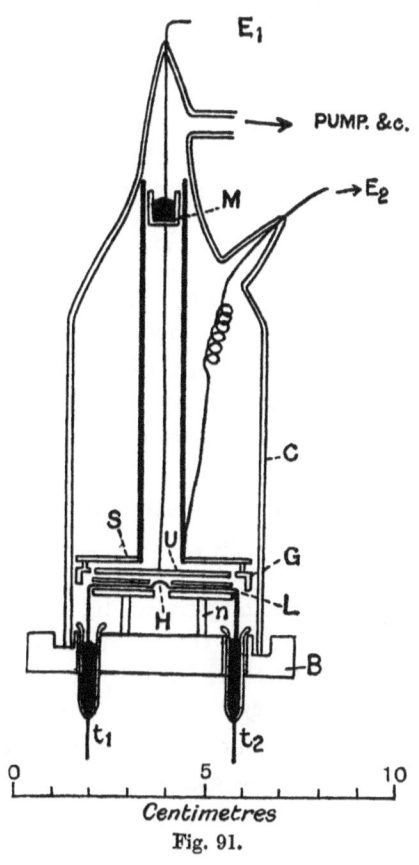

Fig. 91.

Richardson and Brown's[1] first investigation dealt with the u component only. The apparatus used is shown in section in Fig. 91. The electrons were emitted from a small piece H of thin platinum foil heated electrically, and measured by the charge given to the plate U connected to a sensitive electrometer. A retarding potential V_1 was maintained between U and the plate L which was connected so as to be at the same potential as the centre of H and was flush with it. G was a guard ring and S an electrostatic shield. U was covered with platinum to avoid contact

[1] Richardson and Brown, *Phil. Mag.* xvi. p. 353, 1908.

potential differences. If $u_0^2 > \frac{2}{m} eV_1$, an electron starting with normal velocity u_0 will reach U, otherwise it will return to the lower plate. The experiments consisted in finding how the current between the plates varied with V_1. If Maxwell's law holds, the current i between the plates is given by

$$i = Ne \int_{u_0}^{\infty} 2hmu\epsilon^{-hmu^2} du,$$

where $u_0 = \sqrt{2eV_1/m}.$

This gives $i = Ne\epsilon^{-2hV_1 e} = i_0 \epsilon^{-2hV_1 e},$

where i_0 is the current when $V_1 = 0$. Hence

$$\log i/i_0 = -\frac{V_1 e}{k\theta} = -\frac{\nu e V_1}{R\theta},$$

where ν is the number of molecules in a c.c. and R is the gas constant for this amount of gas.

The experiments showed that $\log i$ varied in a linear manner with V_1, and the value of R calculated from the curve agreed well with the accepted value.

Other experiments with platinum coated with lime, the liquid alloy of sodium and potassium, and platinum saturated with hydrogen did agree with the theory, but the experimental evidence was not regarded by the authors as satisfactory.

To determine the distribution of velocity for the components parallel to the surface Richardson[1] used the apparatus shown in Fig. 92. This consists essentially of two slits in parallel metal plates A and B, the one at D containing the emitting platinum strip, the other giving access to the Faraday cylinder T, insulated from the plates, to which the current was measured. The current to the plate B was also measured and the plate and all

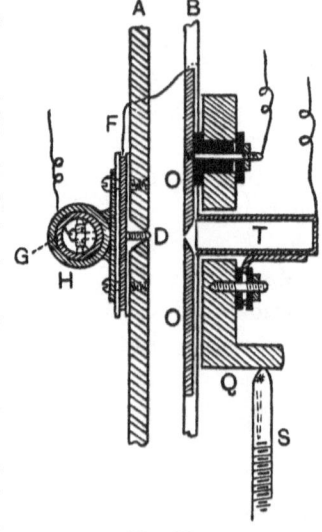

Fig. 92.

[1] Richardson, *Phil. Mag.* xvi. p. 890, 1908; xviii. p. 681, 1909.

connected with it could be moved up and down through known distances by means of the screw S. An *accelerating* potential V_1 is applied between A and B and measurements are taken of the ratio of the current through the slit to that to the plate B for different positions of the slit. For the method of calculating the distribution of velocities from these data, see Richardson, *Emission of Electricity from Hot Bodies*, p. 163. The results show on the whole good agreement with the Maxwell theory. Observations when $V_1 = 0$ are shown in Fig. 93, where the curve is cal-

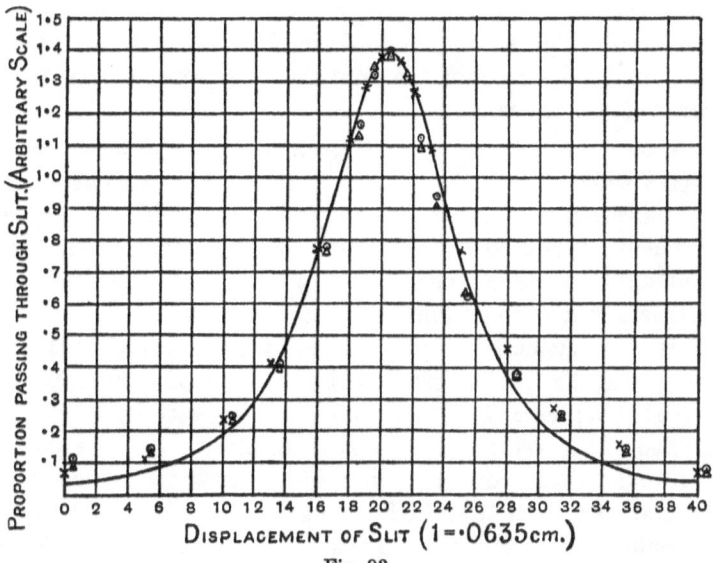

Fig. 93.

culated, and the points were obtained under varying conditions of filament temperature and emission. The discrepancy shown at large distances from the central position is thought to be due to secondary causes such as roughness of the metal surface and the deflection of the moving electrons by gas molecules.

Schottky[1] has also made experiments confirming Maxwell's law for the form of the distribution curve but giving a rather large value for the mean energy. He adopted an improvement in the experimental method by introducing in both the heating circuit and in the circuit measuring the thermionic current, make

[1] Schottky, *Ann. der Phys.* xliv. p. 1011, 1914.

and break switches. These were operated 250 times per second and timed so that when one circuit was closed the other was open. In this way the disturbing effect of the magnetic field of the heating circuit on the motion of the electrons was eliminated. The alternations were so rapid that the temperature of the filament remained sensibly constant.

Doubt was thrown on these results by some experiments conducted by Ting[1] in Richardson's laboratory. He found that with tungsten and platinum, while the distribution was approximately that of Maxwell, it corresponded to an absolute temperature considerably higher than that of the filament, in some cases to nearly twice the amount. A number of experiments have since been made on the point.

Germer[2] used the electrons from a wire and measured the fraction which had sufficient energy to reach a concentric metal cylinder against various opposing voltages. He found good agreement with theory in the case of tungsten. J. H. Jones[3] found also agreement with theory and considers that the chief source of error is contamination of the heated surfaces. Potter[4] working on platinum *in vacuo* and in hydrogen found that in the former case the results agreed with theory, but in the latter gave too high a temperature. Congdon[5] working with tungsten found the theoretical result *in vacuo* and in an atmosphere of argon, but again found that hydrogen gave results corresponding to a higher temperature than that of the filament. Rossiger[6] found Maxwell's law confirmed for the emission from CaO, SrO and BaO. Koller[7] however found that the two latter gave results corresponding to a temperature about 30 per cent. too high. Del Rosario[8] found the theoretical result for tungsten both *in vacuo* and in hydrogen.

The conclusion seems to be that the energy distribution is that corresponding to an electron gas at the temperature of the filament

[1] Ting, quoted by Richardson, *Emission of Electricity*, p. 172.
[2] Germer, *Phys. Rev.* xxv. p. 795, 1925.
[3] Jones, *Proc. Roy. Soc.* cii. p. 734, 1923.
[4] Potter, *Phil. Mag.* xlvi. p. 768, 1923.
[5] Congdon, *Phil. Mag.* xlvii. p. 458, 1924.
[6] Rossiger, *Zeits. f. Phys.* xix. p. 167, 1923.
[7] Koller, *Phys. Rev.* xxv. p. 671, 1925.
[8] Del Rosario, *Phys. Rev.* xxvii. p. 810, 1926.

except perhaps when hydrogen is present. The anomalous emission of platinum in an atmosphere of hydrogen suggests that there may be a real difference in this case, but the great experimental difficulties make it doubtful whether there is really an exception.

In connection with these experiments it should be noticed that the reduction of the energies of all electrons by the same amount, as by passing through a double layer, makes no difference to the velocity distribution, if this is of the Maxwell type, but merely reduces the total number. This can be seen as follows. Write $E = \tfrac{1}{2}mu^2$, the portion of the kinetic energy due to the u component of velocity. Then Maxwell's law can be written

$$N_E dE = N 2h \epsilon^{-2hE} dE.$$

If ϕ represent the loss of energy per electron on passing through the double layer, the number which *emerge* with energies between E and $E + dE$ will be

$$N'_E dE = N_{E+\phi} dE = N 2h \epsilon^{-2h(E+\phi)} dE = N \epsilon^{-2h\phi} 2h \epsilon^{-2hE} dE.$$

Thus $N'_E / N_E = \epsilon^{-2h\phi}$, a constant quantity.

110·2. Richardson was the first to point out that the emission or absorption of electrons must be accompanied by a reversible thermal effect analogous to the latent heat of vaporisation of a gas. In fact, since ϕ_0 represents the difference between the energy of an electron in the metal and one at rest outside, the absorption of heat per electron will be $\phi_0 + 2k\theta$, and the absorption for a current i is $U = \dfrac{i}{e}(\phi_0 + 2k\theta)$. Thus measurements of U can be made to give a value of ϕ_0 which should agree with that found from the variation of thermionic current with temperature. The first experiments made to test this by Wehnelt and Jentzsch[1] using oxide-coated platinum wires did not give satisfactory results. It is now known that such filaments are liable to undergo modification as a result of temperature change which may greatly affect the result. The vacuum also was not perfect, and the bombardment of the positive ions formed in the residual gas led to a change of the cooling into a heating effect at high tempera-

[1] Wehnelt and Jentzsch, *Verh. der Deutsch. Phys. Ges.* x. p. 610, 1908; *Ann. der Phys.* xxviii. p. 537, 1909.

tures. Richardson and Cooke[1] were more successful. They measured the change in resistance of an osmium wire, carrying a constant current, when a thermionic current was allowed to flow. The results were consistent with the theory and gave a mean value of $\phi_0 = 4.7$ equivalent volts. Similar experiments with tungsten gave $\phi_0 = 4.63$ volts, in good agreement with values obtained by the temperature variation method. Wehnelt and Liebrich[2] further investigated the anomalous effects with lime and showed that variations of the thermionic emission occurred with time, and that these were accompanied by corresponding changes in the cooling effect. W. Wilson[3], working with the standard oxide-coated filaments prepared by the Western Electric Company, found complete agreement between the values of ϕ_0 from the two methods. Davisson and Germer[4] have obtained a similar result with oxide-coated (SrO, BaO) platinum wires. They found that the filaments changed their condition with changing temperature, but sufficiently slowly to allow measurements to be made before appreciable change had occurred. Lester[5] has made measurements of ϕ_0 for molybdenum, carbon, tantalum and tungsten, using the cooling effect. Davisson and Germer[6] in a very careful experiment found values of ϕ_0 for tungsten by the two methods of 4.52 (calorimetric) and 4.48 volts (temperature variation of emission).

The inverse effect, the heating of the metal due to absorption of electrons, has also been investigated by Richardson and Cooke[7]. It is rather more complicated than the emission effect, as in general the electrons will arrive at the surface layer with kinetic energy which is partly due to a fall through a potential difference, and not wholly due to thermal agitation as in the converse case. The electrons from two heated osmium filaments were directed by a small potential difference on to thin strips of the metals to be

[1] Richardson and Cooke, *Phil. Mag.* xxv. p. 624, 1915.
[2] Wehnelt and Liebrich, *Verh. der Deutsch. Phys. Ges.* xv. p. 1057, 1913; *Phys. Zeits.* xv. p. 548, 1914.
[3] Cf. Arnold, *Phys. Rev.* xvi. p. 78, 1920.
[4] Davisson and Germer, *Phys. Rev.* xxiv. p. 666, 1924.
[5] Lester, *Phil. Mag.* xxxi. p. 197, 1916.
[6] Davisson and Germer, *Phys. Rev.* xx. p. 300, 1922.
[7] Richardson and Cooke, *Phil. Mag.* xx. p. 173, 1910; xxi. p. 404, 1911.

tested, the effect was detected by the change in resistance of the strip owing to its change of temperature. The heating effect is then plotted against V the applied voltage. This gives a straight line, as it should, and the gain in energy in volts due to the entrance of an electron can be found as the (negative) voltage which would be required to give zero heating effect if the same law continued to hold. If however the metal receiving the electrons is not the same as that emitting them the contact potential difference comes in, and the actual potential difference through which the electron has passed is not the same as that applied by the battery.

Let ϕ_1, ϕ_2 be the values of ϕ_0 for the hot and cold metals respectively, and V_1, V_2 be the potentials at points just outside them when zero potential difference is applied externally. If $V_2 > V_1$, *i.e.* if the contact P.D. accelerates the electrons, they will arrive at the cold metal with mean energy $2k\theta - e(V_2 - V_1)$ and the heat given out on passing into the metal will be

$$2k\theta - e(V_2 - V_1) + \phi_2 \text{ per electron,}$$

(e negative) neglecting small effects due to the kinetic energy of the electrons in the metals, which are very uncertain. But to the order to which ϕ_1, ϕ_2 are independent of θ the investigation on p. 356 shows that $\phi_1 + e(V_2 - V_1) - \phi_2 = 0$. Thus J, the heating effect in equivalent volts, is $2k\theta + \phi_1$, and depends only on the hot metal. On the other hand, if $V_2 < V_1$ the electrons are retarded. This does *not* change their mean energy (see above) but only their number, and the heat liberated is given by $J' = 2k\theta + \phi_2$ depending on the cold metal. This comes to saying that in both cases $J - 2k\theta = \phi$ for the more electronegative element.

Richardson and Cooke's results are not accurate enough to prove the truth of this theory, but they do not disagree with it, and the values obtained for ϕ are in reasonable agreement with those found by the other methods.

Influence of Gases on Thermionic Emission.

111·1. The presence of a gaseous atmosphere may influence the thermionic current in two quite distinct ways. It may modify the actual emission from the solid by altering the surface layer or being dissolved in the metal, or it may alter the current carried by the change in the conditions *outside* the metal, due chiefly to the presence of positive ions formed by collision of the electrons with the gas molecules. If the potentials are insufficient to allow ionisation by collision, or if the gas is very rare, the former is much the more important, and we shall consider it first.

It was early observed that minute traces of certain gases might enormously increase the thermionic emission, *e.g.* platinum in the presence of hydrogen. It was even suggested that the presence of gas was a primary cause of the emission, and that this was an effect of chemical action. This view is now completely abandoned, and the extensive use of filaments, both metallic and oxide-coated, in technical work has shown that it is possible to obtain constant results over long periods of time if the vacuum is good enough. Extreme care must however be taken to remove gas, not only from the volume of the exhausted vessel but also from its walls. The layer of gas clinging to the surface of glass is extremely tenacious, and it is necessary to bake the evacuated bulbs for some hours at a temperature not far from the softening point of the glass in order to remove the gas. Even then it seems doubtful if the removal is complete, as at a high temperature small quantities of gas can diffuse through the glass. In addition prolonged bombardment and heating is required to remove the gas dissolved in the electrodes or other metallic parts or condensed on their surfaces. It is largely for these reasons that the results for the thermionic emission of the same substance obtained by different observers show such large discrepancies. The pressures obtained are of the order of 10^{-9} or 10^{-10} of a mm. Another cause of the irregular results is that the emission varies so rapidly with ϕ; thus a minute speck of impurity may greatly modify the emission of the whole wire.

Platinum.

111·2. Richardson and Wilson have shown that air, nitrogen, water vapour and oxygen at moderate pressures, say 1 mm. or less, have no effect on the emission from platinum if the conditions are such as to prevent impact ionisation.

On the other hand hydrogen enormously increases the emission, sometimes by a factor of 10^5 or more, but the amount of the increase depends greatly on the state of the wire.

The effect of hydrogen on the emission from platinum has been very carefully studied by Richardson and Wilson. For a full account see Richardson, *Emission of Electricity from Hot Bodies*, pp. 117 et seq. The effect appears to be very complicated and is even now not entirely understood. The following are some of the main facts. In all cases the variation of emission with temperature follows a law of the type $i = a\theta^{\frac{1}{2}}\epsilon^{-b/\theta}$ within the limits of experimental accuracy, but the quantities a and b depend on the amount of hydrogen; usually both decrease with increasing pressure of hydrogen, but the effect of b is the more important, so that the result is an increase of current. If the wire has not been previously heated in hydrogen, the values of a and b are functions of the pressure only, though there is a considerable lag before the emission reaches a steady state after a change of pressure. If however the wire has been 'aged' by prolonged heating in hydrogen, it will continue to emit strongly when the pressure of hydrogen is reduced almost to zero, even after it has ceased to emit appreciable amounts of gas. The large emission can however be destroyed by heating to about 1700° C. in a good vacuum, and Wilson has shown that a wire which has retained the power of strong emission still contains a large quantity of hydrogen even after heating in vacuum till gas ceases to come off. It is suggested that this gas is combined with the platinum to form a compound of definite (and very low) dissociation pressure, while in the case of the 'new' wire the gas is in solution to an extent proportional to the outside pressure. In any case the time lag makes it probable that the effect is a volume one, and not merely a change in the layer of gas adsorbed on the platinum. It was pointed out in the second edition of this book that the effect may be analogous to the potential difference in a concentration cell. If two electrolytes are in contact and in one of them there is strong ionisation and in the other weak, there is a difference of potential between them proportional to $R\theta \log p_1/p_2$, where R is the gas constant, and p_1 and p_2 the pressures of the ions in the two solutions. Now the hydrogen dissolved in the platinum will probably be strongly ionised, while that outside is not, so that there will be a difference of potential, and a double layer of electrification at the surface which will modify ϕ. In fact if V is the contact potential difference between pure platinum and platinum containing hydrogen, $\log \frac{i_2}{i_1} = eV/k\theta$. If V contains a term involving θ to the first power we shall get a difference in the factor outside the exponential, so that this accounts for the change in a. It appears from Wilson's experiments that a at constant temperature is proportional to p^z, where z is a proper fraction between 0·5 and 1·0 and diminishes with rising temperature.

Lockrow[1] finds that the 'permanent' effect can be prevented by using a liquid air trap and is of the opinion that it is due to the formation of a film of tap grease on the surface. He also thinks that the 'temporary' effect is due to the action of hydrogen on some impurity. Potter also finds no effect of hydrogen on a wire cleaned by prolonged heating in a high vacuum.

111·3. *Oxides.* Early experiments of Horton and Martyn found that the emission from a Wehnelt cathode was increased by hydrogen, but the former in more recent experiments concludes that this only occurs at considerable pressures, and that at pressures comparable with 0·01 mm. there is little difference in the emissions from lime and Nernst filaments in hydrogen, air, oxygen or nitrogen. Koller[2] found that oxygen decreased the emission from BaO and SrO, while argon, hydrogen, CO and CO_2 increased it.

111·4. *Tungsten.* Langmuir[3] has investigated the emission from tungsten in various gases. In all cases the saturation currents could be represented by $i = a\theta^{\frac{1}{2}}\epsilon^{-b/\theta}$, but the constants were different in different gases. Mercury vapour and the inert gases have no effect. Most other gases increase both a and b above their values for a vacuum. The effect with hydrogen does not vary in a simple manner with the pressure, and Langmuir thinks it may be due chiefly to water vapour either formed by it, or introduced with it. Hydrogen, nitrogen and oxygen are all 'cleaned up' by the tungsten, owing to chemical action resulting in involatile compounds which are deposited on the cold parts of the tube. At certain temperatures the action occurs with nitrogen

[1] Lockrow, *Phys. Rev.* xix. p. 97, 1922.
[2] Koller, *Phys. Rev.* xxv. p. 671, 1925.
[3] Langmuir, *Phys. Rev.* ii. p. 450, 1913; *Phys. Zeits.* xv. p. 516, 1914.

only when the latter is ionised, and the formation of the nitride when impact ionisation sets in, is regarded as the explanation of some peculiarities in the saturation curve.

For all the three gases and for a vacuum the values of $\log a$ and b when plotted against each other give a straight line. This indicates that they all act by the same mechanism, which is contrasted with that in the case of platinum by the different sign in the changes of a and b. Richardson suggests that in platinum we have an effect of positive ions from hydrogen either in the volume of the metal or on the surface layer, while in tungsten there is a corresponding effect of negative ions from oxygen or nitrogen. For a mathematical investigation of the conclusions which can be drawn from the straight line law, see Richardson, p. 135.

111·5. *Other metals.* Wilson has found that the emission from palladium is increased by hydrogen. The following experiment showing a similar effect with sodium was described by the author in the second edition of this book.

A bright surface of sodium was formed in a highly exhausted vessel which contained a well-insulated electroscope; this vessel was placed in a chamber from which all light was carefully excluded and which was so dark that a sensitive photographic plate was not fogged after an exposure of 48 hours; under such circumstances the electroscope very slowly lost a positive charge but retained a negative one. When however a trace of hydrogen was sent into the vessel in the dark chamber, the electroscope began to lose its positive charge much more rapidly; there was no leak if the electroscope was negatively charged. The increased leak from the positively charged electroscope lasted for a few minutes and then disappeared, it could however be renewed by letting in fresh hydrogen and this process could be repeated time after time. This leak was stopped when the electroscope was placed in a magnetic field, showing that it was due to electrons coming from the sodium while the metal was absorbing hydrogen. No leak of the electroscope occurred if oxygen or carbonic acid gas was admitted into the vessel instead of hydrogen.

Relation between the Current and the Potential Difference.

111·6. When the pressure is low, so that the motion of the electrons is unhindered, and no ionisation by collision occurs, the current should be saturated as soon as there is an applied potential difference large enough to balance the opposed contact potential difference, if any, always assuming that the density of the electrons is so small that the effect of their mutual repulsion is negligible. This conclusion is not usually borne out by experiment. In many cases there is a slow increase of current after apparent saturation has been reached, probably due to evolution of gas. The emission from oxides is also difficult to saturate, possibly for the same reason. Again the presence of specks of impurities on the wire may lead to local fields which prevent the current being saturated when it otherwise would have been, and easy saturation has been proposed as a test for the cleanness of a wire. Apart from these effects the potential required to saturate the current from a wire increases with the current when this becomes large, owing to the repulsion exerted by the electrons in the space preventing others from being emitted. In most experiments the wire is heated by an electric current, and the magnetic field due to this influences the path of the electrons so that they never reach more than a certain distance from the cathode depending on the relation between the electric and magnetic fields. If this distance is less than that of the positive electrode they will not convey current. A correction must also be applied for the varying potential along the wire due to its resistance. The method of intermittent heating enables these two last effects to be eliminated, and Schottky[1] found that in this case the current saturated for zero potential difference, after allowing for contact potential, provided the current was weak enough for mutual repulsion to be unimportant.

Langmuir[2] has investigated very thoroughly the case in which the current is limited by the space charge. Some of his experimental results are shown in Fig. 94. The full line represents the saturation current as a function of the temperature, the dotted

---

[1] Schottky, *Ann. der Phys.* xliv. p. 1011, 1914.
[2] Langmuir, *Phys. Rev.* ii. p. 453, 1913.

lines show the current for various fixed voltages as a function of the temperature. Taking any one of these it will be noticed that for low temperatures the current is approximately saturated, as the temperature increases the saturation becomes less complete and for high temperatures the current becomes constant. In this

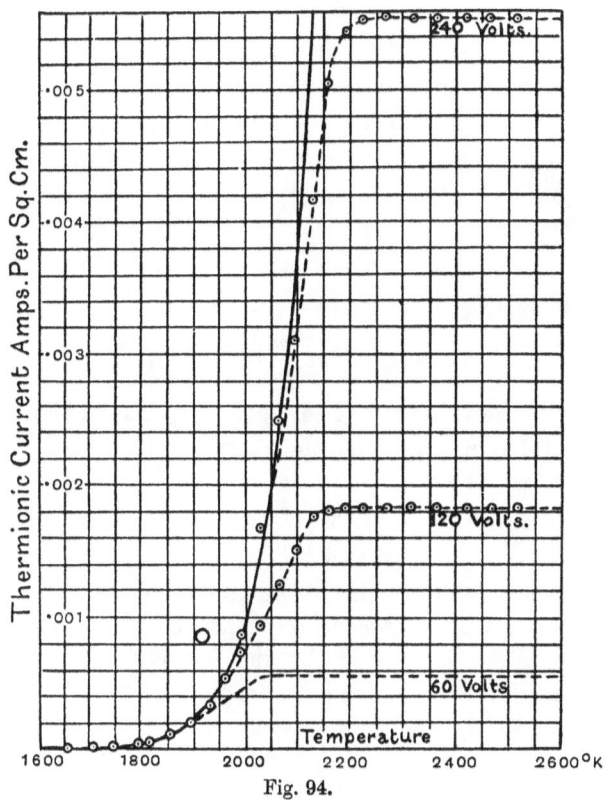

Fig. 94.

state the current is determined by the ability of the field to remove the electrons and not by the number emitted by the wire. Langmuir found that the square of the current under these conditions varies as the cube of the applied voltage.

The following investigation is an extension of one given in the second edition, p. 223. The subject has been investigated by Child and Schottky.

We consider the geometrically simple case of parallel plates. Let x be measured normal to the plates and $x = 0$ at the hot emitting plate, let V be the difference of potential between the hot plate and a point whose coordinate is x, ρ the density of the electricity. Then
$$\frac{d^2V}{dx^2} = -4\pi\rho \quad \ldots\ldots\ldots\ldots\ldots(1).$$

If v be the velocity of the ion at x, v_0 its velocity when starting from the plate, m its mass and e its charge, then
$$\tfrac{1}{2}m(v^2 - v_0^2) = Ve \quad \ldots\ldots\ldots\ldots(2);$$
but since all the ions are of one sign, i the current through unit area is equal to $v\rho$, hence from (1) and (2)
$$\left(v_0^2 + \frac{2e}{m}V\right)\left(\frac{d^2V}{dx^2}\right)^2 = 16\pi^2 i^2 \quad \ldots\ldots\ldots(3);$$
integrating this equation we have, if we write X for dV/dx,
$$X^2 = C + 8\pi i \cdot \frac{m}{e}\left\{v_0^2 + \frac{2e}{m}V\right\}^{\frac{1}{2}} \quad \ldots\ldots\ldots(4).$$

Hence if X is the value at the cold plate, X_0 that at the hot, V the potential difference between the plates, and C the constant of integration, we have
$$X^2 - X_0^2 = 8\pi i \cdot \frac{m}{e}\left[\left\{v_0^2 + \frac{2e}{m}V\right\}^{\frac{1}{2}} - v_0\right] \quad \ldots\ldots(5).$$

Now since all the ions are of the same sign, negative, the curve of V against x will be everywhere concave upward as in Fig. 95. Two cases arise, one shown by the dotted curve, where V is everywhere positive, and the other where V has a minimum as in the full curve. In the former case, if diffusion is neglected the force on an electron is always away from the hot plate, and the current is saturated; any electron which leaves the hot plate will reach the cold one. In the second case only those electrons which are carried past the point A by their initial energy will escape and the rest will return to O. Now the potential at A is of the order $\tfrac{1}{2}mv_0^2$, and the number of electrons for which this exceeds 1 volt is negligible; hence if

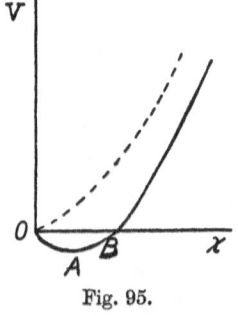

Fig. 95.

the applied potential is of the order of 100 volts we may reckon V from the potential at A without much error, and apply the above equation with A as origin. This makes $X_0 = 0$, and v_0 can be neglected compared with $\frac{2e}{m} V$, so that

$$X^2 = \left(\frac{dV}{dx}\right)^2 = 8\pi i \sqrt{\frac{2mV}{e}} \quad \ldots\ldots\ldots(6).$$

Integrating again and remembering that $\frac{dV}{dx} = 0$ at A, we have, after reduction,

$$i = \frac{V^{\frac{3}{2}}}{9\pi} \sqrt{\frac{2e}{m}} \frac{1}{x'^2} \quad \ldots\ldots\ldots\ldots\ldots(7),$$

where x' is measured from A.

This is Langmuir's result, if x' does not alter appreciably with V. The determination of the position of A is difficult but a solution has been obtained by Schottky[1].

For a given emission of electrons A will be nearer to O the greater the applied voltage and saturation occurs when they coincide. At this point the voltage is given by putting x' equal to d, the distance between the plates, in equation (7).

In general, if V is very large compared with the energy of emission, A will be near to O, x' will be nearly independent of V, and hence i will not vary with the temperature but will follow Langmuir's law. It is easily seen, however, from the general shape of the curve that the ratio of OA to d may greatly exceed the ratio of the potential at A to V, so the error involved in writing d for x' in (7) in the general case may be considerable even though V is of the order of 100 volts.

It should be noticed that the possibility of a steady solution, except in the saturated case, depends on effects which we have so far neglected, namely diffusion and the varying values of v_0; for if it were not for these considerations the electrons would all either pass the point A or not, and the only possible values for the current would be zero and the saturation value.

The general effect of diffusion may be seen from the following considerations. When things have reached a steady state there

---

[1] Schottky, *Phys. Zeits.* xv. p. 526, 1914; *Ann. der Phys.* xliv. p. 1011, 1914.

will be a definite distribution of electrons between the plates and the density of the electrons will vary from point to point; this will produce diffusion currents and thus equation (4), which neglects these currents, is only an approximation. It is a very close approximation at places so far from the hot plate that the drop of potential is considerable in comparison with that required to give an electron an amount of energy equal to that due to thermal agitation, but at smaller distances it ceases to be an approximation at all, and there the currents are due to diffusion, and the equation is

$$i = - De \cdot \frac{dn}{dx} \quad \ldots \ldots \ldots \ldots \ldots (8),$$

where D is the coefficient of diffusion of the electrons and n their density; thus

$$n = n_0 - \frac{i}{De} x \quad \ldots \ldots \ldots \ldots \ldots (9),$$

where n_0 is the density when $x = 0$, and if I is the saturation current

$$\frac{n_0 c \cdot e}{\sqrt{6\pi}} = I - i \quad \ldots \ldots \ldots \ldots \ldots (10)$$

(see p. 212).

The density at the hot plate is thus $\frac{\sqrt{6\pi}}{c} (I - i) ce$, and not infinite as it would be if equation (6) held good right up to the plate, making V vary as $x^{\frac{4}{3}}$. Diffusion cannot carry a current i if $nec/\sqrt{6\pi}$ is less than i. Hence, putting

$$n = \frac{i\sqrt{6\pi}}{ce}$$

in equation (9), we find that the maximum thickness of the region where diffusion prevails is

$$\sqrt{6\pi} (I - 2i) D/ci.$$

Now D/c is of the order λ, where λ is the free path of the electron. Hence the breadth of the diffusion currents will be proportional to $\frac{I - 2i}{i} \lambda$, and thus when the current is only a small fraction of the saturation current, this region may be considerable. The region vanishes when the current is nearly saturated.

Thus we should expect that equation (7) would cease to be a close approximation when i is small, and since the diffusion region depends on λ, the current would be sensitive to small changes in the pressure of the residual gas.

When the currents are carried by diffusion, unless the potential difference across the layer exceeds a value which increases rapidly with I/i the force at the hot plate will be negative and will vanish at a point within the layer.

Relation between Current and Potential Difference when Gas is Present.

111·7. The presence of gas affects the current-potential relation in a variety of ways. At low pressures the most important is that ions formed by collision help to neutralise the space-charge, and so increase the current in the region in which Langmuir's law holds. This effect is very great owing to the large mass of the positive ions compared with that of an electron. If the ratio of mass is M/m the velocity due to the same potential difference is in the ratio $\sqrt{m/M}$ and hence the average time spent in the field is that of $\sqrt{M/m}$. Even for hydrogen ions this is 60:1, and for oxygen four times this. Thus n positive hydrogen ions released per second will (roughly speaking) neutralise the space charge due to a current of $60n$ electrons. If the hot body is a wire the effect may be still larger, for the positive ions may make many orbits round it before being absorbed, and all the time are helping to neutralise electrons.

At large potentials the multiplying effect of ionisation by collision comes in, and we reach a state which is midway between the true thermionic current and the ordinary 'unassisted' discharge in gases. Such a state occurs when a Wehnelt cathode is used in a discharge tube. Such a discharge presents many interesting features and will be considered in the chapter dealing with the high tension discharge. It however may be pointed out here that the existence of ionisation by collision results in the current being never completely saturated.

If the gas is present at a high pressure, the retarding effect of the gas collisions on the electrons comes in; the electrons also will tend to attach themselves to the gas molecules, except in the case of the inert gases, and form heavy ions. As a result the current will be decreased for voltages too small to produce ionisation by collision, for the heavy ions will have smaller velocities and hence will have a greater tendency to accumulate as a space charge. If the pressure is of the order of that of the atmosphere the velocity of an ion depends only on the electric force acting on it, and not to any appreciable extent on its previous history. We can then proceed as follows.

112. Let us consider the case of two parallel plates at right angles to the axis of x, then if only one of the plates is incandescent, or if both are incandescent but the temperature is so low that only positive ions are produced at the surface of the plates, then the ions carrying the current between the plates will be all of one sign and we may apply the results of Art. 50. Hence if X is the electric force, k the velocity of the ion under unit electric force, we have, if i is the current,

$$X \frac{dX}{dx} = \frac{4\pi i}{k};$$

hence if k is independent of x we have

$$X^2 = \frac{8\pi i}{k} x + C.$$

If n is the number of ions per cubic centimeter

$$\frac{dX}{dx} = 4\pi n e,$$

thus

$$n = \frac{i}{ek} \frac{1}{\sqrt{\frac{8\pi i}{k} x + C}}.$$

Thus the density of the ions at the hot plate when $x = 0$ is equal to

$$\frac{i}{ek\sqrt{C}}.$$

Now suppose that the quantity of electricity emitted by the hot plate per second is I, the quantity of electricity passing through unit area of the gas between the plates is i: the difference $I - i$ must equal the charge carried back to the plate by the electrons striking against it. We showed on p. 351 that if n is the number of electrons per unit volume in the gas near the plate, the number striking unit area of the plate per second is

$$\frac{nc}{\sqrt{6\pi}},$$

where c is the velocity of mean square of the electrons. Since

$$n = i/ekC^{\frac{1}{2}},$$

the charge given to the plate by the electrons which strike against it is

$$\frac{ic}{\sqrt{6\pi}kC^{\frac{1}{2}}};$$

as this is equal to $I - i$ we have

$$I - i = \frac{ic}{\sqrt{6\pi k}\sqrt{C}},$$

or

$$C = \frac{c^2}{6\pi k^2} \frac{i^2}{(I-i)^2}.$$

If V is the difference of potential and l is the distance between the plates, $V = \int_0^l X dx$, and since

$$X = \left\{ \frac{8\pi i}{k} x + C \right\}^{\frac{1}{2}},$$

$$V = \frac{k}{12\pi i} \left[\left(\frac{8\pi i l}{k} + C \right)^{\frac{3}{2}} - C^{\frac{3}{2}} \right],$$

or substituting the value previously found for C

$$V = \frac{k}{12\pi i} \left[\left(\frac{8\pi i l}{k} + \frac{c^2}{6\pi k^2} \frac{i^2}{(I-i)^2} \right)^{\frac{3}{2}} - \frac{c^3 i^3}{(6\pi)^{\frac{3}{2}} k^3 (I-i)^3} \right] \ldots(1).$$

This gives the relation between the current and the potential difference. I is the saturation current.

The quantities are supposed to be measured in electrostatic units.

We shall consider two particular cases of this equation; the first is when i is so small compared with I that

$$\frac{c^2}{6\pi k^2} \frac{i^2}{(I-i)^2}$$

is small compared with $\quad \dfrac{8\pi i l}{k}.$

Then equation (1) becomes

$$V = \left(\frac{8\pi i}{k}\right)^{\frac{1}{2}} \tfrac{2}{3} l^{\frac{3}{2}},$$

or
$$i = \frac{9kV^2}{32\pi l^3} \quad \ldots\ldots\ldots\ldots\ldots\ldots\ldots(2).$$

Thus the current increases more rapidly than the potential difference, and diminishes rapidly as the distance between the plates is increased.

This equation has been tested by Rutherford[1]; we cannot however expect the theory to be in very close agreement with the facts, for in deducing equation (1) we have made several assumptions which are not satisfied in practice; in the first place we have assumed that k is independent of x, this will only be true when the temperature is uniform between the plates, it will not be true when one plate is hot and the other cold, for the velocity of the ion depends upon the temperature. Thus H. A. Wilson[2] has shown that in a flame at a temperature of about 2000° C. the velocity of the negative ion under a potential gradient of 1 volt per cm. is about 1000 cm./sec., that of the positive ion under the same gradient 62 cm./sec.; in hot air at a temperature of about 1000° C. the velocity of the negative ion is only about 26 cm./sec., that of the positive about 7·2 cm./sec. McClelland[3] found that the ions from an incandescent wire when they got into the cold air at some distance from the wire travelled with velocities as small as ·04 cm./sec., and that the velocity diminished as the ions got further from the wire, and could be increased again by warming the ions; thus k varies rapidly with the temperature and therefore with x.

The increase of k with the temperature makes the current increase rapidly with the temperature of the hot plate. We see from equation (1) that the current for a constant small difference of potential does not depend upon the amount of ionisation near the plate[4], so that the increase of ionisation at the higher tem-

[1] Rutherford, *Phys. Rev.* xiii. p. 321, 1901.
[2] H. A. Wilson, *Phil. Trans.* A, cxcii. p. 499, 1899.
[3] McClelland, *Phil. Mag.* v. 46, p. 29, 1899.
[4] It must be remembered that equation (1) only applies when the current is small, so that $X = 0$ when $x = 0$; when the current approaches saturation it increases rapidly with the amount of ionisation at the plate.

peratures would not explain the increase of current when the wire gets hotter; a satisfactory explanation of this increase is however afforded by the increase of k with the temperature.

When the temperature of the hot plate is high enough for negative as well as positive ions to exist near the plate, the leak between the hot plate and a cold one will be greater when the hot plate is the negative electrode than when it is the positive: for in the former case the current is carried by negative ions, in the latter by positive, and equation (1) shows that with the same potential difference the current is proportional to the velocity of the ion by which it is carried. Now the velocity of the negative ion is always greater than that of the positive, and the ratio of the velocity of the negative to that of the positive increases rapidly with the temperature; thus the experiments of H. A. Wilson on the leak through gases mixed with the vapours of salt (*l.c.*) show that this ratio at 2000° C. is about 17 while at 1000° C. it is only about 3·5. At ordinary temperatures for the case of ions drawn from the neighbourhood of the hot wire, McClelland's experiments show that this ratio is only about 1·25. The absolute values are still more different. Thus McClelland found for the velocity under a potential gradient of a volt per cm. values ranging from ·006 to ·03 cm./sec., while Wilson at 1000° C. found 26 cm./sec. for the negative and 7·2 cm./sec. for the positive; at 2000° C. the values were respectively 1030 cm./sec. and 62 cm./sec.

The great increase of current produced by changing the sign of a very hot electrode from + to − is a very well-marked phenomenon; one striking example of it is furnished by an old experiment of Hittorf's[1]. In this experiment a bead of salt was placed in a flame between glowing electrodes: the increase in the current was much greater when the bead was placed close to the negative electrode than when it was placed near to the positive. These results, it must be remembered, are only true when the currents are very small compared with their saturation values; the saturation values do not depend upon the velocities of the ions but only upon the number of ions produced in unit time at the surface of the hot metal.

[1] Hittorf, *Pogg. Ann.* Jubelband, p. 430, 1874.

The velocity of an ion under a constant electric force increases as the pressure of the gas diminishes, hence we see from equation (1) that the current when small will increase when the pressure diminishes.

113. The equation (2) will however, except at very low pressures, only hold when the current is an exceedingly small fraction of the saturation current, for in deducing it we have assumed that

$$\frac{c^2}{6\pi k^2} \frac{i^2}{(I-i)^2}$$

is small compared with $\dfrac{8\pi i l}{k}$.

Now c is the velocity of mean square of the electrons, so that if the temperature of the hot plate is 1000° C. c^2 will be of the order 9×10^{12}. k is the velocity of the ion through the gas under unit electrostatic force, *i.e.* 300 volts per centimetre; thus if the air is cold and at atmospheric pressure k will be about 4.5×10^2 and k^2 about 2×10^5; thus c^2/k^2 will be exceedingly large, and it is only when i is very small compared with I that the above condition is fulfilled. If the air is also at 1000°, c^2/k^2 is not so large.

The other case we shall consider is when

$$\frac{c^2}{6\pi k^2} \frac{i^2}{(I-i)^2}$$

is large compared with $\dfrac{8\pi i l}{k}$.

In this case
$$V = \frac{c}{\sqrt{6\pi}} \frac{i}{k(I-i)} l,$$

$$i = \frac{VI}{V + \dfrac{c}{\sqrt{6\pi}} \dfrac{l}{k}}.$$

This equation shows that i now increases less rapidly than the potential difference; it approaches saturation when $\dfrac{V}{l} k$ is large compared with $c/\sqrt{6\pi}$, *i.e.* when the electric field is so strong that the velocity given to an ion by the field is large compared with the velocity of mean square of an electron at the temperature of the hot plate. We see from the numbers given above that it would require a prodigious field to saturate the current from a hot wire through a gas at atmospheric pressure, in fact it would be impossible to saturate the current at all, for the field required to do so is greater than that required to spark through the gas.

113·1. This effect is most likely to be observed in the case of positive emission which occurs at a lower temperature than the negative, so that the ions have small mobility. The following figures found by Richardson[1] for the positive current from a platinum tube 0·2 cm. in diameter, and surrounded at atmospheric pressure by a cold tube 3·2 cm. in diameter, serve to illustrate the point.

Volts on hot tube + ...	0	4	10	20	40	80	400	960
Current ($1 = 1·8 \times 10^{-12}$ amps per sq. cm.)	0	2·6	10	22	32	64	225	390

Table of Constants of Electron Emission.

113·2. To complete our account of electron emission we give a table of the results of a few recent measurements, but with most substances different experimenters get widely different results. Tungsten is probably the most accurately known.

Substance	Φ volts	d degrees K.	A amps per sq. cm.	a amps per sq. cm.	b degrees K.	Φ' volts	Author
Pt	4·24	49,250	27·1	—	—	—	Schlicter
	—	—	—	$4·76 \times 10^{27}$ $7·64 \times 10^{25}$	57,910 53,060	4·987 4·569	Suhrmann
W	4·64	52,600	60·2	—	—	—	Dushman, Rowe, Ewald and Kidner
	4·48	51,860	61	—	—	—	Davisson and Germer
Ta	4·11	47,800	50	—	—	—	Dushman, Rowe
	—	—	—	$7·47 \times 10^{23}$ $2·27 \times 10^{22}$	36,400	3·14	Suhrmann
Mo	4·30	50,000	160	—	—	—	Dushman, Rowe
K	1·12 0·43	12,960 4,980	0·021 $1·26 \times 10^{-11}$	—	—	—	Richardson and Young
C	3·92	45,700	5·9	$1·49 \times 10^{25}$	48,700	4·20	Langmuir
50 % BaO and SrO on Pt	1·79	20,800	3·2	—	—	—	Davisson and Germer

[1] Richardson, *Phil. Trans.* A, ccvii. p. 58, 1908.

The currents are measured in amps per sq. cm. The quantities Φ, d, A refer to a formula of the type $i = A\theta^2 \epsilon^{-\phi_0/k\theta}$, d being the value of ϕ_0/k and Φ being ϕ_0 measured in equivalent volts. The quantities a, b, Φ' refer to a formula of the type $i = a\theta^{-\frac{1}{2}}\epsilon^{-b/\theta}$, and Φ' is the value of kb in equivalent volts.

Of the values given by Suhrmann[1] for Ta and Pt the second refers to a specimen strongly 'degassed,' the first to one less so. The values for Φ and d for Mo and Ta were calculated on the supposition that the constant A had the theoretical value 60·2. The two values for K refer to a two-term formula (see p. 357).

Emission of Positive Electricity by Hot Metals.

114. Hot metals emit positive as well as negative electricity, indeed in many cases until the temperature gets very high the metal emits more positive than negative electricity. The emission of positive electricity can be detected at much lower temperatures than that of negative. Strutt[2] has detected the emission of positive electricity from copper and silver wires at a temperature less than 200° C.

The rate of emission of positive ions from a hot wire depends to a very large extent on the previous history of the wire. On first heating a wire it emits a large quantity of positive electricity; if the wire is kept hot the rate of emission rapidly decays until it falls to a small fraction of its original value. This large initial leak cannot be due to dirt on the surface of the wire, for it occurs with platinum wires which have been boiled in nitric acid. It is not confined to metals, for Owen[3] has shown that it is well marked in a Nernst filament; he found that the filament after heating did not recover its power of giving a large initial leak after exposure to air at atmospheric pressure for several days. The evidence as to the recovery of such a leak in platinum wires after exposure to air is somewhat conflicting, as H. A. Wilson and Richardson, who have each made important investigations on the leak from hot wires, have arrived at opposite conclusions: Wilson thinking that a short exposure to air is sufficient to restore the original

[1] Suhrmann, *Zeits. f. Phys.* xiii. p. 17, 1923.
[2] Strutt, *Phil. Mag.* vi. 4, p. 98, 1902.
[3] Owen, *Phil. Mag.* vi. 8, p. 230, 1904.

leak, while Richardson attributed the increase which he obtained not to the air but to the accidental presence of a complex vapour, probably that of phosphorus, as he has found[1] that the exposure of a platinum wire to a small quantity of phosphorus will for a time produce an enormous increase in the positive leak from the wire.

Richardson found that a wire which had been strongly heated could be kept in a vacuum for more than three months without any recovery of its power to give a large positive leak.

Richardson has made the very interesting discovery[2] that when a wire has by heating lost its power of giving a large positive leak, it can regain the power by being made the cathode for the electric discharge through a gas at low pressure, or even by being placed near such a cathode; in the latter case the recovery of this power does not take place if an obstacle is placed between the cathode and the wire. We cannot tell without further experiments whether this recovery is due to the bombardment of the wire by electrons or positive ions, or to some matter deposited on the wire by the discharge; this point could be tested by placing the wire in a stream of cathode rays in a highly exhausted tube, arranged so that none of the metal torn off from the cathode can reach the wire.

It is important to settle this point because some substances after exposure to cathode rays show what is known as thermo-luminescence, which is in some respects analogous to the behaviour of the wire. The substances showing thermo-luminescence when heated after exposure to cathode rays become luminous; as the heating continues the luminosity gradually fades away and finally disappears, and does not recur until the substance has been again exposed to cathode rays. The luminosity of these substances thus shows analogies with the leak of positive electricity from hot wire.

114·1. In addition Richardson[3] has shown that the power of emitting may be restored by warming with a Bunsen burner the walls of the glass tube in which the wire is mounted. In a particular experiment the current was increased from $2·2 \times 10^{-13}$ amp.

[1] Richardson, *Phil. Mag.* vi. 9, p. 407, 1905.
[2] Richardson, *Phil. Mag.* vi. 8, p. 400, 1904.
[3] Richardson, *Emission of Electricity from Hot Bodies*, p. 203.

to 5×10^{-9} amp. The pressure in the meantime rose only slightly and the increase was apparently not due to this cause.

Other means by which the emission of a wire may be revived are: (1) distillation from another fresh wire, especially if the latter is positively charged; (2) exposure to gases at high pressures (50–100 atmospheres); (3) heating in a gaseous atmosphere or in a Bunsen flame; (4) straining. Richardson found that a manganin wire was revived when subjected to the strain caused by passing a current through it in a varying magnetic field.

In all cases the increased emission rapidly decays when the agency is no longer acting, and the wire is heated *in vacuo*.

The decay of initial emission increases as a rule with rise of temperature and the emission is sensibly constant at the temperature at which it begins to be measurable. The rate of decay is often irregular and there may even be a slight increase during part of the time. The currents from new wires are also very difficult to saturate, the current sometimes increasing almost in proportion to the voltage between 40 and 400 volts. Sometimes there is a *decrease* of current with increasing voltage after about 5 volts. These effects have not been adequately explained.

115·1. Besides this temporary emission, there is a smaller constant emission which takes place when the wire is heated in a gas. The following account of an experiment by Richardson gives an idea of the magnitude of the effects[1].

The wire (platinum) under test was 7 cm. long and 0·01 cm. in diameter. The positive emission on first heating at 804° C. was found to be $1·62 \times 10^{-8}$ amp., the pressure given by the McLeod gauge being 0·00005 mm. This current decayed to one-half its value in 10 minutes and to one-tenth in about an hour. Even after heating *in vacuo* for several hours a day, at temperatures in the neighbourhood of 800° C. for about two weeks the wire still gave small currents under the best available vacuum conditions. Thus at 0·0003 mm. pressure a saturation current of $9·6 \times 10^{-13}$ amp. was obtained at 721° C. when the wire was charged positively. On letting in oxygen to a pressure of 0·045 mm. and keeping the temperature constant the current increased to $1·8 \times 10^{-12}$. It was found that the small current which did not depend on the pressure of the gas gradually disappeared with continued heating, whereas the additional current caused by the gas did not.

[1] Richardson, *loc. cit.* p. 228.

IONISATION BY INCANDESCENT SOLIDS

For all cases in which saturation currents can be measured which change sufficiently slowly with the time for measurements of the temperature variation to be possible, the current can be represented by an expression of the form $a\theta^{\frac{1}{2}}\epsilon^{-b/\theta}$. The value of b is usually less than in the corresponding electron formula, so that the positive current increases less rapidly with increasing temperature than does the negative, and becomes negligible in comparison at high temperatures.

This equation also applies to the emission from a Nernst filament, and to the positive currents which are obtained from a number of salts when heated.

The phenomena of the emission of positive electricity from solids are extremely complicated, much more so than in the case of electron emission. One reason is that, in many cases, the effects are modified profoundly by chemical changes produced by the heating, while in others the emission is due mainly to minute traces of impurities. Richardson gives a good account of a number of experiments that have been made on various substances and under various experimental conditions, and the reader is referred to him for details.

Values of e/m.

116·1. The most powerful method of attack on this complicated problem has been the determination of e/m for the emitted particles. The first method used by one of the authors was an adaptation of that of § 67 to the case of positive ions. For ions from platinum values of the mass were found ranging from 14 to 170 times that of a hydrogen atom, with indications of a few carriers of still greater mass, perhaps platinum dust.

More recent experiments[1] have been made by the method of coincident electric and magnetic fields as used in experiments on positive rays. The ions from heated platinum were found to have a value of e/m corresponding to a weight of 27, when the only gas detectable spectroscopically was CO (molecular weight 28). When the platinum had been heated in hydrogen for some time, the average weight was reduced to 9, indicating the presence of hydrogen atoms or molecules. These results indicate that the

[1] J. J. Thomson, *Proc. Camb. Phil. Soc.* xv. p. 64, 1908.

ions from platinum heated in a fairly good vacuum, and after the initial emission has disappeared, are atoms and molecules of the gases present, with some atoms of the metal itself. As regards the latter point, Jenkins[1] has found evidence of heavy positive ions from tungsten near its melting point, and considers that they consist of atoms of the metal. It seems possible however that they are formed from the vapour of the metal by photoelectric action. The presence of the heavy metallic ions may be connected with the disintegration of the platinum which occurs when it is heated in a gas containing oxygen, and so be indirectly a chemical effect. There is plenty of evidence in the low mobilities found for thermal ions for the existence of heavy particles at high pressures.

The ions corresponding to the large initial emission have been investigated by Richardson[2]. In conjunction with Hulbert he measured the value of e/m for the ions from Pt, Pd, Cu, Ag, Ni, Os, Au, Fe, Ta, W, C, brass, steel and nichrome. The apparatus used was similar to that used for investigating the distribution of velocities parallel to the emitting surface (see Fig. 92). A magnetic field H was applied perpendicular to the plane of the figure, and the position found for the slit which gave the greatest proportion of the ions passing through it. If the displacement of this position from that opposite the heated strip is x, and z is the distance between the plates, $e/m = \dfrac{9Vx^2}{2H^2z^4}$; the values obtained all lay between 21·1 and 30·5, average 26·9, taking oxygen as 16, the only exception being tungsten, which behaved rather erratically and gave 42·1. At first sight these values suggest very strongly that the ions are carbon monoxide or nitrogen (each 28) and perhaps CO_2 for the tungsten. These gases are also among those most commonly present in the residual gas in an evacuated vessel. Richardson however apparently considers that the results are not sufficiently accurate to warrant this conclusion, and experiments made with improved apparatus have led to a different result.

In these experiments great care was taken to prevent the heated strip from bowing when heated, to ensure that H and z

---

[1] Jenkins, *Phil. Mag.* xlvii. p. 1025, 1924.
[2] Richardson and Hulbert, *Phil. Mag.* xx. p. 545, 1910.

were accurately measured and that the electric field was uniform. Some of the observations were taken by a somewhat different method, namely, measuring separately the current carried by the ions to the portions of plate B on the two sides of the slit, and finding for what value of x they were equal. Fig. 96 shows the results obtained for platinum, and for the ions from potassium sulphate. It will be seen that the two agree well with each other and reasonably well with the value to be expected for potassium ions, while after prolonged heating sodium ions seem to be emitted.

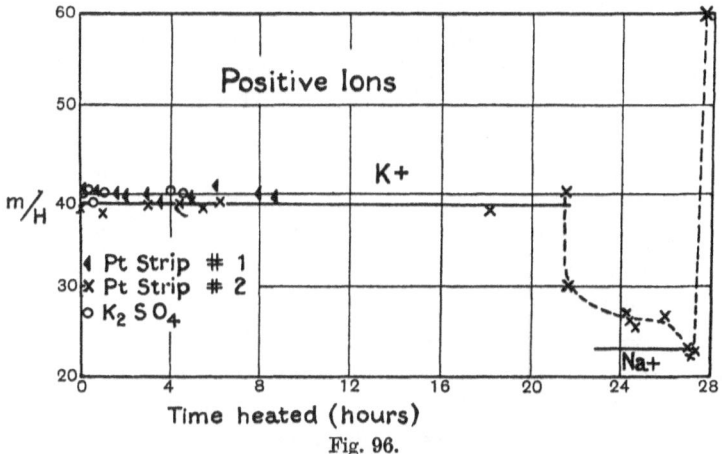

Fig. 96.

Ions from manganin strip and from iron also showed masses of about 40. Since these results occur with metals cleaned by reagents, it is clear that the potassium is not derived simply from dirt but must be present as an impurity in the metal, or, less probably, in the reagents used. Similar results have been found for the ions emitted when the wire is 'revived' by heating or straining. Richardson has calculated that in several cases the total weight of ions emitted in the early emission bears a ratio to the weight of the metal of about $1 : 10^7$.

e/m for Ions from Heated Salts.

117·1. Measurements have also been made of the value of e/m for the positive ions emitted by many salts when heated. In all cases the ions are charged atoms of a metal, but in many cases the great majority are not due to one of the main constituents of the salt, but to an impurity[1]. Richardson[2], using the above method, examined the sulphates of the alkali metals. Lithium sulphate gave at first a single kind of ion of weight 35·9. After 12 hours' heating two kinds of ions appeared of masses 41·8 and 5·5. With continued heating the heavier ions disappeared and after 44 hours were inappreciable. Further heating resulted in the appearance of a new kind of ion of mass 20·6 and the disappearance of the old. By this time the emission was very small and the salt had mostly volatilised. These results are clearly to be explained as due to atoms of different metals coming off at different stages. First potassium originally present as impurity, then the lithium forming the greater part of the whole, and finally sodium from an impurity. The other alkali sulphates did not show more than one kind of ion; the mass found varied slightly with the time of heating but was always close to that of an atom of the metal in question, except in one experiment with caesium which showed at first a value corresponding to rubidium. Sodium fluoride and iodide behaved like the sulphates.

The salts of the alkaline earths have been investigated by Richardson and by Davisson[3]. The ions found in all cases correspond to single charged atoms in spite of the divalent nature of the metals. Salts of Ba and Sr gave ions of the corresponding weights with, in some cases, indications of ions of weight corresponding to K. Calcium and magnesium ions are hard to distinguish from potassium and sodium respectively, but it was concluded that they were present when the corresponding salts were heated. Beryllium salts, on the other hand, gave only ions corresponding in weight to potassium and sodium, and probably due to impurities of these substances. Haloid salts of zinc and cadmium gave mostly doubly charged ions of the metal, but the singly charged ions sometimes occur as well as, in some cases, potassium and sodium.

A large number of salts give emissions apparently due only

---

[1] Garrett, who made the first measurements, by the method of § 67 found about 10 per cent. of the ions from aluminium phosphate to be hydrogen, but this has not been confirmed by subsequent workers.

[2] Richardson, *Phil. Mag.* xx. pp. 981, 999, 1910.

[3] Davisson, *Phil. Mag.* xxiii. p. 121, 1912.

to impurities of sodium or potassium, the latter being the more common. Among these are $FeCl_3$, $AlPO_4$, $Ba_3(PO_4)_2$, $BeSO_4$, $Be(NO_3)_2$, $AgCl$, AgI, $PbCl_2$, $PbBr_2$, $PtCl_2$, AlF_3, $CuCl_2$, the seven last having been investigated by Waterman[1]. He found indications of copper with two charges from $CuCl_2$ and of molybdenum with one charge from the mineral molybdenite.

117·2. Positive ions from salts heated in various ways have been used in recent years in the investigation of the isotopes of the metals they contain. Their use for this purpose of course depends on their being metallic ions. In one method, used by Aston[2], the salt is heated on a strip of platinum and the ions driven off by a large potential, about 20,000 volts, between this and the cathode containing a system of slits used to isolate a beam, which is then analysed by electric and magnetic fields to form a 'mass spectrum.' The current obtained is of the order of 2 milliamps, which is much larger than that due to the unaided thermal emission. On the other hand the pressure used is so low that practically no current passes when the salt is not heated. Possibly there is a reciprocal bombardment of anode and cathode by electrons and positive ions respectively, the impact of each producing the other. Perhaps also the small amounts of gas released from the heated salt help to carry the discharge. By this method ions of the alkali earths have been successfully obtained, but the intensity is very capricious. The discharge seems to concentrate on certain points which cannot be predicted beforehand. The ions when obtained are all singly charged atoms of the metals. For other elements it was found necessary to use the method of anode rays.

These rays are obtained when a discharge is passed through a tube containing sufficient gas to allow it to pass easily, using as an anode a paste of haloid salts made conducting by mixing with graphite. They then appear as a visible beam diverging from the anode. For a fuller account of their properties and production see *Rays of Positive Electricity*, p. 142. These rays consist of charged atoms of the metals of the salt. By analysing these rays by the parabola method one of the authors

[1] Waterman, *Phil. Mag.* xxxiii. p. 225, 1917.
[2] Aston, *Phil. Mag.* xlii. p. 436, 1921.

discovered the presence of two isotopes in lithium[1]. They are given most copiously by the salts of alkalis, but the alkaline earths[2] also give good results. In both cases only singly charged metal atoms were found. Positively charged atoms of iodine and fluorine were also observed, but these were probably due to ionisation of the gas, as they usually did not have energy equal to the full difference of potential between the electrodes. Aston[3] has since obtained these rays for a large number of elements, including Sc, Ti, V, Cr, Mn, Co, Cu, Ga, Ge, Sr, Y and Ag, whose isotopes he has investigated by forming their mass spectra, and at a later date[4] Ba, La, Pr, Nd, Ce, Zr, Cd, Te, Pb, Bi and In. He records one photograph in which doubly charged Ca was observed, but in all cases the singly charged ions were much more numerous.

It is open to question how far these rays can properly be grouped under the heading of thermal ions. They are produced by the bombardment of the anode by the negative ions in the discharge, and it seems to be necessary that a halogen, preferably iodine, should be present. The action of this is to load up the electrons and so produce a large fall of potential of the order of 1000 volts in the immediate vicinity of the anode. The bombardment of the anode by these heavy negative ions must have an intense local heating effect, but whether this alone is sufficient to cause the powerful emission observed is doubtful. When the rays are produced the anode emits a considerable amount of gas, and its surface glows with the spectrum characteristic of the metal, which is also seen in the rays. The rarity of double charges in the case of the alkaline earths is against the process being analogous to the electrolysis of fused salts, but there is undoubtedly violent chemical action going on and this may have an important effect.

Dempster[5] has obtained positive ions for the investigation of isotopes, by bombarding the heated metal with electrons of 30 to 160 volts energy. The metal is heated by a separate circuit as well as by bombardment. This method was successful with Li, Mg, K, Ca and Zn. As the metal was appreciably volatilised

[1] Aston and G. P. Thomson, *Nature*, cvi. p. 827, 1921.
[2] G. P. Thomson, *Phil. Mag.* xlii. p. 857, 1921.
[3] Aston, *Phil. Mag.* xlvii. p. 385, 1924.
[4] Aston, *Phil. Mag.* xlix. p. 1191, 1925.
[5] Dempster, *Phys. Rev.* xviii. p. 415, 1921; xx. p. 631, 1922.

this may simply be a case of ionisation by collision in the gaseous state. In the case of lithium, when freshly heated after exposure to the air, positively charged hydrogen atoms (protons) were obtained at first. The ions all carried a single charge. He has since[1] used lithium that had been allowed to oxidise, and was then bombarded by cathode rays, as a source of protons. Hydrogen molecules, water vapour and nitrogen molecules were also given off as ions. Volmer[2] has measured e/m for the ions from CdI_2, $CdCl_2$, CuI_2, $ZnCl_2$, $PbBr_2$, CaF_2. His method consisted in measuring the magnetic field required to shift a stream of the ions from one electrode to another. He finds in all cases doubly charged atoms of the metals only, but the method does not seem a very sensitive one. It may be mentioned that Kondratjeff[3] using ionisation by electron impacts finds the principal ionisation of $ZnCl_2$ to be into $ZnCl^+$ and Cl^- at about 12 volts. Kunsman[4] has found that a fused mixture of iron oxide with one per cent. of oxide of an alkali or alkaline earth is a good source of ions of the impurity.

Emission from Metals.

117·3. Since the emission from metals is apparently mostly or entirely due to substances other than the metal itself, it can readily be understood that the effects are capricious. The usual rapid decrease with time is presumably due to the removal of the potassium or sodium, which forms the ions from the surface layer of the metal. The steady emission in a gas, which appears as a definite function of the pressure, is probably due to atoms or molecules of the gas ionised in some way by contact with the hot metal. Helium gives an effect, though not a very large one, so the process is not purely chemical. Hydrogen and oxygen both take some time to produce the full effect corresponding to the pressure, at least in the case of platinum, on which most of the experiments have been done. As the result of some experiments on the positive emission from platinum, through which hydrogen is allowed to diffuse, Richardson suggests that the gas in the metal is ionised and a few of the ions escape as such, a view which was also advanced in the early

[1] Dempster, *Proc. Nat. Acad. Sci.* xii. p. 96, 1926.
[2] Volmer, *Zeits. f. Phys.* xxvi. p. 285, 1924.
[3] Kondratjeff, *Zeits. f. Phys.* xxxii. p. 535, 1925.
[4] Barton, Harnwell and Kunsman, *Phys. Rev.* xxvii. p. 738, 1926.

editions of this book. More experiments are needed on the value of e/m for the ions in the 'steady' emission in various gases before any certain conclusions can be reached. For an account of the rather complicated variation of the current with pressure in various gases, see Richardson, pp. 230 *et seq.*

The Kinetic Energy of Positive Ions.

117·4. Richardson[1] has investigated the distribution of velocity among the positive ions, taking the component parallel to the surface and using the same method as for electrons. The results are in agreement with the hypothesis that the distribution was that of Maxwell's law, and the average kinetic energy was that corresponding to the temperature of the hot metal. Brown[2] made investigations on the normal component of velocity, with the same result. These experiments were made on the ions given off by hot platinum, and involve the assumption that the ions carry a single charge. Other experiments by Brown[3] on a number of metals and aluminium phosphate showed considerable divergence from theory in some cases. Some of these may have been due to complications arising from the presence of electrons. The agreement of the earlier results is further evidence that, in most cases at least, the ions are singly charged.

Effect of Vapours on the Rate of Leak.

118. We have seen that in the best vacua we can produce, a metal when first it begins to glow gives off positive electricity and then at considerably higher temperatures negative electricity as well, the rate of emission of negative electricity increasing more rapidly with the temperature than that of the positive, so that at very high temperatures the negative is greatly in excess of the positive. Thus to make a metal emit positive electricity we have to communicate a certain amount of energy to its surface, a larger amount being required to make it give out negative electricity. When the incandescent metal is surrounded by gas at an appreciable pressure we find that the nature of the gas has a very

---

1 Richardson, *Phil. Mag.* xvi. p. 890, 1908.
2 Brown, *Phil. Mag.* xvii. p. 355, 1909.
3 Brown, *Phil. Mag.* xviii. p. 649, 1909.

distinct effect upon the amount of leak. The author[1] has shown that gases such as the vapours of iodine and bromine, chlorine, hydriodic acid gas, hydrobromic acid gas, hydrochloric acid gas, the vapours of potassium iodide, sal-ammoniac, sodium chloride, potassium chloride, which are dissociated by heat conduct electricity on quite a different scale from those which like air, hydrogen or nitrogen do not suffer any dissociation; in the latter case the leak is not greater than could be accounted for by the emission of ions from the electrodes, in the former case it is very much greater, showing that the gas itself is ionised.

The vapours of many metals conduct very well; of the metals I tried, sodium, potassium, thallium, cadmium, bismuth, lead, aluminium, magnesium, tin, zinc, silver and mercury; sodium and potassium had the highest conductivity; while the conductivity of the vapours of mercury, tin, thallium, did not seem any greater than that of air; so that the small conductivity actually observed might have been due to the presence of air and not to the vapour of the metal. See also § 125·5.

118·1. The difficulty in work on these lines is to distinguish with certainty between ions given off from the solids and a true volume ionisation of the vapour. Kalendyk[2] working with CdI_2 vapour at a temperature of about 300° found a definite conductivity at a distance from the solid, and after the vapour had passed through an electric field to remove ions. If the vapour was dry the conductivity had a constant value. If water was present the conductivity was increased but showed time changes. The vapour of KI would conduct when wet but not when dry. The current through the CdI_2 vapour could be represented by the expression $a\epsilon^{-b/\theta}$; in these experiments the temperature of the vapour could be varied without changing that of the salt. Sheard[3] has also found that the vapour from CdI_2 is conducting, but it is very difficult at these comparatively low temperatures to be sure that the salt does not form a solid deposit on the testing electrodes. From the value of b and thermodynamical considerations Kalendyk found for the ionisation potential of the substances tested the

[1] J. J. Thomson, *Phil. Mag.* v. 29, pp. 358, 441, 1890.
[2] Kalendyk, *Proc. Roy. Soc.* xc. p. 634, 1914.
[3] Sheard, *Phil. Mag.* xxv. p. 370, 1913.

following: CdI_2 1·93 volts, ZnI_2 2·72 volts, $ZnBr_2$ 2·97 volts. He also found that PBr_5 and SCl_2, which dissociate with heat, did not conduct, so that dissociation is not necessarily accompanied by conductivity.

In a recent paper Schmidt and Walter[1] find that there is no volume ionisation in CdI_2 and that the only ions emitted from the surface of the salt are positive. This is contrary to Sheard's results and suggests that the effects attributed to volume ionisation are due really to contamination of the electrodes by the salt and subsequent emission of positive ions.

Emission from Salts.

118·2. The emission of electricity from heated salts was observed by Beattie[2], who found that a large number of substances increased the leak across a parallel plate air condenser at a temperature of about 300° C. In an experiment by one of the authors[3] various inorganic salts were spread on a heated porcelain tube and their discharging effect measured on an electroscope with a collector surrounding the tube. The tube was at a red heat and was in air at atmospheric pressure. Chlorides and phosphates were found to discharge positive electricity and the oxides negative; the nitrates discharged positive electricity till converted into oxides, after which they discharged negative. The sign of the charge given off by the salt was opposite to that acquired when rubbed with a pestle in a mortar. Aluminium phosphate gave an abnormally large effect (the salts of the alkali metals were not tried) which decreased with continuous heating but was not much affected by reduction of pressure or change of atmosphere to hydrogen or carbon dioxide. It is now known that this large emission is due to the presence of impurities, and Richardson working with specially prepared aluminium phosphate of very high purity found a very small emission.

In most cases salts, both in a vacuum and in gases, show a more or less complicated change of emission with time which appears to be due, at least in many cases, to chemical changes

[1] Schmidt and Walter, *Ann. der Phys.* lxxii. p. 565, 1923.
[2] Beattie, *Phil. Mag.* v. 48, p. 97, 1899; vi. 1, p. 442, 1901.
[3] J. J. Thomson, *Proc. Camb. Phil. Soc.* xiv. p. 105, 1906.

taking place. The curves often resemble those giving the radiation from radioactive substances where several successive transformations are involved, and it seems probable that in most cases the substance giving the greatest part of the emission is a product formed by the heating. There is often a quick rise to a maximum followed by a slow irregular fall. The presence or absence of water vapour is often important. In the case of aluminium phosphate Richardson found that the rate of decay of emission was most marked when the salt was positively charged. In most cases the emission is affected by the presence of gas, there being often a maximum emission at a certain pressure. The most important fact in connection with the effect of gases is that, as Davisson has shown, the ions emitted are still atoms of the metal of the salt, or of an impurity, and not molecules of the gas. Thus $SrCl_2$ gave Sr ions, $SrSO_4$ gave K or Sr and $AlPO_4$ gave Na ions, as determined by the value of e/m, for all pressures at which the measurements could be made. The gases were CO_2, air and hydrogen. $SrCl_2$ and $SrSO_2$ evolved CO, but no trace was found of ions of mass corresponding to this or to hydrogen.

It may be noticed that the emission from impure aluminium phosphate, which is mostly sodium ions, may exceed that from sodium phosphate initially, though it decays rapidly. In general the effects of small impurities is much larger than one would expect. A great deal of work has been done on the emission from salts, among the authors being Garrett and Willows, Garrett, Schmidt, Sheard, Horton, H. A. Wilson and Richardson (see Richardson, *Emission of Electricity from Hot Bodies*, chap. VIII).

The Distribution of Potential near Glowing Electrodes.

119. We shall confine ourselves to the case when the current passes between two parallel plane electrodes. If one of these be hot and the other cold—too cold to emit either positive or negative ions—the current will be carried entirely by ions of one sign, the electric force will therefore increase continuously from the hot plate to the cold one, and (see p. 376) the distribution of potential will be represented by a curve similar to that in Fig. 97, the lower electrode being the hotter of the two. Similar curves will represent the distribution of potential when both plates are hot provided the temperature of the negative plate is not high enough for negative as well as positive ions to be emitted by the plate, for it is evident that in this case the current has to be carried entirely by positive ions. The shape of the curve will change when both plates are hot enough to emit ions and the negative so hot that negative as well as positive ions are emitted. For, when the field is strong most of the positive ions will be driven from the positive plate and the negative ions from the negative plate; there will be an excess of positive ions at the negative plate, so that in its neighbourhood the potential curve will be concave, and an excess of negative ions at the positive, which

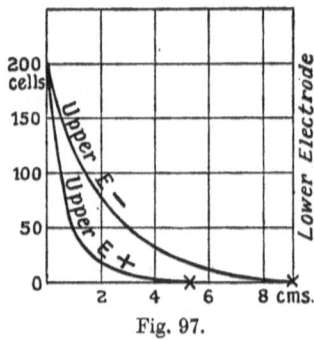
Fig. 97.

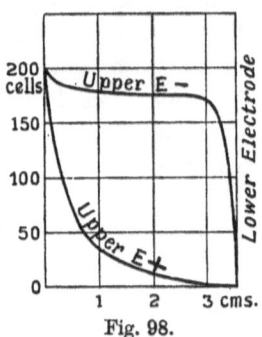
Fig. 98.

will make the potential curve convex. The potential curve will be like the higher curve in Fig. 98, the straight part in the middle showing that except close to the plates there are approximately equal numbers of positive and negative ions present. Curves similar to this have been obtained by H. A. Wilson[1] and Marx[2].

The variation in the current due to varying the sign of the potential difference is shown in some early experiments of H. A. Wilson from which Fig. 99 is taken. This shows the current between two concentric cylindrical electrodes of platinum of diameters 0·75 cm. and 0·3 cm. heated in air at atmospheric pressure. The bulk of the ionisation was probably due to the emission of positive ions from the platinum. In this case the field would be stronger near the inner electrode apart from the effect of space charge. The temperature of the outer electrode

[1] H. A. Wilson, *Phil. Trans.* A, cxcii. p. 499, 1899.
[2] Marx, *Ann. der Phys.* ii. p. 768, 1900.

was 1080° C. in this experiment, that of the inner considerably less.

When the hot plates are made of different materials, Pettinelli and Marolli[1] have shown that the magnitude of the current depends upon which metal is used as the cathode, thus with electrodes of carbon and iron the current when the carbon was cathode was three or four times the current when the iron was

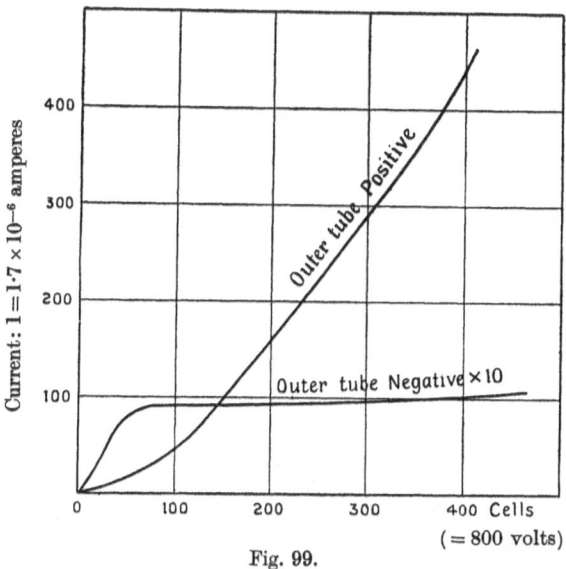

Fig. 99.

cathode; they state that the current is greatest when the more porous substance is used as the cathode. These effects are much more marked at high than at low temperatures; it is probable that they do not commence until the temperature is high enough to produce negative ions.

Mobilities of Ions from Salts.

120·1. While the ions when first formed are seen from the experiments of § 117·1 to be charged atoms, this is not the case after they have existed for a short time in gas at a considerable pressure.

Garrett and Willows[2] found for the ions from the haloid salts of zinc heated to about 360° C. mobilities of the positive ions of

[1] Pettinelli and Marolli, *Atti della Accad. dei Lincei*, v. p. 136, 1896.
[2] Garrett and Willows, *Phil. Mag.* viii. p. 452, 1904.

about ·006 in cm. per sec. per volt cm.$^{-1}$. In these experiments the mobilities were measured by the method used by McClelland and at a temperature much lower than that of the salt. This value is of course much less than that of X-ray ions, so it is assumed that the ions have become loaded up by gathering round them molecules of salt or gas.

Moreau[1] has made a series of experiments on the mobilities of the ions obtained by blowing vaporised salt solutions through a tube heated to 800°–900° C. The mobilities were measured after the mixture had cooled down to 170° C. or less. They were intermediate between those of X-ray ions and those of the heavy Langevin ions and decreased as the gas cooled.

Garrett[2] has used Method VIII to find k for the ions from various salts (ZnI_2, BiI_3, PbI_2, CdI_2). In these experiments the air in which the mobilities were measured was at the temperature of the salt, 215° C. The pressures varied from 10 to 90 mm. of mercury. At 40 mm. pressure the mobilities of the positive ions were about ·03 except in the case of CdI_2, where they were about ·06. The mobilities of the negative ions at the same pressure were about ·4 for ZnI_2 and also for CaI_2, BaI_2; these latter gave no positive ions when dry. At lower temperatures, down to about 190° C., the mobilities of the positive ions decreased. The product of pressure and mobility was approximately constant in the case of positive ions from BiI_3 and for negative ions. This shows that no great change in the structure of the ions occurs over the range of pressures used.

Todd[3], however, working at pressures down to 0·1 mm. found a marked increase in the value of the product below about 1 mm. in the case of air and about 10 mm. in the case of hydrogen. He worked with the positive ions from aluminium phosphate; this salt is hardly volatile at the temperatures used, so the loading up of the ions is probably due to gas molecules rather than salt vapour. He found the product of mobility and pressure was about the same as for X-ray ions. For a discussion of the problems connected with the mobility of ions see Chap. III.

[1] Moreau, *Ann. de Chem. et de Phys.* viii. p. 201, 1906.
[2] Garrett, *Phil. Mag.* xiii. p. 739, 1907.
[3] Todd, *Phil. Mag.* xxii. p. 791, 1911.

CHAPTER X

IONISATION IN GASES FROM FLAMES

121. It has been known for more than a century that gases from flames are conductors of electricity: a well-known application of this fact—the discharge of electricity from the surface of a non-conductor by passing a flame over it—was used by Volta in his experiments on Contact Electricity. We shall not attempt to give any historical account of the earlier experiments on this subject, because the conditions in these experiments were generally such that the interpretation of the results obtained is always exceedingly difficult and often ambiguous: the reason of this is very obvious—to investigate the electrical conditions of the flame wires are generally introduced, these become incandescent and so at once add to the electrical phenomena in the flame the very complicated effects we have been discussing in the last chapter.

The gases which come from the flame, even when they have got some distance away from it and have been cooled by the surrounding air, possess for some time considerable conductivity, and will discharge an insulated conductor placed within their reach. The conductivity can be entirely taken out of the gas by making it pass through a strong electric field; this field abstracts the ions from the gas, driving them against the electrodes so that when the gas emerges from the field, although its chemical composition is unaltered, its conducting power is gone. This result shows too that no uncharged radioactive substances, such as emanate from thorium and some other substances, are produced in the flame; these would not be taken out by the field, so that if they existed the conductivity of the gas would not be destroyed by the field. If not driven out of the gas by an electric field the ions are fairly long lived. Thus in some experiments Giese noticed that the gas retained appreciable conductivity 6 or 7 minutes after it had left the flame. The ions stick to any dust there may be in the air and then move very slowly so that their rate of recombination becomes exceedingly slow. McClelland[1] has

[1] McClelland, *Phil. Mag.* v. 46, p. 29, 1898.

shown that the velocity of the ions under a given electric force decreases very much as they recede from the flame; thus close to the flame the velocity under the force of a volt per centimetre was ·23 cm./sec., while some distance away from it the velocity was only ·04 cm./sec.

In order that a conductor should be discharged by a flame it is not necessary that it should be placed where the gases from the flame would naturally strike it—thus for example it will be discharged if placed underneath a Bunsen flame. The explanation of this is that the electric field due to the charged conductor drags out of the flame and up to the conductor ions of opposite sign to the charge.

This ionised gas is produced by flames of coal gas whether luminous or not, by the oxy-hydrogen flame, by the alcohol flame of a spirit lamp, by a flame of carbonic oxide; it is not however produced in very low temperature flames such as the pale lambent flame of ether. Thus to produce the ionised gas high temperature as well as chemical combination is required. That chemical combination alone is insufficient to produce ionisation is shown by the case of hydrogen and chlorine which do not conduct even when combining under ultra-violet light[1]. Braun[2] has shown that in the explosive wave produced in the combination of certain gases there is ionisation, but in this case there is also very high temperature.

In the coal-gas flame the part where the gas comes in contact with the air and where there is most combustion is positively electrified, while the interior of the flame is negatively electrified; this accounts for the effect produced by holding a negatively electrified body near the flame, the luminous part turns to the negative body, and if this is near, stretches out until it comes into contact with it; if the flame be placed between two oppositely charged plates the bright outer portion of the flame is attracted towards the negative plate while the inner portion moves, but less markedly, towards the positive plate. This effect is illustrated by Fig. 100 taken from a paper by Neureneuf[3]. In some experi-

[1] J. J. Thomson, *Proc. Camb. Phil. Soc.* xi. p. 90, 1901.
[2] Braun, *Zeitschrift für Physikalische Chemie*, xiii. p. 155, 1894.
[3] Neureneuf, *Annales de Chim. et de Phys.* v. 2, p. 473, 1874.

ments made by Holtz[1], one of which is figured in Fig. 101, the flame was divided by the electric field between the plates into two sheets; the reader will find many other interesting experiments on the effect of an electric field on the shape of flames in the papers by Neureneuf and Holtz. It appears from these results that in the bright portion of the flame where combustion is taking place there is an excess of positive electricity, while in the unburnt coal gas there is an excess of negative, a fact discovered a long time ago by Pouillet[2]. If the hydrogen and oxygen were ionised by the heat, then since negative ions of oxygen combine with positive ions of hydrogen to form water, the negative oxygen ions and the positive hydrogen ones would get used up, and there would

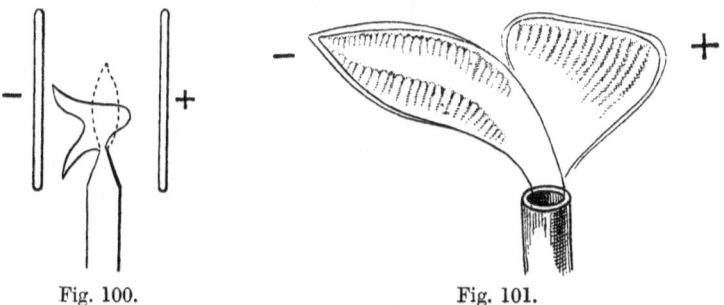

Fig. 100. Fig. 101.

be an excess of positive electricity in the oxygen and of negative in the hydrogen. It is possible too that at a temperature corresponding to that of vivid incandescence in a solid the molecules of a gas may like those of a solid give out electrons, on this account there would be a tendency for the hotter parts of the flame to be positively, the colder negatively, electrified. When as in luminous flames we have small particles of solid carbon raised to the temperature of vivid incandescence the electrical effects are complicated by those due to incandescent solids, which as we have seen in the last chapter are very considerable.

When two wires connected together through a sensitive galvanometer are placed in different parts of the flame currents flow through the galvanometer; suppose one of the wires is placed in the cool inner portion of the flame where there is an excess of

[1] Holtz, *Carl. Répert.* xvii. p. 269, 1881.
[2] Pouillet, *Ann. de Chim. et de Phys.* xxxv. p. 410, 1827.

negative electricity, while the other wire is placed at the outside of the flame where there is an excess of positive electricity; there will, neglecting any ionisation due to the wire, be a current from the hot outer portion of the flame to the cool inner portion through the galvanometer: the wire in the outer portion will however certainly be raised to incandescence; if its temperature keeps so low that only positive ions are produced at its surface, then there will on this account be a current of electricity from the hot to the cool part of the flame through the flame and thus in the opposite direction to the previous current. If however the wire got so hot that it emitted more negative than positive ions the effect of the incandescence of the wires would be to increase instead of diminishing the current due to the flame itself. Thus we see that these currents will vary in a complex way with the temperature. For an account of the currents which can thus be tapped from a flame and for other electrical properties of flames we must refer the reader to the papers of Erman[1], Hankel[2], Hittorf[3], Braun[4], Herwig[5], and especially of Giese[6], who was the first to suggest that the conduction of electricity through flames and hot gases was due to the motion of charged ions distributed through the gases: there is a very complete account of these researches in Wiedemann's *Elektricität*, Bd. IV. B, chap. 4.

121·1. Lenard has shown that the streak of coloured vapour formed about a bead of a salt placed in the flames is deflected by a horizontal electric field as though it had a positive charge. Andrade[7], who has further investigated this effect, finds that the deflection of the streak is greater than that of the flame, though Wilson[8] has since cast doubt on this point. If the field was continued for some time, Andrade found that a deposit was formed on the negative plate (placed out of the flame), which gave the characteristic colour of the metal when heated in a flame. This deposit was not

---

[1] Erman, *Gilbert. Ann.* xi. p. 150, 1802; xxii. p. 14, 1806.
[2] Hankel, *Pogg. Ann.* lxxxi. p. 213, 1850; cviii. p. 146, 1859.
[3] Hittorf, *Pogg. Ann.* cxxxvi. p. 197, 1869; Jubelbd. p. 430, 1874.
[4] Braun, *Pogg. Ann.* cliv. p. 481, 1875.
[5] Herwig, *Wied. Ann.* i. p. 516, 1877.
[6] Giese, *Wied. Ann.* xvii. pp. 1, 236, 519, 1882; xxxviii. p. 403, 1889.
[7] Andrade, *Phil. Mag.* xxiii. p. 865, 1912; also xxiv. p. 15, 1912.
[8] Wilson, *Phil. Trans.* A, ccxvi. p. 63, 1915.

formed if the plates were uncharged, or on the positive plate except to a slight extent with sodium. This shows that positive ions of the metal are present in these cases.

In some cases a deflection of part of the streak towards the positive plate was observed, and at a pressure of three atmospheres it divided into two equally bright portions, one deflected each way. This recalls the effect shown in Fig. 101 and suggests that the movement of the streak is caused in a similar way to that of the flame (see § 126 below).

Conduction of Electricity through Flames.

122. The passage of electricity through flames has been investigated by Arrhenius[1], H. A. Wilson[2], Marx[3], Starke[4], Moreau[5], Stark[6], Tufts[7], Tufts and Stark[8] and later writers. The most important phenomena of flame conduction are as follows.

Distribution of Electric Intensity between the Electrodes.

There is a very intense electric field close to the negative electrode and a weak uniform field between the electrodes; the field near the positive electrode, although not nearly so intense as that close to the negative, is stronger than that at some distance from either electrode. The distribution of electric intensity is of the type shown in Fig. 102.

Fig. 103 represents the distribution of electric potential measured by H. A. Wilson between electrodes 18 cm. apart in a long flame from a quartz tube burner. The difference of potential between the electrodes was 550 volts and it will be noticed that a drop of 450 volts occurs quite close to the cathode.

If X is the electric intensity at a point x, q the amount of ionisation per unit volume, k_1, k_2 the velocities of the positive and

[1] Arrhenius, *Wied. Ann.* xlii. p. 18, 1891.
[2] H. A. Wilson, *Phil. Trans.* A, cxcii. p. 499, 1899; *Proceedings Physical Society.*
[3] Marx, *Ann. d. Phys.* ii. pp. 768, 798, 1900; *Verh. d. D. Phys. Ges.* v. p. 441, 1903.
[4] Starke, *Verh. d. D. Phys. Ges.* v. 364, 1903; vi. p. 33, 1904.
[5] Moreau, *Ann. de Chimie et de Physique*, vii. 30, p. 1, 1903.
[6] Stark, *Physik. Zeitschr.* v. p. 83, 1904.
[7] Tufts, *Physik. Zeitschr.* v. p. 76, 1904.
[8] Tufts and Stark, *Physik. Zeitschr.* v. p. 248, 1904.

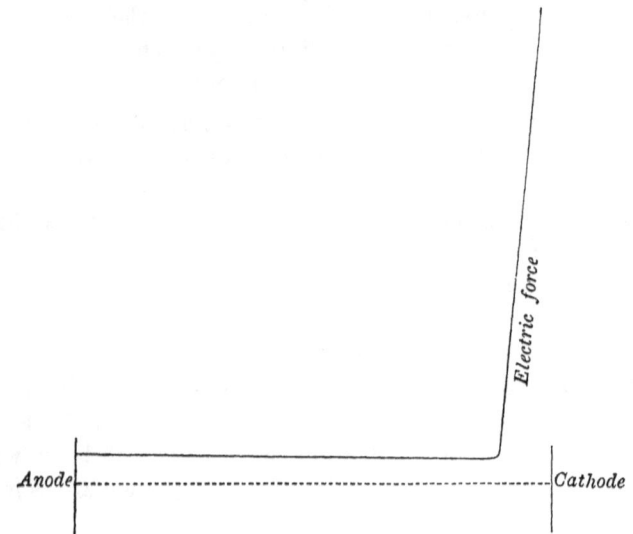

Fig. 102

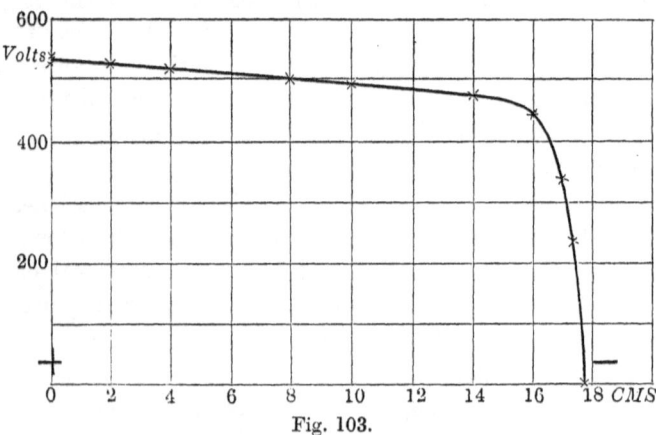

Fig. 103.

negative ions under unit force, m, n the numbers of positive and negative ions per unit volume, we have by equation (7), p. 195,

$$\frac{d^2X^2}{dx^2} = 8\pi e\,(q - anm)\left(\frac{1}{k_1} + \frac{1}{k_2}\right).$$

Since X is constant along the flame $\frac{d^2X^2}{dx^2}$ vanishes, hence

$$q = anm.$$

Thus the ionisation balances the recombination; as recombination of the ions is certainly taking place in the flame it follows that there must be ionisation throughout the flame.

122·1. It was shown in §§ 44 and 45 that the whole fall of potential in a case like this, where there is a region of constant force, can be regarded as made up of three parts, the fall of potential near each electrode and that in the uniform field. Write

$$V = V_1 + V_2 + V_0,$$

where V_1 is the fall of potential at the anode, V_2 at the cathode and V_0 in the uniform part of the field. By § 44,

$$V_1/V_2 = \frac{k_1^2}{k_2^2},$$

and since the results show that the potential fall at the cathode is much greater than at the anode, it follows that the mobility k_2 of the negative ions is much greater than k_1, that of the positive. Under the conditions of these experiments

$$V_2 = \tfrac{1}{2} \frac{a^{\frac{1}{2}}}{q^{\frac{3}{2}}} \cdot \frac{i^2}{e^2} \cdot \frac{k_2}{(k_1 + k_2)^2} \cdot \beta_2^{\frac{1}{2}} \quad \text{(see p. 201)}.$$

Substituting for β_2 and neglecting k_1/k_2 we find

$$V_2 = \frac{\sqrt{\pi} i^2}{q^{\frac{3}{2}} \cdot e^{\frac{1}{2}} \cdot k_1^{\frac{1}{2}}}.$$

If l is the distance between the plates,

$$V_0 = \left(\frac{a}{q}\right)^{\frac{1}{2}} \frac{i}{e (k_1 + k_2)} \left(l - \frac{i}{qe}\right) \quad \text{(see p. 198)},$$

the term i/qe being the width of the region occupied by the electrode falls of potential. The term V_1 varies as i^2 but is unimportant.

We see at once that V is of the form $Ai^2 + Bi$; Wilson[1] has shown that this result is confirmed experimentally with considerable accuracy. Fig. 104 shows the type of apparatus used, the flame being formed from twenty-five small brass tubes, to each of which a fused quartz tube is attached with a short piece of india-rubber tubing. The electrodes E are platinum discs, and their distance apart can be adjusted. In this case the current is horizontal and the vertical velocity of the flame makes very little

[1] H. A. Wilson, *Electrical Properties of Flames*, p. 62, 1912.

difference. Experiments have sometimes been made with horizontal gauze electrodes, through which the flame streamed. In this case the theory must be modified, especially if the lower electrode is negative, and there is then a supply of negative ions through the cathode. The value of the constant B is on this theory approximately

$$\left(\frac{a}{q}\right)^{\frac{1}{2}} \frac{l}{ek_2} = \frac{l}{nek_2},$$

where n is the number of ions of each kind per c.c. The potential gradient is therefore i/nek_2, and this portion of the flame obeys Ohm's law, the specific conductivity being nek_2. Wilson[1] and

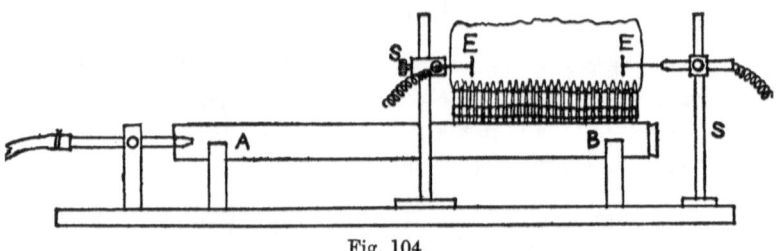

Fig. 104.

Gold[2] have verified this over a considerable range of current. The specific resistance varies greatly with the mixture being burnt. Wilson found in one case about 2×10^6 ohms cm. Gold's experiments gave about $1 \cdot 4 \times 10^5$ ohms cm. The width of the layer in which the cathode fall of potential occurs is taken on this theory as

$$\frac{k_2}{k_1 + k_2} \ i/qe \ \text{or} \ i/qe \ \text{approximately}.$$

The theory assumes that q is constant; if the electrodes are massive pieces of metal they will considerably cool the flame near them. Now, as pointed out in the earlier editions of this book, the process of ionisation is analogous to the dissociation of a diatomic gas into atoms, and the expression for the amount of this dissociation contains a factor $\epsilon^{-a/\theta}$, where θ is the absolute temperature. Thus a comparatively slight cooling of the gas near the cathode would produce a great diminution in q. This will increase V_2 and also

[1] H. A. Wilson, *Phil. Mag.* Oct. 1905.
[2] Gold, *Proc. Roy. Soc.* A, lxxix. 1907.

the thickness of the region of non-uniform electric force. By adjusting the temperatures of the electrodes the fall of potential near them can be controlled, and in particular Marx[1] has shown that by keeping the positive electrode cool, the potential drop there may be made greater than at the cathode.

If the cathode emits any considerable number of electrons, these will carry part of the current and so reduce V_2. The extent of this will depend on the ratio of i to i_0, the current carried by these electrons. When $i = i_0$ there is no cathode fall. Gold[2] however has found that the fall in potential remains even when the current is reduced to a very small value, so that for ordinary currents the effect is unimportant. This was for metal electrodes, for a cathode covered with lime Tufts[3] found that the cathode fall was greatly reduced. This is doubtless due to the copious thermionic emission from incandescent lime.

Conductivity of Gases containing Salt Vapours.

123. When the vapours of salts are introduced into a flame the conductivity between metallic terminals is very greatly increased, and the electrical properties are simpler and more regular than in pure flames; the laws of the flow of electricity through these salt-laden flames have been investigated by Arrhenius[4] and H. A. Wilson[5]. The method—devised by Arrhenius and adopted by Wilson—of introducing the salt into the flame was as follows: a dilute solution of the salt was sprayed into exceedingly fine drops by a Gouy sprayer, the spray got well mixed with the coal gas on its way to the burner, and in the flame the water evaporated and the salt vaporised. The amount of salt supplied to the flame in unit time was estimated by determining the rate at which a bead of salt, introduced into an equal and similar flame so as to produce the same coloration as that produced by the spray in the original flame, burnt away. The salts used were chiefly the haloid and oxy-salts of the alkali metals and earths. The conductivity due to the salt was determined by subtracting from

[1] Marx, *Ann. der Phys.* iv. p. 2, 1900.
[2] Gold, *Proc. Roy. Soc.* A, lxxix. 1907.
[3] Tufts, *Phys. Zeits.* v. p. 76, 1904.
[4] Arrhenius, *Wied. Ann.* xlii. p. 18, 1891.
[5] H. A. Wilson, *Phil. Trans.* A, cxcii. p. 499, 1899.

the current observed when the salt was in the flame the current with the same electromotive force in the pure flame. It was found that when the concentration of the solutions is small, equivalent solutions[1] of all salts of the same metal impart the same conductivity to the flame. With large concentration this is no longer the case, the oxy-salts giving greater conductivity than the haloid salts. According to Arrhenius all the salts in the flame are converted into hydroxides, so that whatever salts are used, the metal in the flame always occurs in the same form. The relation between the current and the electromotive force is represented by Fig. 105 taken from Wilson's paper. In a later paper Wilson has shown that these curves are parabolas.

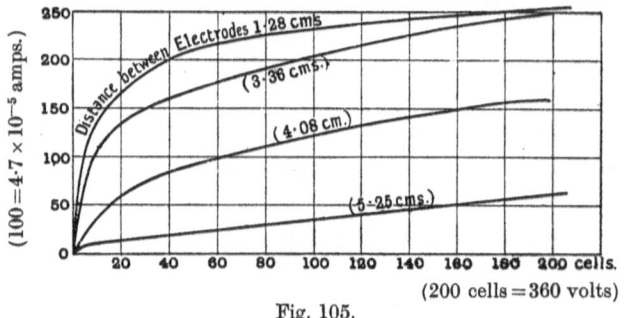

Fig. 105.

When the salt vapour is put into the flame by means of a bead of salt on a platinum wire, very little effect on the current is produced when the bead is inserted in the body of the flame, whereas a very large increase in the current occurs when the bead is put close to the cathode. This does not however necessarily mean that the salt vapour is not ionised except close to the cathode, for the velocity of the negative ion is very much greater than that of the positive. When this is the case the distribution of the electric force is such that increased ionisation produces little effect except close to the cathode. This is because the greater part of V is V_2, the fall at the cathode, which depends on the value of q in the region near the cathode, and will not be affected unless the salt gets near the cathode. It then decreases rapidly, varying

---

[1] Equivalent solutions are those in which the weight of salt per litre is proportional to the molecular weight of the salt.

as $q^{-\frac{3}{2}}$. That there is increased ionisation when the salt is placed in the body of the flame is shown very clearly by an experiment made by H. A. Wilson; he measured the potential difference between two neighbouring points a and b in a pure flame and found that the electric force was about 1·6 volts per cm., he then introduced between a and b a bead of salt and found that though the current was not appreciably altered the electric force between a and b had fallen to a very small fraction of a volt per cm.; since the electric force is equal to

$$\frac{i}{k_2}\sqrt{\frac{a}{qe^2}},$$

this result would indicate that the value of q in the salted flame must be some hundreds of times its value in the unsalted.

We may write the equation for V in the form

$$V = \left(\frac{i}{q^{\frac{1}{2}}}\right)^2 \frac{1}{q^{\frac{1}{2}}}\left(\frac{\sqrt{\pi}}{e^{\frac{1}{2}}k_1^{\frac{1}{2}}} - \frac{a^{\frac{1}{2}}}{e^2 k_2}\right) + \left(\frac{i}{q^{\frac{1}{2}}}\right)\frac{a^{\frac{1}{2}}l}{ek_2}.$$

We see that when V is constant $i/q^{\frac{1}{2}}$ must increase as q increases, and that for large values of i, i will vary approximately as $q^{\frac{3}{4}}$. Since $i/q^{\frac{1}{2}}$ increases with q we see that for a constant difference of potential the electric force in the body of the flame will increase with q, while the fall of potential at the cathode will diminish. Thus the potential gradient in the uniform part of the flame will be steeper in a salted flame than in a pure one, while the cathode fall will be less; we see too that the thickness of the layer near the cathode where the electric force is variable is less in a salted than in a pure flame, so that the uniform part of the field comes closer to the cathode.

123·1. Fig. 106 shows the distribution of potential between the electrodes with potassium carbonate on the negative electrode. Comparing this with Fig. 103 we see that the changes are those predicted by the theory.

Wilson has shown that, if some potassium carbonate is put on one of two electrodes in a flame, the difference in resistance in the two directions will rectify an alternating current passed through the flame. When the flame is completely filled with salt vapour the distribution of potential between the electrodes becomes again similar to that in a flame without salt, but the current is much greater for a given potential difference.

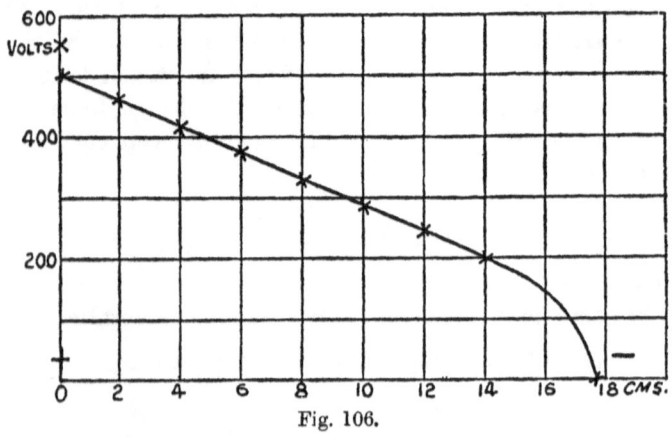

Fig. 106.

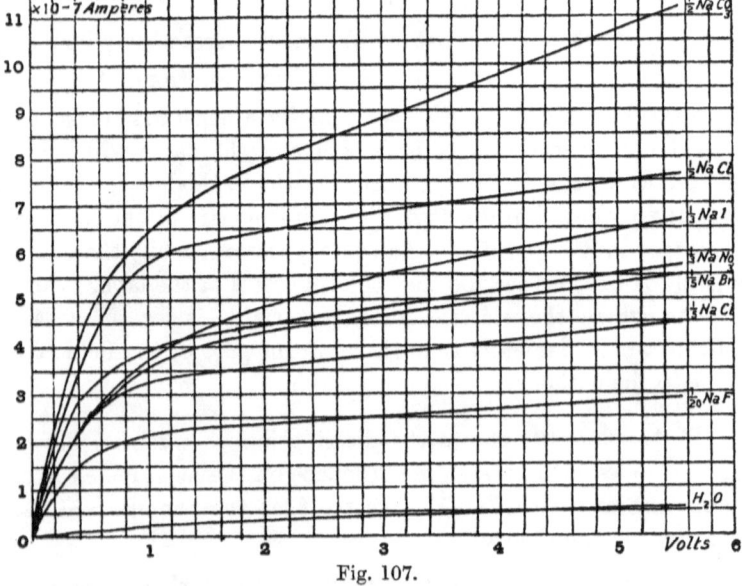

Fig. 107.

In the early measurements of conductivity the whole flame was filled with salt and the total change in current measured. Thus Arrhenius found that $i = A\sqrt{KV}$ approximately. Curves given in Fig. 107 taken from a paper by Smithells, Dawson and Wilson[1] show the variation of the current with the voltage, and the strength K of the solution, for a series of salts. In these experiments the chief effect was at the cathode. Recent measurements have aimed at determining the conductivity in the region of constant potential gradient. This gives nek_2, and as k_2 is probably independent of the presence of the salt, the ions being in all cases electrons, it is a measure of the number of ions present per c.c.

The Conductivity given to the Flame by the Salts of the different Alkali Metals under the same condition as to temperature, potential difference and concentration.

124. The Caesium salts conduct the best, and then follow in order the salts of Rubidium, Potassium, Sodium, Lithium, and Hydrogen. The order of the conductivities is thus the same as that of the atomic weights of the metals; the difference between the metals is very large, as is shown by the following table given by H. A. Wilson:

	Chlorides			Nitrates		
Potential difference	5·60	·795	·237	5·60	·795	·237
	Current			Current		
Caesium	123	60·5	22·2	303	115	36·6
Rubidium	41·4	26·4	11·3	213	82·4	25·9
Potassium	21·0	13·4	5·75	68·4	29·3	9·35
Sodium	3·49	2·45	1·15	3·88	2·67	1·32
Lithium	1·29	·87	·41	1·47	·99	·53
Hydrogen	·75	...	·27			

The above refers to the total change in current between concentric cylindrical electrodes, and includes the effect on the electrode falls of potential and on the conductivity in the region of uniform field (see below).

[1] *Phil. Trans.* A, cxciii. p. 89, 1899.

On the Variation of Conductivity with the strength of the Solution and the nature of the Salt.

125·1. In experiments on these points the conductivity is determined from the potential difference observed between two probes inserted in the flame in the region of uniform temperature gradient, and from the current density through the flame. The use of probes may lead to errors which have been considered in detail by Zachmann[1]. Experiments by Andrade[2] and by Wilson[3] agree that for the salts of a given metal the conductivity and luminosity of the flame are proportional.

Wilson[3] has compared his experiments with the theory that the ionisation in the flame can be regarded as analogous to the dissociation of a diatomic molecule, the products of dissociation being an electron and a positively charged atom or molecule. We shall see later on that there is good reason to believe that the positive ion is a metal atom, at least in the case of the alkali metal salts, and we will treat the problem from this point of view, assuming that the salt is wholly dissociated chemically in the flame. Let the concentration of the solution sprayed be k, and let $m = kG$ be the number of gram atoms of metal per c.c. of the flame; G can be found from measurements of the cross-section of the flame, of its upward velocity, and of the amount of solution sprayed per sec. The flame molecules will also be ionised, though to a less extent. Let p^+, p^-, p, p_0^+ and p_0 be the partial pressures respectively due to the positive metallic ions, the electrons, the neutral metal atoms, the positive flame ions, and the neutral flame molecules. Then, assuming that the ionisation of the flame itself can be regarded as due to one constituent of pressure p_0, we have $K_0 = p_0^+ p^-/p_0$ and $K = p^+ p^-/p$, where K and K_0 are equilibrium constants for the ionisation of metallic atoms and of the ionised constituent of the flame respectively. If c is the conductivity (assumed to be due entirely to the electrons),

$$c = nek_2 = k_2 Nep^-/R\theta,$$

[1] Zachmann, *Ann. der Phys.* lxxiv. p. 461, 1924.
[2] Andrade, *Phil. Mag.* xxiv. p. 15, 1912.
[3] H. A. Wilson, *Phil. Trans.* A, ccxvi. p. 63, 1915.

where N is Avogadro's number and R is the gas constant for a gram molecule. Now
$$p^- = p^+ + p_0^+,$$
also
$$p + p^+ = mR\theta = p^+p^-/K + p^+,$$
so
$$p^+ = kGR\theta/(1 + p^-/K).$$
Again
$$p_0^+ = p_0 K_0/p^-.$$
Hence
$$p^- = \left[\frac{kGR\theta}{1 + p^-/K} + \frac{p_0 K_0}{p^-}\right],$$
and
$$\{(p^-)^2 - p_0 K_0\}(1 + p^-/K) = kGR\theta p^-.$$

Let c_0, p_0^- be the values of c, p^- for the flame without salt, *i.e.* when $k = 0$, then $p_0^- = \sqrt{p_0 K_0}$ and $p^- = c\sqrt{p_0 K_0}/c_0$. Hence
$$(c^2 - c_0^2)(1 + c\sqrt{p_0 K_0}/c_0 K) = c \cdot c_0 kGR\theta/\sqrt{p_0 K_0}$$
or
$$\frac{kc \cdot c_0}{c^2 - c_0^2} = \frac{\sqrt{p_0 K_0}}{RG\theta} + \frac{c p_0 K_0}{c_0 K R G \theta} = b + ac/c_0,$$

where b and a are constants for any given temperature. Wilson finds good agreement with this equation for caesium chloride, taking $b = 10^{-3}$ and $a = 10^{-4}$, k being measured in grams per litre. The proportion of atoms ionised can also be found. It is
$$p^+/(p + p^+) = 1/(1 + p^-/K) = b/(b + ac/c_0).$$

For very small concentrations of salt this gives about 90 per cent. ionised, for a solution of 8 grams per litre 3·5 per cent. were ionised under the conditions of these experiments.

Wilson also examined the behaviour of RbCl, NaCl, K_2CO_3 and found good agreement with these also. The values of b were proportional to the weights containing equal numbers of metal atoms (*e.g.* for K_2CO_3, *half* the molecular weight), k being measured in grams per litre. He also found that solutions of different salts of the same metal having equal numbers of atoms per c.c. produced equal effects on the conductivity. This held for large concentrations as well as small, so the contrary result mentioned on p. 408 must be due to a difference in the action of the salts on the electrodes causing differences in the cathode fall of potential. This, together with the law of variation of b, is strong evidence that the salts undergo a change preparatory to being ionised, so that the same system is ionised in each case.

For large conductivities $c^2 \propto k$. In this case the conductivity is wholly due to the salt, and the ratio c^2/k gives a measure of the ease of ionisation of the metal, being proportional to K. The following table, taken from Wilson's paper, shows the reciprocal of this quantity, namely the relative concentrations (caesium = 1) required to give equal conductivities, and the percentage ionised when the concentration is very small.

Metal	Atomic Weight	Concentration	Per cent. ionised
Na	23	626·0	1·6
K	39	7·25	58
Rb	85	4·48	69
Cs	133	1·00	91

The flame was a Bunsen flame burning a mixture of gas and air, the temperature is not stated, but was probably about 2000° absolute.

Barnes[1] has confirmed the formula in the case of caesium chloride, and Bennett[2] in the case of caesium and rubidium chlorides. Bryan[3], however, found appreciable discrepancies and considered that a formula of the type $k^{\cdot 435} = Ac + B$ fitted his results better.

The following table, taken from Zachmann's[4] paper, shows the value of the absolute conductivity, in some cases, of flames sprayed with sodium vapour.

Nature of flame	Metal atoms per cm.³	Specific conductivity in mhos.	Temp. abs.
Pure flame	—	0·5–0·8 × 10⁻⁶	—
Illuminating gas flame (1/100 normal solution)	6·28 × 10¹²	10–15 × 10⁻⁶	2000°
Illuminating gas flame (100/100 normal solution)	6·28 × 10¹⁴	100–140 × 10⁻⁶	2000°
Alcohol flame (100/100 normal solution)	7·56 × 10¹⁴	200 × 10⁻⁶	2050°

[1] Barnes, *Phys. Rev.* xxiii. p. 178, 1924.
[2] Bennett, *Phil. Mag.* iii. p. 127, 1927.
[3] Bryan, *Phys. Rev.* xviii. p. 275, 1921.
[4] Zachmann, *Ann. der Phys.* lxxiv. p. 461, 1924.

125·2. If two salts are introduced simultaneously, the conductivity produced will be less than the sum of the two acting separately, for the electrons due to each will tend to diminish the dissociation of the other, just as the presence of a common ion 'suppresses' the ionisation of an electrolyte. Wilson shows that on the above theory, and if the conductivities are large, $c^2 = c_1^2 + c_2^2$, where c_1 and c_2 are the conductivities for the two salts separately. He finds this result confirmed by experiment, as also does Zachmann (*loc. cit.*). Bryan finds that his experiments agree with the formula

$$c^{2\cdot 3} = c_1^{2\cdot 3} + c_2^{2\cdot 3}.$$

It must be remembered that the above theory depends on the salt being wholly transformed into whatever bodies are directly ionised. It is possible that this transformation is not complete in some cases at the temperatures used. This may explain the different results found by Smithells, Dawson and Wilson for different salts of the same metal when the concentrations were large, but they are probably due to electrode effects. For solutions of salts of the alkaline earths Bryan[1] found $c^2 - c_0^2 = Ak^{\frac{2}{3}}$, which he shows can be accounted for if it is supposed that in the flame these salts are transformed into the involatile oxides, which then form small solid particles and emit electrons. The same formula also holds for solutions of $AlCl_3$ and H_3BO_3, which however give only a slight increase of conductivity.

125·3. In most of these experiments the velocity of the flame is measured by introducing puffs of salt-laden air at regular intervals and viewing these stroboscopically. The temperature is determined by comparing the light from the flame containing a little sodium with that from a black body. The latter is viewed through the flame in a spectroscope and its temperature adjusted till the D lines merge in brightness with the continuous spectrum. The temperature of the black body is then determined with an optical pyrometer.

Variation of the Conductivity with the Temperature of the Flame.

125·4. The theory of conductivity in flames has recently been much strengthened by the application of thermodynamics. Saha[2]

[1] Bryan, *Phys. Rev.* xviii. p. 275, 1921.
[2] Saha, *Proc. Roy. Soc.* A, xcix. p. 135, 1921, and other papers.

has applied Nernst's equation of the 'reaction isobar' to ionisation by heat. If K is the equilibrium constant as defined above, the equation is
$$\log_{10} K = -\frac{U}{4 \cdot 571 \theta} + \Sigma nc_p \frac{\log_{10} \theta}{R} + \Sigma nC.$$
Here U is the heat of reaction per gram molecule, θ the absolute temperature, R the gas constant for a gram molecule and the summations refer to the molecules taking part in the chemical equation, those appearing on opposite sides of the equation having opposite signs. Finally the quantities C are the so-called 'chemical constants' of the different molecular species. In this case the equation is $M \rightleftharpoons M^+ + e - U$, where M represents a neutral metal atom and M^+ the same singly ionised, $K = p^+p^-/p$, the pressures being measured in atmospheres. The specific heats and chemical constants of M and M^+ cancel out. Those for the electrons are calculated by regarding the swarm of free electrons as a monatomic gas. Thus $c_p = 5/2R$, and Saha uses for C the Sackur-Tetrode-Stern relation $C = -1 \cdot 6 + \frac{3}{2} \log M'$, where M' is the molecular weight, in this case $5 \cdot 5 \times 10^{-4}$. Hence $C = -6 \cdot 5$. It is convenient to express U in terms of the ionisation potential V measured in volts, and the equation becomes
$$\log_{10} K = -\frac{5050 V}{\theta} + 2 \cdot 5 \log_{10} \theta - 6 \cdot 5.$$
Saha originally used this equation to calculate the ionisation in the sun and stars, and his work and that of Russell, R. H. Fowler and Milne have been very effective in throwing light on many points in solar and stellar physics.

Under laboratory conditions the number of substances which show this effect is limited to those for which the energy of ionisation is not a large multiple of the mean energy of a molecule at the available temperatures; in fact the expression for the number of ions will contain a factor $\epsilon^{-\frac{Ve}{k\theta}}$, as in the corresponding case of emission from solids, k being Boltzmann's constant, and this factor soon becomes negligibly small if $Ve/k\theta$ gets large. In addition, ϕ_0, the work done to remove an electron is less for most solids than the ionisation potential of most gases, so that if the gases are heated by contact with a solid the emission from the latter will swamp the ionisation in the gas. For this reason it is in flames that the theory can best be tested, and of these, those containing salts of the alkali metals are best as the ionisation potentials of these metals are the smallest known. The increasing effectiveness of the alkalis with increasing atomic weight is explained by the corresponding decrease in ionisation potential. Thus V for Cs is 3·9 volts, for Rb 4·1, K 4·3, Na 5·1, and Li 5·4. Those of Ba (5·2), Sr (5·7) and Ca (6·1) are also fairly low, and these elements have a considerable effect on the conductivity of a flame.

Noyes and Wilson[1] were able to show that the values of K found from experiments on conductivity were closely parallel to those calculated from Saha's equation. Experiments directly to test Saha's formula have been made by Barnes[2] and by Bennett[3]. Barnes finds very good agreement between values of θ, calculated by determining K by experiments on the conductivity and using Saha's equation, and those observed directly. In several cases the error was less than 25°. He used a solution of CsCl sprayed into the flame, and determined the constants a and b in the equation

$$\frac{kcc_0}{c^2 - c_0^2} = b + \frac{ac}{c_0}.$$

The value of K is then given by $K = (b^2/a)\, GR\theta$ (see § 125·1). Bennett used a similar method with RbCl, except that he made experiments at several temperatures, while Barnes' were all done at about 2000° abs. Bennett found good agreement at the higher temperatures but considerable differences below 1800° abs. The method assumes complete dissociation of the salt into atoms, and it is possible that this is not the case at the lower temperatures. The success of the theory is however ample to prove that it is the metallic atoms which are being ionised. Bennett also tried to calculate V for the molecules of the flame from the variation of K_0 with temperature, using Saha's equation. The result, 1·8 volts, is obviously wrong, as this would make the gases of the flame more easily ionisable than the metal atoms, which would therefore not increase its conductivity. A rough calculation, using his data and comparing the conductivity of the unsalted flame at 1720° abs. with that of one containing salt at the same temperature, gives $V = 7\cdot 4$ volts. As there are many kinds of molecules in the flame

[1] Noyes and Wilson, *Proc. Nat. Acad. Sci.* viii. p. 303, 1922.
[2] *Loc. cit.*   [3] *Loc. cit.*

this will be a mean, but strongly weighted in favour of the lower values. The failure of the temperature variation method is probably due to the fact that the composition of the flame was altered in order to produce the required changes in temperature.

125·5. Langmuir and Kingdon[1] have shown that caesium vapour in contact with a heated filament becomes ionised in accordance with Saha's theory. At 1200° abs. the results are in good agreement with the equation, at high temperatures the rate of production of positive ions is limited by the number of atoms striking the filament, practically all that do so being ionised. At low temperatures a layer of adsorbed ions is formed on the filament which causes a great increase in the electronic emission. The caesium can regain electrons from the filament and so the emission of positive ions is decreased. Killian[2] has obtained similar results with potassium and rubidium, and uses the results at high temperatures to determine their vapour pressures. The positive ions can be detected by their effect in partially neutralising the space charge.

Saha, Sur and Mazundar[3] have recently made experiments to test the thermal ionisation of metallic vapours, taking great care to avoid the possibility of the conductivity observed being due to emission from the walls of the apparatus. Using temperatures up to 2000° C. they found a small effect with Mg, but with Na and K the effect was so large as to require a milliameter for its measurement. The results are in general agreement with Saha's theory.

Velocity of the Ions.

126. The velocity of the ions in flames containing salt vapours has been determined by H. A. Wilson[4], who used a method of which the principle is as follows. Suppose that in a flame we have two electrodes, one vertically over the other, and that we introduce a bead of salt just underneath the upper electrode; the vapour from this bead will be carried along by the upward rush of gases in the flame, and unless the ions in the salt vapour are

---

[1] Langmuir and Kingdon, *Proc. Roy. Soc.* A, cvii. p. 61, 1925.
[2] Killian, *Phys. Rev.* xxvii. p. 578.
[3] Saha, Sur and Mazundar, *Zeits. f. Phys.* xl. p. 648.
[4] H. A. Wilson, *Phil. Trans.* A, cxcii. p. 499, 1899.

driven downwards by the electric field between the electrodes, none of them will reach the lower electrode. If however the ions from the salt do not reach the electrode the current between the electrodes will be unaffected by the presence of the salt. Thus when the potential difference between the electrodes is small the current will not be increased by the introduction of the salt, but as soon as the electric force between the electrodes is sufficient to drive one of the ions against the blast in the flame, the current will be increased by the bead of salt. This is illustrated by the curves in Fig. 108 taken from Wilson's paper; we see that when

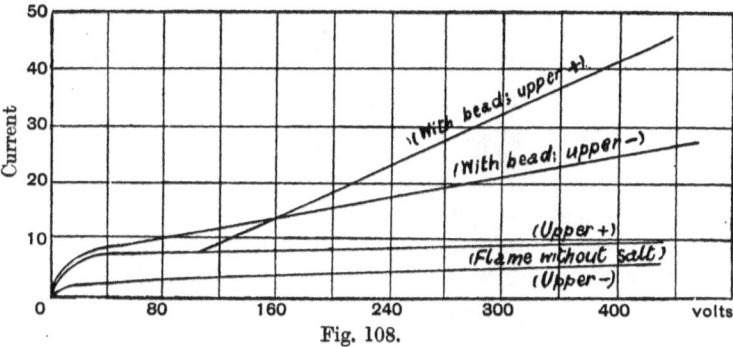

Fig. 108.

the upper electrode was positive the current was not increased by the bead until the potential difference between the electrodes was about 100 volts, while for greater differences of potential the bead produced a substantial increase in the current. Thus when there was a difference of 100 volts between the electrodes, the smallest electric force in the space traversed by the ion must be just sufficient to give to the positive ion a downward velocity equal to the upward velocity of the gas in the flame. Since the electric field is not uniform between the electrodes (see p. 404), it is necessary to measure the distribution of potential between the electrodes in order to determine the minimum electric force; when this and the upward velocity of the gas in the flame are known we can determine the velocity of the ions in a flame under a given electric force. By this and similar methods Wilson deduced the following values for the velocities of the ions under an electric force of a volt per centimetre.

In a flame whose temperature was estimated to be about 2000° C., the velocity of the negative ion, whatever salts were put in the flame, was about 1000 cm./sec.

The velocities of the positive ions of salts of Caesium, Rubidium, Potassium, Sodium, and Lithium were all equal, and were about 62 cm./sec.

In a stream of hot air whose temperature was estimated at 1000° C. the following results were obtained for the velocities under a potential gradient of 1 volt per cm.

Negative ions ... ... 26 cm./sec.
Positive ions of salts of Li, Na, K, Rb, and Cs 7·2 cm./sec.
Positive ions of salts of Ba, Sr, and Ca ... 3·8 cm./sec.

The absolute numbers must be regarded as only approximately true, the relative values are probably much more accurate.

The velocities are very much less at 1000° C. than they are at 2000° C., but we notice that while the negative ion at the lower temperature moves at only 1/40 of its pace at the higher, the velocity of the positive ion is by the same fall in temperature only reduced to about 1/8·5 of its value.

These determinations of the velocity throw some light on the character of the ions; for suppose e is the charge of electricity on the ion, X the electric force acting upon it, the mechanical force acting on the ion is equal to Xe; if λ is the mean free path of the ion, v its velocity of translation, then the time between two collisions is λ/v, and in this time the force acting upon it will give it a velocity in the direction of the force equal to $Xe\lambda/vm$, where m is the mass of the ion; the average velocity parallel to X due to the electric force will therefore be $Xe\lambda/2vm$, and this will be the velocity with which the ion will, under the electric force, move through the gas. The equal velocity of all negative ions from whatever source they may be derived might at first sight seem to indicate that, as Arrhenius supposed, all the salts were converted to hydroxides in the flame, and that the negative ion was in every case the radicle OH: let us calculate what on this supposition would be the velocity of the negative ion at a temperature of 2000° C. We do not know the free path of OH through a mixture of coal gas and air, but as the free path of the molecule H_2 through hydrogen at 0° C. and at atmospheric pressure is $1\cdot8 \times 10^{-5}$ cm., and the free path of O_2 through oxygen under the same circumstances is $1\cdot06 \times 10^{-5}$ cm., we may as a rough approximation take for the mean free path of OH through the mixture the value $1\cdot4 \times 10^{-5}$ cm. at 0° C.; at 2000° C. λ the mean free path would be this value multiplied by 2273/273, *i.e.* $1\cdot2 \times 10^{-4}$. To get the value of v we remember that mv^2 is the same for all gases at the same temperature, while at different temperatures it is proportional to the absolute temperature. For O_2 at 0° C. $v = 4\cdot25 \times 10^4$ cm./sec., hence for OH at 0° C. $v = 5\cdot6 \times 10^4$ cm./sec., and for OH at 2000° C. $v = 1\cdot6 \times 10^5$: e/m for OH is equal to $1\cdot1 \times 10^3$, hence substituting these values in the expression $Xe\lambda/2vm$ and putting $X = 10^3$ we find for the velocity under the potential gradient of one volt per cm. 37 cm./sec.: the actual velocity is as we have seen 1000 cm./sec.: hence we conclude that the radicle OH cannot be the carrier of the negative charges. The great velocity of the negative ions at these high temperatures points to the conclusion that the negative ions start as electrons and gradually get loaded by molecules condensing round them; at temperatures as high as 2000° the time they exist as free electrons is an appreciable fraction of their life; while they are free electrons they have an exceedingly large velocity, so that though this is enormously reduced when they become the nucleus of a cluster, their average velocity is very considerable. At low temperatures condensation takes place much sooner, so that the average velocity is lower.

The fact that under an electric field the velocities of the positive ions of all the salts of the univalent metals are the same, shows that these too become the nucleus of a group whose size only depends upon the charge on the positive ion; since the velocities of the positive ions for the divalent metals while equal among themselves are less than those of the monovalent metals, we conclude that the divalent ions become the centres of clusters more complex than those which collect round the monovalent ions.

Marx and Moreau also made measurements of the mobilities and found values of the same order as those found by Wilson.

126·1. The above gave the state of knowledge on this subject at the time of the second edition of this book. While these older determinations of mobilities were enough to show that the negative ions are electrons, and the positive atoms, most of the methods used involved serious assumptions.

Andrade[1] has made a number of measurements of the mobility of the positive ions, using two methods. One, due to Lenard, consists in measuring the angle of deflection of a streak of salt in the flame, caused by a transverse electric field. If the electric force was X and the upward velocity of the flame U, then the tangent of the angle of deflection was assumed to be Xk_1/U. The values obtained were very low, ·04 to ·166 for Sr at one atmosphere in cm./sec. per volt/cm. While these may perhaps be explained by supposing that in the highly conducting streak with its many electrons the positive ions only remain charged for a small fraction of the time, it seems doubtful if the movement of the streak can be regarded as simply due to the mobility of the ions it contains. When a current of density i passes normally across a surface separating two gases containing different numbers of ions, the electric forces X, X' in the two gases are given by

$$i/e = n(k_1 + k_2)X = n'(k_1' + k_2')X',$$

where n, n' are the number of ions of either kind in the two gases. Thus X is in general not equal to X', and there will be an electrified layer formed on the surface of separation of density $(X - X')/4\pi$. This layer will be pulled by the electric field and will take the medium with it. It is probably this that accounts for the dividing of the streak into two which Andrade observed at high pressures.

Andrade's second method was to measure the current to a narrow strip in the plane of the negative electrode; both electrodes were outside the flame. The ions from the salt streak are moved sideways by the electric field, and upwards by the flame, and the height above the bead of salt at which the strip had to be placed in order to receive ions of the salt, gives a measure of the distance which the flame has carried the ions up in the course of their path to the electrode. If the upward velocity of the flame is known, this gives the time of passage and hence the velocity due to the

[1] Andrade, *Phil. Mag.* xxiii. p. 865, 1912.

electric field. For positive ions from $SrCl_2$, Andrade found by this method mobilities of 2·5 to 1·8.

H. A. Wilson[1] has since revised his work on the mobilities of the positive ions, and has come to the conclusion that his previous values were too high, and were probably due to particles of salt being carried back into the flame lower down by currents of air. To prevent this he fitted a cylindrical shield round the lower part of the flame (Fig. 109), and found that it required a much greater potential difference than before to make the presence of the bead of salt in the upper part of the flame appreciably affect the current between the electrodes. With this arrangement Wilson found mobilities of about 1 cm./sec. per volt/cm., but says that there may have been a few ions with velocities about equal to those of Andrade. As in the earlier experiments there was no appreciable difference in the mobilities of ions from different metals. He also tried Andrade's first method, with the modification that his field was vertical, and he used puffs of vaporised salt solution whose velocity was found stroboscopically. He could detect no change of velocity when a strong field was thrown on and off. It seems probable that the ions which are drawn away from the salted part of the flame in this method, are not numerous enough to give any appreciable colour, except when, as sometimes happened, the field was strong enough for an arc to form. Both these methods may be expected to give, not the true mobility of the ion, but the mobility multiplied by the fraction of the time during which a particular atom is an ion. As this fraction will be different for different atoms, it can easily be seen that the methods are not likely to lead to any very sharp determination even of the apparent mobility.

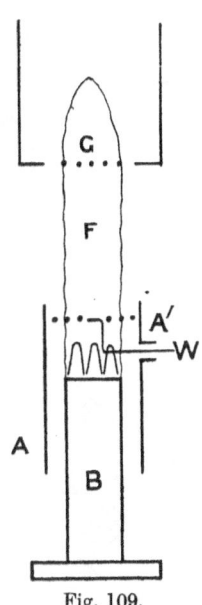

Fig. 109.

Wilson's original determination of the mobility of the negative ion, by the same method as that used for the positive, was very rough, as when the upper electrode is negative the current is

[1] H. A. Wilson, *Phil. Trans.* A, ccxvi. p. 63, 1915.

greatly increased by the reduction in the cathode fall of potential irrespective of any motion of the negative ions. Now that the number of ions per c.c. in the region of uniform electric gradient can be regarded as known, the conductivity here, which is equal to nek_2, can be used to find k_2.

Measurements on these lines have been made by Barnes and Bennett (*loc. cit.*). Barnes finds values between 20,000 and 8000, the latter value being the more reliable. Bennett gets smaller values, from 2700 at 1920° abs. (about the temperature used by Barnes) down to 1800 at 1570° abs. Two other methods of measuring the mobility of the negative ions depend on the results of the next two paragraphs.

Electrical Conductivity for rapidly alternating Currents.

126·2. Wilson and Gold[1] have investigated the electrical properties of flames for rapidly alternating currents. The method was to form a Wheatstone's bridge of three condensers, and a flame between concentric cylindrical electrodes. This system was

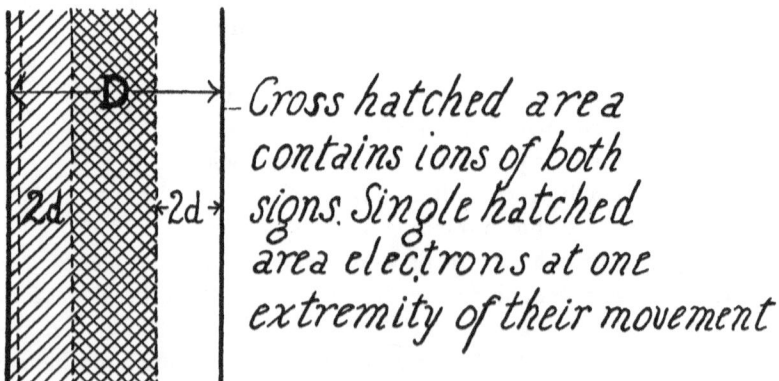

Fig. 110.

then excited by high frequency current from a Tesla coil. A galvanometer and electrolytic rectifier were connected across the bridge. By adjusting the capacities of the condensers it was found possible to get an almost perfect balance of the bridge for all frequencies, showing that the flame behaved as a fourth condenser

[1] H. A. Wilson and Gold, *Phil. Mag.* xi. p. 484, 1906.

whose capacity could thus be determined. The action of the flame may be regarded as follows. Owing to the much greater mobility of the electrons than the positive ions, the latter will hardly move in the alternating field, while the former will oscillate with an amplitude d. If $4d$ is less than the distance D between the electrodes there will be a region of width $D - 4d$, where the flame is permanently neutral, while a swarm of electrons of width $D - 2d$ will oscillate backwards and forwards. Any electron formed outside this latter region will speedily be removed to an electrode. When the electrons are symmetrically placed there is no force tending to move them, but when they are nearer one electrode than the other, the positive ions left free tend to pull them back. This is analogous to the opposing potential difference set up by charging an ordinary condenser.

On the assumption that the mass of the electrons and the resistance to their motion through the flame can both be neglected, Wilson finds that the increased capacity C due to the flame is $\sqrt{\dfrac{\rho}{8\pi V_0}}$ per unit area, where ρ is the density of charge of the positive ions, and V_0 is the amplitude of the varying external electromotive force. In fact, if t_1 and t_2 are the widths of the regions next the plates free from electrons at any instant, the potential difference

$$V = 2\pi\rho\,(t_1{}^2 - t_2{}^2) = 4\pi\rho\,(t_1 - t_2)\,d,$$

since $t_1 + t_2 = 2d$. Hence

$$\frac{dV}{dt} = 8\pi\rho \cdot \frac{dt_1}{dt} \cdot d = 8\pi \cdot i \cdot d,$$

where i is the current per unit area. But $C\dfrac{dV}{dt} = i$, and therefore $C = \dfrac{1}{8\pi d}$. The value of d is given by $V_0 = 8\pi\rho \cdot d^2$, which leads at once to Wilson's result. In agreement with this theory the capacity is found to vary inversely as the square root of the potential difference applied to the electrodes. It is also found to be approximately proportional to the square root of the conductivities of the flames for steady currents and low voltages. As the latter are probably approximately proportional to the number of ions per c.c., the variation of the capacity with ρ is in agreement with the theory.

The capacities can be used to determine ρ and the results are in reasonable agreement, in the case of salted flames, with what might be expected from the density of the salt. Gold[1] has used this method to find ρ for a salt-free flame, and hence, by measurement of the potential gradient in the uniform region, to find the mobility of the electrons. He found $k_2 = 5300$ cm./sec. per volt/cm.

The method has been brought up to date by Bryan and Wilson[2], who measure the capacity of the flame by finding its effect on the frequency of a circuit containing inductance of which it forms part. The results fit the theory as well as can be expected in view of the assumptions involved. For salted flames they give values of k_2 which vary with the root mean square electric force used. As this varies from 9 to 64 volts/cm., k_2 varies from 20,000 to 8000 approximately. This variation is in agreement with Loeb's theory (see §§ 37·6 and 39·4) and indeed the values found fit Loeb's curve for nitrogen at atmospheric pressure very well; though in view of the great difference in temperature this must probably be regarded as a coincidence only. There is a marked tendency for k to decrease in the case of flames containing very large amounts of salt. This may be due to the electrons getting attached to neutral metal atoms during part of their existence. In these experiments ρ was of the order of 1 E.S.U. per c.c. and the frequency usually 6×10^5 per sec.

Transverse Electromotive Force produced by a Magnetic Field acting on a Flame carrying a Current.

127. If an electric current is flowing through a flame parallel to the direction x, and a magnetic force at right angles to this direction, say parallel to the direction y, is applied to the flame, a transverse electromotive force is produced which is at right angles to both x and y. This electromotive force has been detected and measured by Marx[3]. The general explanation of this effect, which is analogous to the 'Hall' effect in metals, is easy; the calculation of its magnitude, except in a few special cases, is however beset by difficulties.

[1] Gold, *Proc. Roy. Soc.* A, lxxix. p. 43, 1907.
[2] Bryan, *Phys. Rev.* xxiii. p. 189, and Bryan and Wilson, next paper.
[3] Marx, *Ann. der Phys.* ii. p. 798, 1900.

As there is a current parallel to x flowing through the flame, the average direction of the positive ions will be along, say, the positive direction of x, that of the negative ions in the opposite direction. Let V be the average velocity of the positive ions, V' that of the negative; if these are moving in a magnetic field where the magnetic force H is parallel to y, they will be subject to mechanical forces tending to move them in the same direction, this direction being parallel to z, at right angles to both x and y. The magnitudes of the mechanical forces acting on the positive and negative ions are respectively HeV and HeV', where e is the charge on an ion. The displacement of the ions under these forces will (if V is not equal to V') produce a current of electricity through the flame parallel to z; if however the ions cannot escape in this direction the current will soon stop, as the accumulation of ions will produce a back pressure and an electrostatic field which will balance the effect of the mechanical forces arising from the magnetic field.

We shall now proceed to deduce the equations which give the disturbance produced by the magnetic field; these equations are not limited to the case of flames, but apply to all cases of the conduction of electricity through a gas containing ions.

Let the direction of the primary current, *i.e.* the current before the magnetic field is applied, be taken as the axis of x, let the magnetic force act downwards at right angles to the plane of the paper: then the force on the ions will be in the plane of the paper and at right angles to the axis of x; we shall take the axis of z in this direction.

Let H be the intensity of the magnetic force,

X, Z the components of the electric force parallel to the axes of x and z respectively,

u, v the velocities of the positive and negative ions under unit electric force,

p_1, p_2 the pressures at any point due to the positive and negative ions respectively,

m, n the numbers of positive and negative ions per cubic centimetre at any point.

We shall assume that these ions behave like a perfect gas, so that $p_1 = Rm$, $p_2 = Rn$, where R is a constant proportional to the absolute temperature.

Let us consider first the positive ions; their velocity parallel to the axis of x is Xu, hence the mechanical force on an ion parallel to z due to the magnetic field is $euXH$; the force on the ion due to the electric field is Ze, and the force on the ions in unit volume due to the variation in the pressure at different points in the field is $-dp_1/dz$, hence the total force parallel to z on the positive ions in unit volume is equal to

$$-\frac{dp_1}{dz} + me\,(uXH + Z),$$

and the number crossing in unit time one square centimetre of surface at right angles to z is equal to

$$\frac{u}{e}\left\{-\frac{dp_1}{dz} + me\,(uXH + Z)\right\};$$

similarly the flux parallel to x is equal to

$$\frac{u}{e}\left\{-\frac{dp_1}{dx} + me\,(-uZH + X)\right\};$$

if we neglect terms depending upon H^2 the term uZH may be omitted, and the flux parallel to x is then

$$\frac{u}{e}\left\{-\frac{dp_1}{dx} + meX\right\}.$$

Similarly the flux of the negative ions parallel to z is equal to

$$\frac{v}{e}\left\{-\frac{dp_2}{dz} + ne\,(vXH - Z)\right\},$$

and the flux parallel to x to

$$\frac{v}{e}\left(-\frac{dp_2}{dx} - neX\right).$$

Let q be the number of ions produced in one cubic centimetre of the gas in one second, αnm the number of ions which recombine in one second in unit volume; then by the equation of continuity we have, when things are in a steady state,

$$\frac{u}{e}\frac{d}{dz}\left\{-\frac{dp_1}{dz} + me\,(uXH + Z)\right\} + \frac{u}{e}\frac{d}{dx}\left(-\frac{dp_1}{dx} + meX\right) = q - \alpha mn,$$

$$\frac{v}{e}\frac{d}{dz}\left\{-\frac{dp_2}{dz} + ne\,(vXH - Z)\right\} + \frac{v}{e}\frac{d}{dx}\left(-\frac{dp_2}{dx} - neX\right) = q - \alpha mn;$$

we have also, using electromagnetic units,
$$\frac{dX}{dx} + \frac{dZ}{dz} = 4\pi e (m-n) V^2,$$
where V is the velocity of light,

and
$$\frac{dX}{dz} - \frac{dZ}{dx} = 0.$$

Since $p_1 = Rm$, $p_2 = Rn$, we have as many equations as there are variables, p_1, p_2, m, n, X, Z. The solution will however depend very greatly upon the boundary conditions; thus one solution is $Z = 0$, p_1 and p_2 constant, and X independent of z and the same as when the magnetic force is zero: this, however, involves a transverse flux of positive ions equal to mu^2XH and of negative ions equal to nv^2XH, and is not consistent with a steady state unless there is some means for this transverse stream to escape. If there is no way of escape for the transverse streams of ions the flux of the ions parallel to z must vanish at the boundaries of the gas. Let us suppose that it vanishes throughout the gas, then we have

$$-\frac{dp_1}{dz} + me(uXH + Z) = 0 \quad \ldots\ldots\ldots(1),$$

$$-\frac{dp_2}{dz} + ne(vXH - Z) = 0 \quad \ldots\ldots\ldots(2).$$

Putting $p_1 = Rm$, $p_2 = Rn$ and $(m - n)e = \rho$ we get from (1) and (2)

$$\frac{R}{e}\frac{d\rho}{dz} = eXH(mu - nv) + Ze(m + n) \quad \ldots\ldots(3);$$

and since
$$4\pi \frac{d\rho}{dz} = \frac{1}{V^2}\left(\frac{d^2Z}{dx^2} + \frac{d^2Z}{dz^2}\right),$$

(3) becomes

$$\frac{R}{4\pi eV^2}\left(\frac{d^2Z}{dx^2} + \frac{d^2Z}{dz^2}\right) = eXH(mu - nv) + Ze(m + n) \ldots(4),$$

an equation to find Z. In the terms on the right-hand side, we may put for X, m, n the values when $H = 0$, if we are content to neglect terms in H^2.

Since $V^2 = 9 \times 10^{20}$, $e = 1\cdot 59 \times 10^{-20}$ (in electromagnetic units), $R = 3\cdot 7 \times 10^{-14}$, for a gas at $0°$ C., we see that (4) may be written

$$2 \times 10^{-16}\left(\frac{d^2Z}{dx^2} + \frac{d^2Z}{dz^2}\right) = eXH(mu - nv) + Ze(m + n).$$

If the sum of the partial pressures due to the positive and negative ions were 1 atmosphere, $e(m+n)$ would be about ·4, hence we see that if the pressure of the ions is large compared with 10^{-15} atmospheres and if Z does not vary exceedingly rapidly with x, a very approximate solution of (4) will be

$$Z = \frac{XH(nv-mu)}{m+n} \quad \ldots \ldots \ldots \ldots (5).$$

This may be written

$$Z = \frac{H(i_n - i_p)}{e(m+n)},$$

where i_n and i_p are respectively the currents carried by the negative and positive ions.

At a place where there is no free electricity $m = n$; in this case (5) becomes

$$Z = \tfrac{1}{2}XH(v-u).$$

This is the formula usually employed, but we see from the preceding work that it is only applicable in a very special case.

No account has been taken in the above calculation of the velocity of the gases in the flame. This can be allowed for as follows. Let the quantities X, Z, H be supposed measured with reference to the moving flame gases, then the above theory will hold, but the electric and magnetic fields with respect to the observer will be X', Y', Z', H_x', H_y', H_z'. To a first approximation, if the components of the flame velocity are U_x, U_y, U_z, we have

$$X' = X + HU_z, \quad V^2 H_x' = U_y Z,$$
$$Y' = Y = 0, \quad V^2 H_y' = V^2 H - U_x Z + U_z X,$$
$$Z' = Z - HU_x, \quad V^2 H_z' = -U_y X.$$

The velocity of the flame is of the order 300 cm./sec., $V^2 = 9 \times 10^{20}$, H is of the order 10^2 and X of the order 10^9, Z being usually smaller.

The equations reduce to $X' = X$, $Y' = 0$, $Z' = Z - HU_x$, $H_y' = H$, H_x' and H_z' negligible. Thus the effect of the motion of the flame is to make the observed cross force less than that calculated above by HU_x. This is usually a small correction unless X is small. If the main current is horizontal $U_x = 0$ and the original expression for Z holds good.

When solutions of KCl of various strengths were sprayed into a flame Marx[1] found values of Z/XH varying from $10\cdot18 \times 10^{-6}$ for the pure flame to $3\cdot7 \times 10^{-6}$ when a saturated solution of KCl was sprayed into it, the sign of the result showing that the velocity of the negative ion is greater than that of the positive. If we apply the preceding formula we find, on the supposition that the measurements were made in a part of the flame where there was no free electricity, that the difference between the velocities of the negative and positive ions under an electric force of one volt per centimetre, *i.e.* 10^8 units, would vary from 2036 cm./sec. for the pure flame to 740 cm./sec. for the flame containing the concentrated solution; the value 940 found by H. A. Wilson by direct experiment is between these limits.

Wilson has pointed out that Marx's result is likely to be too low, as the ions could escape round the edges of the flame to the main electrodes.

If the electric and magnetic forces are considerable there will, when there is no escape for the transverse flow of ions, be very considerable variations in the number of ions in the gas; for putting $p_1 = Rm$, $p_2 = Rn$, we get from equations (1) and (2),

$$R \frac{d}{dz} \log mn = eXH(u+v),$$

or
$$mn = C\epsilon^{eX(u+v)z/R},$$

where C is a constant. To see what variation this implies let us take the case of air ionised by X-rays, the pressure being $1/1000$ of an atmosphere; then since $u+v$ at atmospheric pressure is 3×10^{-8} cm./sec., at the assumed pressure it will be 3×10^{-5}, and if X is 10 volts a centimetre, *i.e.* 10^9, and $H = 10^2$, then since $e/R = 4 \times 10^{-7}$, we see that

$$mn = C\epsilon^{1\cdot2z};$$

thus in the space of a centimetre parallel to z, mn will about triple in value: this variation in the number of ions will affect the distribution of the current parallel to x, the current will be greatest where there are most ions and will therefore no longer be independent of z: this variation in the current may affect the distribution of potential between the electrodes and thus introduce fresh sources of disturbance into the problem.

[1] Marx, *Ann. der Phys.* ii. p. 798, 1900.

In the case when there are only ions of one sign present, say the negative, there is a very simple solution of the preceding equations, for we see that $Z = HXv$, p_2 constant and X the same as when there is no magnetic force, satisfies these equations.

127·1. Some later measurements of k_2 have been attempted by this method. Wilson[1] found a value of about 10,000, and later one of 2450. Watt[2] has made a recent determination using a method similar to that used by Wilson in the latter experiment. The main current passed horizontally along the flame which was of considerable length, and burnt between the poles (10 cm. square) of a large electromagnet with field horizontal. To measure the Hall effect two probes were placed in the flame at a constant distance apart, so that the line joining them could be rotated about a horizontal axis, parallel to the magnetic field. The probes were connected to the quadrants of a quadrant electrometer, and turned until there was no difference of potential between them when the magnetic field was off. When the field was on, the angle θ through which they had to be turned, to again have the same potential, was taken as a measure of the effect, and the mobility calculated from the formula $k_2 = \dfrac{\tan \theta}{H}$. The results varied with the electric force at the point. For unsalted flames k_2 varied from 2660 for 5 volts/cm. to 1570 for 51 volts/cm., for salted flames it varied from 2640 for ·76 volts/cm. to 1580 for 32 volts/cm. For equal fields the mobility was slightly higher for unsalted flames. It will be noticed that, writing $\tan \theta = Z/H$, the theory developed above would give $k_2 = \dfrac{2 \tan \theta}{H}$. The difference appears to be due to the neglect of the terms $\dfrac{dp_1}{dz}, \dfrac{dp_2}{dz}$ in the expression for the force on the ions. If this correction is made the results come rather nearer to those obtained by the other methods, but they are still lower than the average. It is assumed that the top and bottom of the flame are insulated; with so many ions about it seems very doubtful if this would be strictly true.

[1] H. A. Wilson, *Electrical Properties of Flames*, p. 115.
[2] Watt, *Phys. Rev.* xxv. p. 69, 1925.

127·2. Wilson has made some measurements on the increased resistance of a flame due to a transverse magnetic field. He found that it varied in a parabolic manner with H, there being a certain unsymmetry probably due to the effect of the magnetic field on the upward motion of the flame. Watt also found an increase of resistance of the order of 2 per cent. He found that the appearance of the flame was greatly modified by the field of 5000 gauss which he used. With one direction of the field the upward motion of the flame was almost stopped, while with the other it was unchanged or increased. The large forces between the ions and flame gases, of which this behaviour is a sign, must considerably modify the theory, and introduce serious difficulties of a hydrodynamical nature into an already complicated problem. Heaps[1] has taken account of the effect of the changed motion of the flame gases due to the forces on the ions, and finds an expression for the change in resistance, and believes that Wilson's observations can be accounted for in this way.

It will be seen from the above account that the value of k_2 for the ions in a flame is still very uncertain, though it seems probable that the results which give 20,000 cm./sec. per volt/cm. for small values of the field, and about 8000 for stronger fields, are the most reliable. The absolute temperature of the flames was about 2000° C.

Maximum Current that can be carried by the Vapour of a Salt.

128. H. A. Wilson[2] has made an exceedingly important set of experiments on the maximum current that can be carried by a given amount of salt vapour; in these experiments the solution containing the salt vapour was not sprayed into a flame, but into air heated by passing through a long platinum tube raised to bright yellow heat by a furnace; a smaller central tube was placed along the axis of the outer tube and the current between the inner and outer tubes measured. When solutions of the strength 1/10th normal were sprayed and the temperature of the tubes raised and the potential difference increased, a stage was reached when neither an increase in the temperature nor in the

[1] Heaps, *Phys. Rev.* xxiv. p. 652, 1924.
[2] H. A. Wilson, *Phil. Mag.* vi. 4, p. 207, 1902.

potential difference produced any increase in the current. Wilson measured this limiting current and found that it was equal to the current which when passing through an aqueous solution of the salt would electrolyse in one second the same quantity of salt as was sprayed in that time into the hot air; thus if the salt had been supplied to water at the same rate as it was supplied to the hot air, the maximum current that could be sent through the aqueous solution would be the same as that which could be sent through the air; this was proved for the following salts of the alkali metals: $CsCl$, $CsCO_3$, RbI, $RbCl$, Rb_2CO_3, KI, KBr, KF, K_2CO_3, NaI, $NaBr$, $NaCl$, Na_2CO_3, LiI, $LiBr$, $LiCl$, Li_2CO_3.

CHAPTER XI

IONISATION BY LIGHT. PHOTO-ELECTRIC EFFECTS

129. THE discovery by Hertz[1] in 1887 that the incidence of ultra-violet light on a spark gap facilitated the passage of the spark, led immediately to a series of investigations by Hallwachs[2], Hoor[3], Righi[4] and Stoletow[5] on the effect of light, and especially of ultra-violet light, on charged bodies. It was proved by these investigations that a newly cleaned surface of zinc, if charged with negative electricity, rapidly loses this charge however small it may be when ultra-violet light falls upon the surface; while if the surface is uncharged to begin with, it acquires a positive charge when exposed to the light, the negative electrification going out into the gas by which the metal is surrounded; this positive electrification can be much increased by directing a strong air-blast against the surface. If however the zinc surface is positively electrified it suffers no loss of charge when exposed to the light: this result has been questioned, but a very careful examination of the phenomenon by Elster and Geitel[6] has shown that the loss observed under certain circumstances is due to the discharge by the light reflected from the zinc surface of negative electrification on neighbouring conductors induced by the positive charge, the negative electricity under the influence of the electric field moving up to the positively electrified surface.

The ultra-violet light to produce these effects may be obtained from an arc lamp, or by burning magnesium, or by sparking with an induction coil between zinc or cadmium terminals, the light from which is very rich in ultra-violet rays. Sunlight is not rich in ultra-violet rays, as these have been absorbed by the atmosphere, and it does not produce nearly so large an effect as the arc-light.

[1] Hertz, *Wied. Ann.* xxxi. p. 983, 1887.
[2] Hallwachs, *Wied. Ann.* xxxiii. p. 301, 1888.
[3] Hoor, *Repertorium der Physik*, xxv. p. 91, 1889.
[4] Righi, *C. R.* cvi. p. 1349; cvii. p. 559, 1888.
[5] Stoletow, *C. R.* cvi. pp. 1149, 1593; cvii. p. 91; cviii. p. 1241; *Physikalische Revue*, Bd. i., 1892.
[6] Elster and Geitel, *Wied. Ann.* xxxviii. pp. 40, 497, 1889; xli. p. 161, 1890; xlii. p. 564, 1891; xliii. p. 225, 1892; lii. p. 433, 1894; lv. p. 684, 1895.

Elster and Geitel, who have investigated with great success the effects produced by light on electrified bodies, have shown that the more electropositive metals lose negative charges even when exposed to ordinary daylight. They found that amalgams of sodium or potassium enclosed in a glass vessel lose a negative charge in the daylight, though the glass would stop any small quantity of ultra-violet light that might be left in the light after its passage through the atmosphere. When sodium or potassium by themselves instead of their amalgams were used, or, what is more convenient for many purposes, the liquid alloy formed by mixing these metals in the proportion of their combining weights, they found that the negative electricity was discharged by the light from a petroleum lamp; while with the still more electropositive metal rubidium the negative electricity could be discharged by the light from a glass rod just heated to redness. They found, however, that the eye was more sensitive to the radiation than the rubidium, for no discharge could be detected until after the radiation from the glass rod was visible. I have found however that electropositive metals like rubidium give off negative electricity in the dark, especially if a trace of hydrogen is present.

Elster and Geitel arrange the metals in the following order with respect to their power of discharging negative electricity:

        Rubidium.
        Potassium.
        Alloy of Potassium and Sodium.
        Sodium.
        Lithium.
        Magnesium.
        Thallium.
        Zinc.

For copper, platinum, lead, iron, cadmium, carbon, and mercury the effects with ordinary light are too small to be measurable. The order of the metals for this effect is the same as in Volta's series for contact-electricity, the most electropositive metals giving the largest photo-electric effect. Many substances besides metals discharge negative electricity under the action of ultra-violet light: lists of these substances will be found in papers by G. C. Schmidt[1]

[1] G. C. Schmidt, *Wied. Ann.* lxiv. p. 708, 1898.

and O. Knoblauch[1]. Among the more active photo-electric solids are, fluor-spar, the various coloured varieties of which vary greatly in the degree to which they possess this property; the sulphides of antimony, lead, arsenic, manganese, silver, and tin (the sulphates do not possess this property); hydroxide of tin, iodide of lead, many aniline dyes in the solid state.

129.1. It is now believed that all substances would show the effect with light of sufficiently short wave length, and the above results merely indicate substances which show it for light in the visible or close ultra-violet. For a more recent account of the photo-electric effect in solids and liquids other than metals, see Hughes' *Photo-electricity*, chaps. VII and VIII.

Nature of the Ions produced by the action of Ultra-violet Light on Metals.

130. The experiments made by the author and Lenard (see §§ 67, 68) show that in high vacua metals when illuminated with ultra-violet light give out electrons, *i.e.* bodies whose mass is only about $\frac{1}{1800}$ of that of the hydrogen atom; when however the metal is surrounded by gas the electrons soon strike against the molecules, get attached to them and have to drag them along with them as they move under the action of the electric field. The velocity of the negative ions through different gases has been measured by Rutherford (see p. 102), who showed that the velocity of the ion did not depend upon the nature of the metal on which the light fell, but that it did depend on the nature of the gas through which the ion had to travel, and that the velocity through any gas of the negative ion produced by ultra-violet light was very approximately the same as that of the ion produced by X-rays through the same gas.

The diminution of the photo-electric effect produced by a transverse magnetic field when the pressure of the gas is low, which was discovered by Elster and Geitel[2], has already been discussed on p. 247.

The photo-electric effect seems to disappear immediately the light is cut off. Stoletow[3], who made a series of experiments on

[1] O. Knoblauch, *Zeits. f. Physikalische Chemie*, xxix. p. 527, 1899.
[2] Elster and Geitel, *Wied. Ann.* xli. p. 166, 1890.
[3] Stoletow, *Aktinoelektrische Untersuchungen, Physikalische Revue*, i. p. 725, 1892.

this point, could not obtain any evidence that there was any finite interval between the incidence of the light and the attainment of the full photo-electric effect, or between the eclipse of the light and the total cessation of the effect, and he showed that the interval must at any rate be less than $\frac{1}{1000}$ of a second.

130·1. Marx and Lichtenecker[1] showed that if a sensitive potassium cell was illuminated by light from a revolving mirror, the photo-electric effect was independent of the speed of the mirror, even when the time of illumination was reduced to 10^{-7} sec. and the intensity of illumination was only ·56 erg per sq. cm. per sec. The effect was proportional to the intensity of the light.

Lawrence and Beam[2] have shown that the time lag, if any, is less than 3×10^{-9} sec. both at starting and stopping the light.

Energy of Emission of Photo-electrons.

131. Lenard showed that an illuminated plate continued to emit electrons when maintained at a small positive potential, showing that the electrons were emitted with a definite energy. As the potential of the plate was increased the number of electrons gradually diminished, indicating that they were not all emitted with the same velocity. Part, at any rate, of this difference is due to the fact, shown by Ladenburg[3], that the electrons come from a layer of finite thickness; thus those which come from the deeper parts of the layer will, when they emerge, have made more collisions with the molecules of the metal than those which come from the shallower parts, and so will have had their velocities more reduced before they emerge from the metal.

Lenard[4] also made the very important discovery that the maximum velocity of projection of the electrons is independent of the intensity of the incident light. The *number* emitted is proportional to the intensity, but the velocity of each electron depends only on the nature of the emitting surface if the frequency of the light is kept the same. This is a result of the most fundamental importance, as it shows that the energy gained by the electrons cannot be directly derived from the electric force which,

[1] Marx and Lichtenecker, *Ann. der Phys.* xli. p. 124, 1913.
[2] Lawrence and Beam, *Phys. Rev.* xxix. p. 904, 1927.
[3] Ladenburg, *Ann. der Phys.* xii. p. 558, 1903.
[4] Lenard, *Ann. der Phys.* viii. p. 149, 1902.

on Maxwell's theory, exists in the incident beam of light. This result has been amply confirmed by a number of experimenters.

131·1. The maximum energy of emission can be determined by measuring the positive potential to which the plate will rise on illumination. If this is V, then $Ve = \frac{1}{2}mv^2$, where v is the velocity of the fastest electrons. It was early shown that the maximum energy increased with the frequency of the light used, the material of the plate being kept the same. It was at first uncertain whether the energy or the velocity was a linear function of the frequency, the earlier experiments not being sufficiently accurate to distinguish with certainty between these alternatives.

Hughes[1] and Richardson and Compton[2] were however able to show that the former law was correct. Hughes was able to obtain much more consistent results than the earlier experiments by using surfaces formed by the distillation *in vacuo* of the substance under investigation, which was made to condense on to the plate which was to be illuminated. In this way he was able to get rid of many of the difficulties due to surface films of gas and other impurities. The relation between the maximum energy and the frequency can be written in the form $\frac{1}{2}mv^2 = k\nu - V_0 e$, where V_0 is a potential characteristic of the substance. Einstein suggested that k was equal to h, Planck's constant ($6 \cdot 55 \times 10^{-27}$).

The earlier results gave a value of k from 10 to 20 per cent. less than this, but by a very careful series of experiments Millikan[3] was able to show that the agreement held to less than 1 per cent. The theory of the experiments is as follows. The illuminated surface is surrounded by a considerably larger conductor which we may call the 'case,' and the photo-electric current between the surface and the case is measured as a function of the accelerating or retarding potential, typical results being shown in Fig. 111. Account has to be taken of the fact that the actual potential difference between plate and case differs from the applied potential by the contact potential difference between the substances of which the plate and case are made. When this correction is made the curves take the form shown in the lower figure, showing that

[1] Hughes, *Phil. Trans.* A, ccxii. p. 205, 1912.
[2] Richardson and Compton, *Phil. Mag.* xxiv. p. 577, 1912.
[3] Millikan, *Phys. Rev.* vii. p. 355; 1916.

the energies of emission vary between zero and a maximum which varies with the frequency of the light. For any substance there will be a minimum frequency given by $h\nu = V_0 e$, below which no emission will take place. This is known as the 'photo-electric threshold.' If the maximum emission energies are plotted against

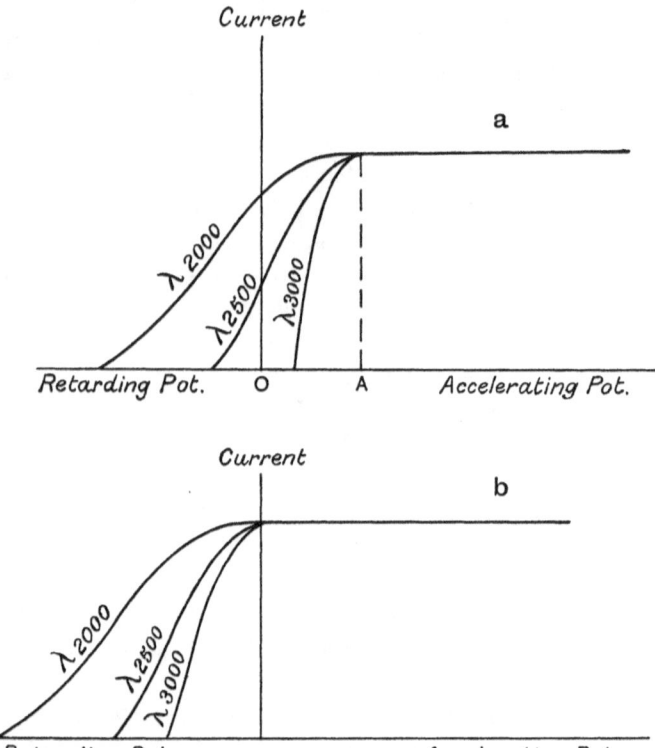

Fig. 111.

the frequency of the light, the result is a straight line whose slope gives h and whose intercept on the frequency axis gives the threshold. Following Kadesch[1], Millikan used oxidised copper as the material of the case, the threshold for this under the conditions of experiment was about $\lambda\,2536$ and the wave lengths used were all greater than this, so that no complication arose from electrons emitted by the case. Great care was taken to

[1] Kadesch, *Phys. Rev.* iii. p. 367, 1914.

ensure that the light used was not contaminated by the presence of small quantities of light of higher frequency than that which the monochromator was supposed to be isolating. The substances used were Li, Na and K. The surfaces used were prepared freshly by cutting *in vacuo*. He obtained values of h of 6.569×10^{-27} from experiments on sodium and 6.584×10^{-27} from experiments on lithium. The final value was given as 6.56×10^{-27}. Sabine made experiments on Zn, Cd and Cu using very short wave lengths in a vacuum spectroscope and combining the results obtained with those of other observers for longer wave lengths. Though obviously not a very satisfactory method, this gave results in fair agreement with the theory.

Robinson[1] found that, using thin films of platinum, the maximum velocity of emission was about 12 per cent. greater on the side from which the light emerged than on that on which it was incident. In view of the agreement with theory of Millikan's experiments, which were made with electrons coming from the side on which the light was incident, this result seems rather improbable though confirmed by Stuhlman[2]; Piersol[3] has recently found that there is no difference between incident and emergent velocities if sufficient precautions are taken against stray light. Herold[4] extends this to the complete distribution of velocities among the electrons. Chien Cha[5] however finds a difference with films of spluttered platinum though not with films formed by evaporation. He attributes the effect in the former case to excluded gas. Piersol however worked with spluttered films. It is clear that there is no effect of fundamental importance.

131·2. The curves of Fig. 111 can be used to determine the proportion of electrons emitted with a given energy. Certain precautions are however required. Whatever the shape of the illuminated plate and the case the current will cease when the work done by the electron in moving from plate to case is just greater than the maximum energy of emission, but if the paths of some of the electrons are at an angle to the electric field they

[1] Robinson, *Phil. Mag.* xxv. p. 115, 1913.
[2] Stuhlman, *Phys. Rev.* iii. p. 195, 1914.
[3] Piersol, *Phys. Rev.* xxiii. p. 144, 1924.
[4] Herold, *Ann. der Phys.* lxxxv. p. 587, 1928.
[5] Chien Cha, *Phil. Mag.* xlix. p. 262, 1925.

will be prevented from reaching the case by a smaller potential than would be required to bring them completely to rest. In Richardson and Compton's experiments the illuminated plate was small and was surrounded by a large spherical case, so that the field between the two was radial and in the direction of motion of the electrons. In this case the ordinate of one of the curves in Fig. 111 measures the number of electrons emitted with an energy greater than the corresponding retarding potential, and the slope of the curve gives the number per unit range of energy at the energy considered. Richardson and Compton found that the slope plotted against the energy gave a symmetrical curve with a well-defined maximum. Ramsauer[1] has developed another method of obtaining the energy, or rather velocity, distribution curve. He deflects the electrons by means of a uniform magnetic field through a semicircle and measures the charge given to a Faraday cylinder. His results are similar to those of Richardson and Compton. This method has also been used by Klemperer[2] and by Wolf[3], and the former has pointed out some possible sources of error. The latter finds that the energy corresponding to the maximum of the curve is a linear function of the frequency, the slope being not much different from h. He used the liquid Na–K alloy. Ramsauer also found straight lines for Zn, Au and brass but with a different slope. Kluge[4] and Bennewitz[5] have found that with changes in the gas content of the metal the maximum and mean energies change together. Herold[6] found closely similar velocity distribution curves for a number of metals. The effect of changing the gas content was mostly a shift of threshold. It is still uncertain whether the variation in energy of the electrons can be wholly accounted for by energy losses during their passage through the material, or whether their initial energies differ. The former would seem more probable on theoretical grounds, but some experiments of Compton and Ross[7] on the escape of photo-electrons

---

[1] Ramsauer, *Ann. der Phys.* xlv. pp. 961 and 1121, 1914.
[2] Klemperer, *Zeits. f. Phys.* xvi. p. 280, 1923.
[3] Wolf, *Ann. der Phys.* lxxxiii. p. 1001, 1927.
[4] Kluge, *Ann. der Phys.* lxxxii. p. 432, 1927.
[5] Bennewitz, *Ann. der Phys.* lxxxiii. p. 913, 1927.
[6] Herold, *Ann. der Phys.* lxxxv. p. 587, 1928.
[7] Compton and Ross, *Phys. Rev.* xiii. p. 374, 1919.

from thin films suggest that they lose their energy completely at a single collision, if at all, in which case the experimental curves give the actual distribution of the energy as originally received by the electrons. Herold has examined the velocity distribution for thin films of platinum. He finds no change with thickness down to $5{\cdot}5\mu\mu$, but a film of $1{\cdot}9\mu\mu$ showed a diminution in the relative number of very slow electrons.

It should be noticed that the form of the curves when corrected for contact difference of potential shows that no further electrons are produced by an accelerating field.

Millikan has shown that his results lead to interesting conclusions concerning the work required to extract the electrons from their normal positions. Suppose that for one metal V is the potential required to be applied between plate and case to stop all the electrons. Then

$$(V + K)\,e = h\nu - V_0 e,$$

where K is the contact potential difference between plate and case. Let dashes refer to the same experiment with a different plate but the same case and frequency of light. Then subtracting, we find

$$(V' - V) + (K' - K) = V_0 - V_0' = \frac{h}{e}(\nu_0 - \nu_0'),$$

where ν_0, ν_0' are the threshold frequencies. Now it is found that $V' = V$, provided the measurements are made quickly one after the other so that the surface film of the case has no time to change. Hence

$$K' - K = \frac{h}{e}(\nu_0 - \nu_0').$$

But for metals at least, in which free electrons are present, $e(K' - K)$ is the difference in the work done in removing the free electron from the metal to infinity. Hence the energy of the photo-electron when free in the metal but before escaping from the surface differs from $h\nu$ only by a constant, if at all. This suggests strongly that the photo-electrons from metals were originally free in the metal. If this is the case, $h\nu_0$ is the work required to remove a free electron from the metal and should therefore be the same as the corresponding quantity determined from thermionic measurements. The annexed table shows the extent of the agreement. Unfortunately both quantities are uncertain, being so much affected by minute surface changes, but the agreement is as good as can be expected for the metals, while for CuO it is reasonable to suppose that the photo-electrons are not originally free and that work has to be done to make them so.

Metal	Photo-electric threshold		Thermionic volts
	λ_0	V_0	
Cs	7500 (Cornelius)	1·50	1·38 (Dushman (quoted))
K	7000 (Richardson and Young)	1·60	{1·12 (Richardson and Young) / 0·43}
Ca	4000 (Hamer)	2·80	2·24 (Dushman (quoted))
Th	3465 (Hamer)	3·24	2·94 (Dushman (quoted))
Mo	2854 (Roy)	4·33	4·30 (Dushman)
Ta	3148 (Roy)	3·93	4·11 (Dushman)
Pt	2800 (Richardson and Compton)	4·40	4·24 (Schlicter)
W	2734 (Roy)	4·52	{4·48 (Davisson and Germer) / 4·64 (Dushman)}
C	2615 (Hamer)	4·29	3·92 (Langmuir)
CuO	{2500 (Millikan) / 2570 (Barton)}	{5·0 / 4·80}	1·9 (Richardson)

131·3. The photo-electric threshold of metals varies very greatly with the extent to which they are freed from gas by heating *in vacuo*. Thus Kluge and Bennewitz find variations of half a volt to a volt or even more in the maximum energy of photo-electrons from Pt and Pd according to treatment, the measurements in all cases being *in vacuo*. Du Bridge[1] finds 1958 ± 15 Å. for the threshold of Pt after very thorough outgassing. Kazda[2] and Dunn[3] find 2735 Å. for the clean surface of running mercury.

In view of the very variable results obtained by different observers for the photo-electric threshold of solids, and the extent to which it is influenced by minute surface changes, it seems doubtful if any of the results so far obtained accurately represent the value for a pure gas-free surface. Indeed several observers,

[1] Du Bridge, *Phys. Rev.* xxix. p. 451, 1927.
[2] Kazda, *Phys. Rev.* xxvi. p. 643, 1925.
[3] Dunn, *Phys. Rev.* xxix. p. 693, 1927.

especially in Germany, have suggested that the whole effect is due to the presence of gas, and have shown that the current can in some cases be reduced to 1 per cent. by repeated distillation *in vacuo*. It seems probable, however, that this is merely an extreme case of a shift of the threshold due to surface layers. The experimental results are very contradictory (see Hughes' *Report on Photo-electricity*, p. 102; also Hallwachs, *Phys. Zeits.* xxi. p. 561, 1920). Richardson[1] finds evidence that the surfaces of sodium and potassium behave as though they were covered with patches for which the work of extraction of the electrons is abnormally low, these patches usually covering only a small area of the surface but changing in size with the treatment of the surface (see § 109·1).

131·4. As an indication of the values of the threshold wave length to be expected under ordinary conditions (surfaces sandpapered but not otherwise treated), the following values obtained by Hamer[2] are given. The measurements were made *in vacuo*:

Metal	Ag	Al	Bi	C (graphite)	Ca	Cd	Cu	Fe
λ Å.	3390	3590	2980	2615	4000	3130	2665	2870

Metal	Ni	Pb	Pt	Se	Sn	Th	W	Zn
λ Å.	3050	2980	2780	2670	3180	3460	2615	3420

131·5. It should be pointed out that probably the bulk of the four or five volts required to remove an electron from a metal is accounted for by the attraction between the electron when just outside the surface and its image in the conducting surface. As pointed out in the second edition of this book, this involves an amount of work $\frac{e^2}{4r}$, where r is a distance which measures the extent to which the actual surface differs from a mathematical conducting plane, and is thus of the order of the radius of an atom. If $r = 10^{-8}$ the effect is represented by a P.D. of 3·6 volts. It is worthy of note in this connection that Hughes found that for elements of the same valency, V_0 decreased as the atomic volume increased.

131·6. Richardson[3] has developed a statistical theory of the

[1] Richardson and Young, *Proc. Roy. Soc.* A, cvii. p. 377, 1925.
[2] Hamer, *Optical Soc. of Amer. J. and Rev. Sci. Inst.* ix. p. 251, 1924.
[3] Richardson, *Phil. Mag.* xxiii. p. 594, 1912; xxiv. p. 574, 1912; xxvii. p. 476, 1914.

photo-electric effect based on the idea that, in a constant temperature enclosure, the photo-electrons emitted by the action of the black body radiation in the enclosure would increase in numbers till the number returning to the substance by collision with it just balanced the number emitted by the radiation. This leads to the result that the electrons emitted by light of a given frequency ν have a mean energy $h\nu - w_0$, and that their number has a maximum at about $\nu = \tfrac{3}{2}\nu_0$. The saturation photo-electric current should have a value $i = AT^2 \epsilon^{-\frac{w_0}{kT}}$, the same law of variation with temperature as the thermionic effect. W. Wilson[1] and Roy[2] have verified this for a number of metals when a surface is illuminated by light from a constant temperature enclosure. Richardson went on to consider whether the thermionic effect itself could be considered as the 'auto-photo-electric' effect due to the radiation of the body acting photo-electrically on the body itself. It appears however that, while the law of temperature variation is the same, the magnitude would be only of the order 10^{-7} of the observed thermionic effect. In this case it seems doubtful how far the theoretical basis of the preceding results is reliable, as the thermo-dynamical argument is only strictly applicable if the two effects are identical.

Photo-electric Properties of Gases.

131·7. If light is able to cause the emission of electrons from the molecules of a gas, we shall clearly find an increase in conductivity, the residues of the molecules forming positive ions, and the electrons, either free or attached to neutral molecules, forming the negative ones. Early experiments on conductivity in gases caused by light gave rather contradictory results. It is now known that in most cases gases require light of shorter wave length to give a photo-electric effect than the majority of solids, and in consequence there is great danger of spurious effects due to electrons emitted by dust particles or by the walls of the apparatus. The former case is characterised by the fact that the positive ions have a very small mobility; thus Lenard, who illuminated air by the light from sparks between the aluminium terminals of an induction coil, found positive ions with a mobility of only ·0015 cm./sec.

[1] W. Wilson, *Proc. Roy. Soc.* A, xciii. p. 359, 1917.
[2] Roy, *Proc. Roy. Soc.* A, cxii. p. 599, 1926.

Bloch was able to show that these heavy ions were not formed in air filtered to remove dust particles. Lenard however obtained slight ionisation in filtered gases very close to the spark, the light from which passed through quartz. The best way of avoiding the difficulty of the effect of the walls is to use a stream of gas which flows from the place of illumination to the ionising chamber. The presence of positive ions in this gas may then be taken as evidence of ionisation, if precautions have been taken to avoid dust. An additional difficulty is that of finding some substance transparent to very short wave lengths to act as a window through which to illuminate the gas. Even fluorite begins to absorb strongly in the region under consideration, and it has the additional disadvantage that different specimens vary greatly in transparency.

On theoretical grounds one would expect the minimum frequency for ionisation to be connected with the ionisation potential, as found by Franck and Hertz's method, or from spectral considerations, by the formula $h\nu = Ve$. There is however the possibility that the molecule might be ionised in stages, the first stage being the removal of an electron from a normal to an abnormal orbit, and the next the completion of ionisation by the removal of the electron by a second quantum of suitable radiation or some other source of energy. Since however the electron only remains for a short time in the abnormal state, this would usually occur only with intense light.

Ionisation of Air and Inorganic Gases.

131·71. The first experimenter to detect a strong ionisation in air was Hughes[1] who made use of the apparatus shown in Fig. 112. The source of light was a discharge in hydrogen at a pressure of about 1 mm. The filtered air streams up against the window F, and then passes into the chamber C, in which the conductivity is tested. The presence of positive ions was taken as the test of ionisation by the light. When F was a quartz plate, 3 mm. thick, no positive ions could be detected, though the apparatus was sensitive enough to detect a current of the order of 10^{-15} amp. On substituting a certain plate of fluorite for the quartz a current

[1] Hughes, *Proc. Camb. Phil. Soc.* xv. p. 483, 1910.

was obtained from the positive ions of the order of 10^{-11}. It was found that only one plate of fluorite gave this effect, the others tried being presumably less transparent. According to Lyman the transparency limit for quartz is $\lambda\,1450$. Hughes determined the transparency limits of some of his fluorite plates by measuring the maximum velocity of emission of electrons

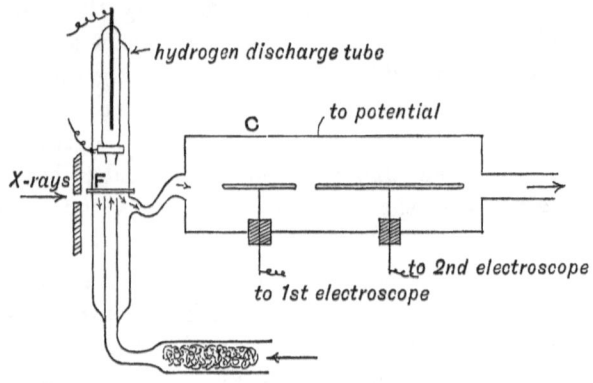

Fig. 112.

emitted from a suitable substance by light transmitted through them. He found they varied very much in transparency. One which transmitted down to $\lambda\,1350$ did not transmit the ionising radiation. Unfortunately the one plate which did allow ionisation was broken before its transparency could be measured. Hughes concludes that air is ionised by a wave length between $\lambda\,1250$ and $\lambda\,1350$. Hughes also used his apparatus to compare the mobility of the positive ions produced by light with those produced by X-rays. This could be done by measuring the ratio of the currents received by the two electrodes when the gas streamed past at a constant rate. He found that the mobilities for photo-electric and X-ray ions were the same.

Palmer[1] also found ionisation in air, oxygen and nitrogen from light which had passed through fluorite. He found that passing the light through an absorbing layer of oxygen before entering the ionisation chamber reduced but did not entirely destroy the effect. His maximum current was of the order of 10^{-11} amp. and he found nitrogen to give about four times the

[1] Palmer, *Phys. Rev.* xxxii. p. 1, 1911.

effect of oxygen. Hydrogen gave a very slight effect, possibly due to impurities.

Lenard and Ramsauer[1] made a very careful investigation of the ionisation in a number of gases. The source of light was a very intense spark between aluminium terminals and under favourable conditions very strong ionisation was obtained. They found that minute traces of impurities, both gaseous and solid, had an enormous effect. It was necessary to pass the air through liquid air to freeze out impurities, and to bake the glass tubing of the apparatus to get rid of occluded gases. They found that under these conditions air could be ionised by light that passed through a fluorite plate but not by the light transmitted by a quartz plate. Hydrogen from a cylinder gave an effect about equal to that in air.

They found that carbon dioxide and ammonia could be ionised by light passing through quartz, thus confirming a result found by Thomson. Traces of ammonia and carbon dioxide in air were found to increase the ionisation for light through fluorite by factors of 44 and 14 respectively.

Ludlam[2] found that light transmitted through fluorite produced no ionisation in chlorine, though it did in air. Henry and Whiddington found that the light from a carbon arc transmitted through quartz produced no ionisation in iodine.

Ludlam and West[3], using light from an aluminium spark transmitted through fluorite, found ionisation in iodine vapour at room temperature, obtaining currents of the order 10^{-12} amp. They found no ionisation in nitrogen, in contradiction to Palmer, but the absorption of 1 cm. of air was sufficient to destroy the effect in the iodine. It is not certain whether it is the molecules or atoms which are ionised, but the latter would normally be present only in very small numbers at this temperature. There is also the possibility of cumulative action as in Compton and Smyth's experiments (p. 452).

Mohler[4] has been able to show the photo-ionisation of argon and neon due to light emitted by the same gas under electron

[1] Lenard and Ramsauer, *Sitz. d. Heidelberg. Akad. d. Wiss.*, 1910–11, quoted by Hughes.
[2] Ludlam, *Phil. Mag.* xxiii. p. 757, 1912.
[3] Ludlam and West, *Proc. Roy. Soc. Edin.* xlv. p. 34, 1925.
[4] Mohler, *Phys. Rev.* xxviii. p. 46, 1926; *Proc. Nat. Acad. Sci.* xii. p. 494, 1926.

bombardment in another part of the vessel, so that the radiation did not have to pass through any solid substance. In this way he has discovered several critical potentials. The ionisation was detected by the effect of the positive ions on the electron emission from a hot filament (see p. 451). Hydrogen gave no ionisation in similar circumstances.

Organic Vapours.

131·8. Stark[1] found marked ionisation in various organic vapours for the light from a mercury arc transmitted through quartz. He used a somewhat different method of distinguishing between true ionisation of the gas and the leak from the electrodes. He varied the pressure of the vapour and considered that ionisation of the gas occurred when the current increased with the pressure. The voltages used were of course too small to produce ionisation by collision, never exceeding 4 volts per cm. He found ionisation in anthracene, diphenylmethane, α-naphthylamine and diphenylamine. All these vapours fluoresce, and Stark considered that the ionisation and fluorescence were connected and that the same electron took part in both effects. Serkof, working on the same lines, found that aniline could also be ionised under the above conditions. Hughes[2] tried carbon disulphide, zinc ethyl and tin tetrachloride, but found no trace of ionisation from the light of a mercury lamp. The shortest wave length acting under these conditions is λ 1849.

Metallic Vapours.

131·9. Steubing[3] found that the light from a mercury lamp could ionise mercury vapour, but it seems somewhat doubtful if sufficient precautions were taken to distinguish between ionisation and surface effects. Anderson[4] investigated the effect of visible light on potassium vapour, and found effects possibly due to a thin layer of solid condensed on the electrodes. Similar results were obtained by Gilbreath[5] for light filtered through glass.

[1] Stark, *Phys. Zeits.* x. p. 787, 1909.
[2] Hughes, *Proc. Camb. Phil. Soc.* xvi. p. 375, 1911.
[3] Steubing, *Phys. Zeits.* x. p. 787, 1909.
[4] Anderson, *Phys. Rev.* i. p. 233, 1913.
[5] Gilbreath, *Phys. Rev.* x. p. 166, 1917.

Kunz and Williams[1] however showed that caesium vapour was only ionised by light of less than about λ 3184 which corresponds to the ionisation potential 3·9 volts. Since the ionisation potential for potassium is 4·3 volts corresponding to λ 2857, this suggests that Anderson and Gilbreath did not really get ionisation, but see below.

Kunz and Williams in a second paper find that the threshold for caesium vapour is 3180 Å., in very good agreement with the series limit at 3184 Å. The difficulty in these experiments is that the true gaseous ionisation, as distinct from spurious effects on the walls, is so small that it is difficult to observe with monochromatic light. Thus in their last paper their monochromator had to be set to give a breadth of 80 Å., and most other observers have not been able to use a monochromator at all, and been obliged to rely on filters. Foote and Mohler[2] have partly avoided this difficulty by the use of a very ingenious device due to Kingdon[3]. If a hot wire is emitting electrons and the potential is such that the current is limited by the space charge, the presence of a few positive ions will increase the current by partly neutralising the charge. The production of one heavy positive ion will neutralise the emission of many electrons because of its slower speed, and also because at low pressures its chance of striking the fine wire cathode, and so being absorbed, is very small. Thus such an apparatus is a very sensitive indicator of positive ions while little affected by the electrons produced by stray light striking the walls of the apparatus. Using this device, the above authors found that they could use a monochromator with a fine slit. The ionisation showed a sharp peak which occurred at the theoretical limit 3184 Å. within the errors of experiment, but there was a considerable effect up to 3400 Å. The authors attribute this to ionisation by two stages, the first produced by the light, the second by heat collisions or possibly infra-red radiation.

Further experiments by Mohler, Foote and Chenault[4] show that ionisation can be caused by the resonance radiation, *i.e.* the

[1] Kunz and Williams, *Phys. Rev.* xv. p. 550, 1920; xxii. p. 456, 1923.
[2] Foote and Mohler, *Phys. Rev.* xxvi. p. 195, 1925.
[3] Kingdon, *Phys. Rev.* xxi. p. 408, 1923.
[4] Mohler, Foote and Chenault, *Phys. Rev.* xxvii. p. 37, 1926.

light of the frequency of a line of the principal series. The limit of the series showed as the largest of a number of maxima, corresponding to each line of the series, when photo-electric effect was plotted against wave-length. The remaining energy required for ionisation is believed to be supplied by the energy of thermal agitation of the molecules, transferred to the electron by collisions. Foote[1] has also shown that mercury vapour can be ionised by the mercury line 2537. The ionisation is not simply proportional to the intensity of the light.

Samuel[2] finds for potassium that light up to 2804 Å. ionises but not up to 2893 Å. He used filters. Williamson[3], also with filters, found the limit between 2800 and 3100 Å. Lawrence[4], who used a monochromatic illuminator, found that ionisation of potassium did not occur for wave-lengths above 2610. This is difficult to reconcile with the theoretical value λ 2857. Lawrence suggests that the ionisation is due to molecules, but it does not seem clear why the more numerous atoms should not produce an effect. The effectiveness of the radiation increased as the wave-length diminished (but see § 131·92).

Ionisation by Stages.

131·91. The idea of ionisation in successive stages has been invoked recently by various writers (Richardson and Bazzoni, Millikan, Van der Byl and Compton) to account for the phenomenon of the low voltage arc, in which an arc is formed with a potential difference less than the ionisation potential of the gas. In this case it was suggested that one, or both, of the stages is caused by an electron impact. It has been shown by Compton and Smyth[5] that the ionising potential of fluorescing iodine vapour is only 6·8 volts, while for the normal vapour it is 9·4 volts. The frequency required to produce fluorescence corresponded to 2·3 volts. Presumably a mixture of this light with that corresponding to 6·8 volts, *i.e.* λ 1815 would cause the ionisation of iodine if it was

[1] Foote, *Phys. Rev.* xxix. p. 609, 1927.
[2] Samuel, *Zeits. f. Phys.* xxix. p. 209, 1924.
[3] Williamson, *Phys. Rev.* xxi. p. 107, 1923.
[4] Lawrence, *Phil. Mag.* l. p. 345, 1925.
[5] Compton and Smyth, *Phys. Rev.* xvi. p. 501, 1920.

intense enough. A more recent paper by Compton[1] leads to the conclusion that in most cases successive ionisation is more likely to be caused by the successive absorption of two quanta of radiation than by electron impacts.

The first critical potential of nitrogen is about 8·2 volts and of oxygen 7·9 volts, while the next are at 16·9 and 15·5 respectively. It is not however quite certain whether the lower values are ionisation or resonance potentials. They correspond to $\lambda\,1510$ and $\lambda\,1570$. In view of the fact that Hughes found no ionisation at $\lambda\,1350$, these lower values cannot be effective unless the determinations are rather badly wrong. On the other hand, if the effect is due to the next critical at 15·5 volts (oxygen) the wavelength would be $\lambda\,800$, and it hardly seems probable that any fluorite plate would transmit this, though Lenard and Ramsauer thought they had evidence of light of about $\lambda\,900$. The matter needs further investigation with a view to finding as accurately as possible the threshold wave-length for the ionisation of air. Ludlam and West's work (§ 131·71) suggests that it is the oxygen rather than the nitrogen that is ionised.

131·92. Some recent measurements have been made of the 'efficiency' of photo-electric ionisation, *i.e.* the number of ions produced per erg of light. If $I_\nu d\nu$ is the intensity of radiation between the limits of frequency ν and $\nu + d\nu$, then if $B_\nu I_\nu d\nu dt$ is the chance of an atom absorbing the radiation and emitting a photo-electron, B_ν may be called the photo-electric efficiency. Mohler, Foote and Chenault[2] made relative measurements for caesium vapour. They found $B_\nu = B_{\nu_1} \epsilon^{-a(\lambda_1 - \lambda)}$, where a is a constant and ν_1, λ_1 refer to the frequency corresponding to the ionisation potential. A variation as a simple power of λ, suggested by Milne and Kramers on theoretical grounds, will not fit these results. Little[3] has made some absolute measurements. Light of $\lambda\,3184$ of intensity one erg per sq. cm. releases 4×10^{-10} electrons per atom. This number falls to $1·2 \times 10^{-10}$ at about $\lambda\,2600$ and then increases.

[1] Compton, *Phys. Rev.* xx. p. 293, 1922.
[2] Mohler, Foote and Chenault, *Phys. Rev.* xxvii. p. 37, 1926.
[3] Little, *Phys. Rev.* xxx. pp. 109 and 963, 1927.

Photo-electric Effects involve an Absorption of Light.

132. Stoletow[1] at an early stage in the history of this subject called attention to the connection between the photo-electric effects and the absorption of the ultra-violet light; he pointed out that water which does not give photo-electric effects does not absorb many of the visible or ultra-violet rays, while solutions such as those of methyl-green or violet, which are photo-electric, show strong absorption. Hallwachs[2], who investigated the subject in greater detail, found in all the photo-electric liquids which he tried strong absorption for the ultra-violet light, but that strong absorption was not always accompanied by photo-electric effects; thus for example the aqueous solution of fuchsine is photo-electric, while the alcoholic solution is not, and yet the alcoholic solution absorbs more ultra-violet light than the aqueous one.

132·1. Differences in absorption appear to account for the asymmetry observed in some cases between the number of photo-electrons emitted by a thin film of metal spluttered on to a quartz plate when it is illuminated by light from opposite sides. Kleeman[3] and Stuhlman[4] both found that under these conditions a platinum film emitted 17 per cent. more electrons when illuminated on its free side than when illuminated through the quartz. Stuhlman also got similar results with other metals. Partzsch and Hallwachs[5] however were able to show both by measurement and theoretical considerations that more light entered the film from the quartz side than from the air side with equal illumination. The theoretical ratios were in close agreement with those found by Stuhlman for the various metals tested.

In the case of electrons ejected from atoms by X-rays, which is of course an analogous process, it is known that there is a tendency for them to be emitted in a forward direction, in fact the majority appear to start out at an angle between 70° and 80° to the direction of the X-rays. This is attributed to the momentum $h\nu/c$ associated with the quantum of radiation. On this view the angle

[1] Stoletow, *Physikalische Revue*, Bd. i. 1892.
[2] Hallwachs, *Wied. Ann.* xxxvii. p. 666, 1889.
[3] Kleeman, *Proc. Roy. Soc.* A, lxxxiv. p. 92, 1910.
[4] Stuhlman, *Phil. Mag.* xx. p. 331, 1910; xxii. p. 854, 1911.
[5] Partzsch and Hallwachs, *Ann. der Phys.* xli. p. 247, 1913.

would be $\cot^{-1} c/v$, where v is the velocity given to the electron and c that of light. In the case of visible radiation the angle would be practically 90° and there would be no appreciable tendency for the electrons to be carried forward with the light. For a theoretical treatment on the basis of wave mechanics, see Wentzel, *Zeits. f. Phys.* xl. p. 574, and xli. p. 828, 1927.

Stuhlman[1] has more recently made some elaborate experiments on the variation of photo-electric effect with the thickness of the film in the case of Pt and Ag. He found a maximum effect for a thickness (of order 10^{-6} cm.) which varied with the wave-length. Compton and Ross[2] found more complicated curves with two maxima at thicknesses appreciably less than the above.

Predwoditelew and Netchaewa[3] have shown that in the case of fuchsine and other allied dyes, the photo-electric emission from a thin film undergoes remarkable periodic variations with the thickness, four or five maxima being observed. Akulov[4] has shown that in the case of fuchsine these can be accounted for by the stationary waves formed in the thin film on the classical theory. He states that similar considerations can account for the existence and position of the maximum observed by Stuhlman.

132·2. A number of workers have shown that given the other conditions constant the number of electrons ejected in unit time is proportional to the intensity of the light. Apparent exceptions are due to changes produced by the light in the surface layer, or to the charging up of non-conducting parts of the apparatus by the emitted electrons.

Arnold and Ives[5] find that there is a lag in the increase of current produced by illuminating an oxide-coated filament emitting thermo-electrons. This cannot be accounted for by the heating produced by the light, and is perhaps a surface change analogous to those which cause photo-electric fatigue.

Lenard found for a surface of soot that proportionality existed

[1] Stuhlman, *Phys. Rev.* xiii. p. 374, 1919.
[2] Compton and Ross, *Phys. Rev.* xiii. p. 109, 1919.
[3] Predwoditelew and Netchaewa, *Zeits. f. Phys.* xxix. p. 332, 1924.
[4] Akulov, *Zeits. f. Phys.* xli. p. 920, 1927.
[5] Arnold and Ives, *Proc. Nat. Acad. Sci.* vii. p. 323, 1921.

over a range of $1:10^6$. Richtmeyer[1] using sodium *in vacuo* found strict proportionality over a range of intensity from 600 to ·007 foot candles. Besides earlier experiments with potassium, Elster and Geitel[2] found that the law held down to an illumination of 10^{-7} ergs per sq. cm. per sec. In a still later paper they extended this result down to 3×10^{-9} ergs per sq. cm. per sec. The fact that the photo-electric effect starts instantaneously even with small energies has an important bearing on the theory, as it shows that the energy of the electrons cannot have been accumulated from the incident beam by the resonance of some portion of an atomic system. On the classical theory a free electron can absorb the energy from radiation over an area of the order of a square wave-length. Elster and Geitel[2] obtained a measurable effect with blue light of intensity 3×10^{-7} ergs per sq. cm. per sec. acting on a potassium surface. The energy which the electron could absorb is thus

$$(4 \times 10^{-5})^2 . 3 \times 10^{-7} = 4 \cdot 8 \times 10^{-16} \text{ ergs per sec.}$$

The energy it receives is $h\nu = 4 \cdot 9 \times 10^{-12}$ ergs; thus the light would have to act for 10^4 secs. before the first electron could be emitted. Thus either the energy is derived from the atom and released by a trigger action or the energy in the wave front is concentrated into small bundles, one of which may be absorbed by a single electron. The second alternative is that now generally accepted, both on account of the difficulty of devising a suitable trigger mechanism, and because of the success of the Bohr theory of spectra which involves analogous assumptions.

132·3. Another point of considerable importance is the relation between the total energy absorbed from the light and the sum of the energies of the emitted electrons. Taking only the electrons which actually escape, the former is always many times the latter; in one special case Pohl and Pringsheim[3] mention that a potassium surface under certain conditions emits electrons whose total energy is 2 per cent. to 3 per cent. of the light absorbed. This is under conditions when the selective effect is important. While it is perfectly possible that the balance is accounted for by the electrons

---

[1] Richtmeyer, *Phys. Rev.* xxix. p. 71, 1909; also xxx. p. 385, 1910.
[2] Elster and Geitel, *Phys. Zeits.* xiii. p. 468, 1912.
[3] Pohl and Pringsheim, *Verh. d. Deutsch. Phys. Ges.* xv. p. 173, 1913.

which do not escape from the metal, there is no experimental proof that anything like the whole energy of the absorbed radiation is used in producing photo-electrons according to the quantum law.

Connection between the Rate of Leak in Gases and the Strength of Electric Field.

133. The first measurements on this subject were made by Stoletow[1], who used the following arrangement: the light from an arc lamp passed through a hole in a metal screen, and after passing through a perforated plate C fell upon a parallel metal plate D; these plates were connected together through a battery, the negative pole of the battery being connected with D, the plate illuminated by the light. The current passing between the plates was measured by a very sensitive galvanometer. By means of this arrangement Stoletow measured the relation between the current and the potential difference between the plates, making experiments with the plates at distances apart varying from about 2·5 millimetres to 100 millimetres; the results of these experiments, in which the gas between the plates was air at atmospheric pressure, are represented by the curves of Fig. 113; the abscissæ represent the potential differences between the plates, the unit being 1·43 volts (the electromotive force of a Clark's cell); the ordinates represent the current passing between the plates, the unit being $8·6 \times 10^{-11}$ amperes; the symbol on the curve, for example $x + 25$, indicates that the distance between the plates was $x + 25$ millimetres, where x is a small distance, about 1·5 mm., that was not very accurately determined; the diameter of the plates was 22 mm. An inspection of the curves shows that when the distance between the plates is small and the electromotive force large the current increases much more slowly than the electromotive force; it is, however, evidently far from saturation; while when the plates are separated by distances greater than 25 mm. there is no approach to saturation. The curves corresponding to the greater distances between the plates show that under small electromotive forces the current increases more rapidly than the potential difference. As far as the measurements represented in the figure go, i is approximately the same at all distances d, provided V is proportional to d,

[1] Stoletow, *Journal de Physique*, ii. 9, p. 468, 1890.

V being the potential difference, *i.e.* i is a function of the mean value of the electric force between the plates; this law, as Stoletow showed in a later paper[1], does not apply for any great range of potential differences—at lower pressures especially, the departures from it are soon very apparent.

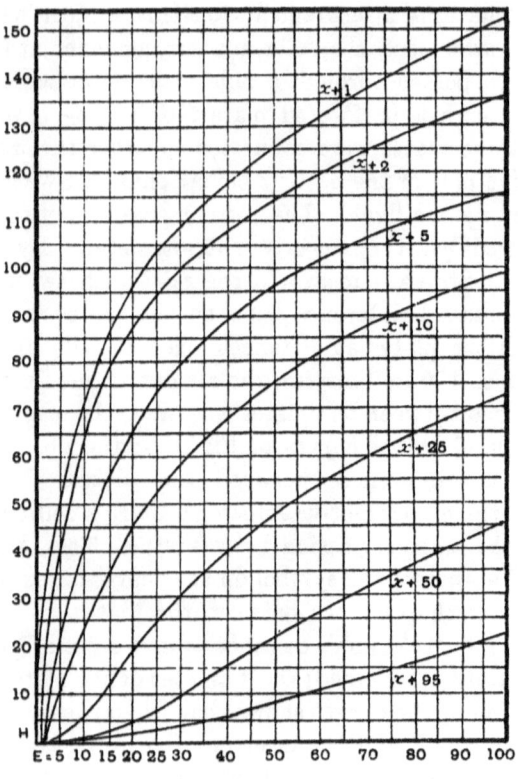

Fig. 113.

Since in this case the ions are all of one kind we may (see p. 376) apply the equation

$$X^2 = X_0^2 + \frac{8\pi i x}{k},$$

where k is the velocity of the ion under unit electric force, i the

---

[1] Stoletow, *Journal de Physique*, ii. 9, p. 469, 1890.

intensity of the current, X_0 and X the values of the electric force at the plate and at a point distant x from it.

To form an estimate of the variation in the electric field which is produced by the presence of the negative ions between the plates, let us take one of Stoletow's experiments in which under an electric field of 150 volts per cm. the current was $3 \cdot 3 \times 10^{-11}$ amperes. The velocity of the negative ions produced by a field of 1 volt per centimetre has been shown by Rutherford to be about 1·5 cm./sec. Hence using electrostatic units, X and X_0 being the values of X at places a centimetre apart, we get, putting $i = 10^{-1}$, $k = 4 \cdot 5 \times 10^2$, $X + X_0 = 1$, in the preceding equation, $X - X_0 = 1/180$ or a little less than 2 volts per cm., thus the

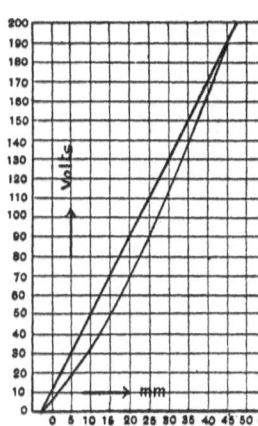

Fig. 114.

variation in the strength of the field is comparatively small. Stoletow, who determined the intensity of the field between two parallel plates one of which was illuminated by ultra-violet light, was not able to detect any variation in the intensity. Schweidler[1], who investigated this point at a later period, found that the distribution of potential between the plates when the ultra-violet light was in action, was not quite uniform; his results are shown in Fig. 114, where the curved line represents the distribution of potential when the light was shining, the straight one when it was not. The curvature of the potential curve in the light is all in one direction, indicating the presence of an excess of negative ions in every part of the region between the plates. The variation in the intensity of the field between the plates has also been observed and measured by Buisson[2] and used by him to determine the velocity of the negative ions; he finds that under a potential gradient of a volt per cm. this velocity is about 2·2 cm./sec.

Schweidler[3] has also made experiments on the relation between

[1] Schweidler, *Wien. Ber.* cvii. p. 881, 1898.
[2] Buisson, *Comptes Rendus*, cxxvii. p. 224, 1898.
[3] Schweidler, *Wien. Ber.* cviii. p. 273, 1899.

the current and the strength of the electric field over a wider range than in Stoletow's experiments: his results for air at atmospheric pressure are shown by the curve (Fig. 115). It will be noticed that when the strength of the field approaches the value

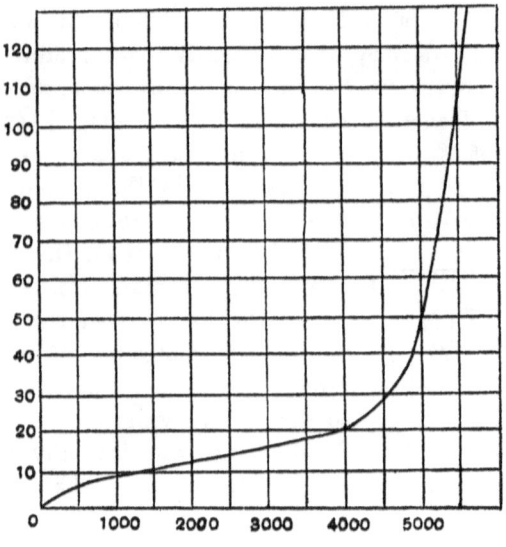

Fig. 115.

5730 volts, which is the strength required to produce a spark in the dark across the plates which were 3 mm. apart, there is a very great increase in the current.

This rapid increase of the photo-electric effect in the neighbourhood of the sparking potential was first observed by Kreusler[1]. The relations between the leak from plates of iron, aluminium, copper, zinc, silver and amalgamated copper, and the strength of field are represented in the curves given in Fig. 116; the abscissæ measured from 0 represent the difference between the electromotive force applied and that required to produce discharge in the dark. The increase in the leak is so great that it cannot be adequately represented in a moderately sized figure; a better idea in the case of the zinc plate can be derived from the following table given by

[1] Kreusler, *Ann. der Phys.* vi. p. 398, 1901.

133] IONISATION BY LIGHT. PHOTO-ELECTRIC EFFECTS 461

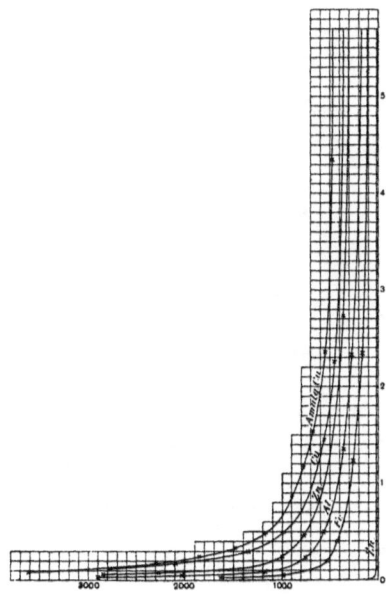

Fig. 116.

Kreusler. V is the potential difference and i the current; the potential required to produce a spark was 4060.

V	i ($1 = 10^{-10}$ amp.)	V	i ($1 = 10^{-10}$ amp.)	V	i ($1 = 10^{-10}$ amp.)
4040	136·39	3050	0·19	3300	0·36
3970	25·67	2540	0·09	3440	0·58
3780	5·88	1760	0·06	3640	1·36
3700	2·40	1170	0·05	3710	1·98
3590	1·39	1760	0·06	3760	3·88
3440	0·70	2530	0·08	3970	21·09
3300	0·40	3060	0·17	4040	80·51

These figures also show evidence of an effect often observed when using ultra-violet light—the decrease of sensibility with the time; thus of the two readings taken with the greatest potential difference the later one was very appreciably less than the earlier one. This 'fatigue' of the plates is probably due to a chemical change in the surface layer caused by the light.

The increase in the rate of leak when the electric field approaches a certain strength is also very evident when the gas is at lower pressures. The effect of altering the pressure of the gas was

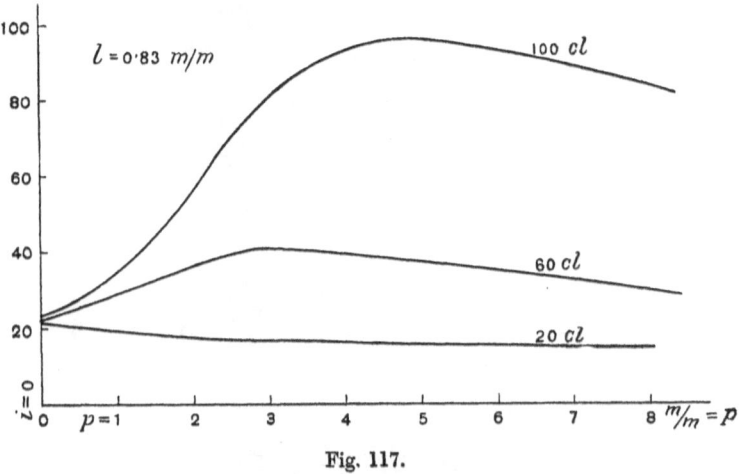

Fig. 117.

first investigated by Stoletow[1], and subsequently by Schweidler[2] and Lenard[3]. Stoletow showed that as the pressure was diminished, starting from atmospheric pressure, the current slightly increased, the change in the current being small compared with that in the pressure; on carrying the reduction of pressure still further, a stage was reached (if the strength of the field was not too small) when the current increased rapidly as the pressure diminished; this went on until the current reached a maximum value, after which it began to decline, but at the lowest obtainable pressures it had a finite value which was independent of the strength of the electric field.

The variation of the current with the pressure when the potential difference remains constant is exhibited in the curves (copied from Stoletow's paper) shown in Fig. 117; the distance between the plates was ·83 millimetre and the figures on the curves indicate the potential difference expressed in terms of Clark's cells (1 Clark's cell = 1·43 volts).

[1] Stoletow, *Journal de Physique*, ii. 9, p. 468, 1890.
[2] Schweidler, *Wien. Ber.* cviii. p. 273, 1899.
[3] Lenard, *Ann. der Phys.* ii. p. 359, 1900.

The values of the current at a series of pressures, when the distance between the plates was 3·71 mm. and the potential difference about 90 volts, are shown in the following table:

Pressure in millimetres	Current	Pressure in millimetres	Current	Pressure in millimetres	Current
754	8·46	2·48	74·7	0·105	65·8
152	13·6	1·01	105·8	0·0147	53·8
21	26·4	0·64	108·2	0·0047	50·7
8·8	32·2	0·52	102·4	0·0031	49·5
3·3	48·9	0·275	82·6		

We see by an inspection of the curves in Fig. 117 that the pressure at which the current is a maximum increases with the electric force between the plates: Stoletow has shown that p_m, the pressure at which the current is a maximum, is proportional to E/d, where d is the distance and E the potential difference between the plates; this law may also be expressed by saying that if λ is the mean free path of a molecule at the pressure for maximum current, when the electric force is X, then $X\lambda$ is constant. The curves in Fig. 117 show that at very low pressures the current is independent of the strength of the electric field, *i.e.* is saturated. This is also well shown by the following numbers taken from Lenard's paper. V is the potential difference in volts and i the current; the vacuum was the best obtainable, the pressure being less than ·002 mm. of mercury.

V	i		V	i	
45000	24·5 × 10⁻¹⁰ Coulomb/sec.		500	23·4 × 10⁻¹⁰ Coulomb/sec.	
25000	26·6	,, ,,	120	21·9	,, ,,
8900	22·5	,, ,,	14	19·9	,, ,,
4100	24·8	,, ,,	9	15·9	,, ,,
3110	24·5	,, ,,	1	7	,, ,,
1300	24·5	,, ,,	0	4	,, ,,

Stoletow has shown that when the pressure is of the order of p_m, the current does not depend merely upon the value of E/d, where E is the potential difference and d the distance between the plates, for with a constant value of E/d the current at these pressures increases rapidly with the distance between the plates.

Schweidler[1] has given curves representing the relation between the current and the potential difference at several pressures. Similar curves have been obtained by Varley[2] at the

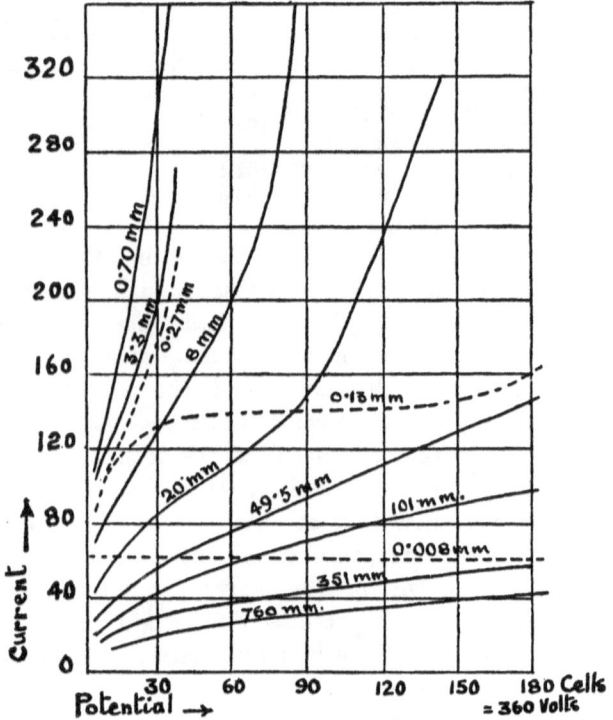

Fig. 118. Distance between the electrodes 3·5 mm. Gas hydrogen.

Cavendish Laboratory, some of these are reproduced in Figs. 118 and 119. The curves show three distinct stages; the first when the electric force is weak, then the current increases rapidly with the electric force; the rate of increase gradually dies away as the electric force increases, and the second stage is reached when the current only varies slowly, at some pressures hardly at all, with the electric field; with still larger electric forces a third stage is reached when the current increases rapidly with the electric force and also with the distance between the electrodes.

[1] Schweidler, *Wien. Ber.* cviii. p. 273, 1899.
[2] Varley, *Phil. Trans.* A, ccii. p. 439, 1904.

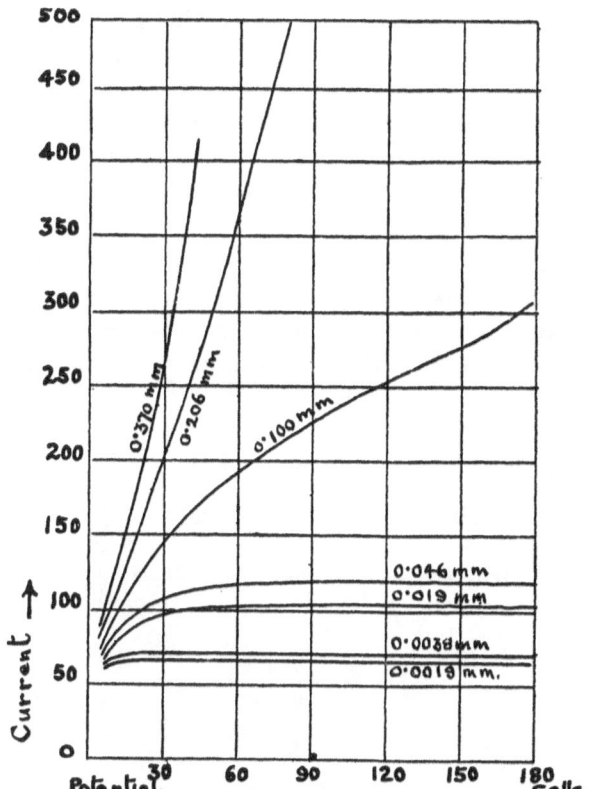

Fig. 119. Distance between the electrodes 3·5 mm. Gas air.

Theoretical Considerations relating to the Connection between the Current and the Strength of the Electric Field.

134. It will be convenient to confine our attention in the first place to electric fields which are weak compared with those required to produce discharge in the dark. The view we take of the action of the ultra-violet light is that under the action of this light the metal emits from each unit area in unit time a certain number of electrons; that these electrons soon, when gas surrounds the metal, get attached to one or more molecules of the gas and form negative ions.

The negative ions accumulate in the space between the plates and, since like the molecules of a gas they are moving about in all directions, some of them will strike against the negative plate and give up their negative charges to it: the steady state will be reached when the negative electricity given out by the plate struck by the ultra-violet light, minus the quantity of electricity given up to the plate by the ions striking against it, is equal to the current through the gas.

To express this condition mathematically let us take the case of two parallel plates one of which is illuminated by ultra-violet light; let X be the electric intensity at right angles to the plates, the illuminated plate being at the lower potential, n the density of the negative ions between the plates[1], k the velocity of the negative ion under unit electric force, e the charge on an ion, i the current through unit area, then
$$i = nekX.$$

The number of negative ions which in unit time strike against unit area of the plate is (see p. 351) equal to $\dfrac{cn}{\sqrt{6\pi}}$, where c is the average velocity of translation of the negative ions. If I is the number of electrons emitted by unit area of the plate in unit time, then
$$Ie = \frac{cne}{\sqrt{6\pi}} + i,$$

or since
$$i = nekX,$$
$$Ie = \frac{ci}{\sqrt{6\pi}kX} + i,$$

or
$$i = \frac{Ie\sqrt{6\pi}kX}{c + \sqrt{6\pi}kX}.$$

When X is so small that kX is small compared with c, i.e. when the velocity of the ion due to the electric field is small compared with the mean velocity of translation of the ion,
$$i = Ie\sqrt{6\pi}kX \frac{1}{c}.$$

Thus when the electric force is small the current obeys Ohm's law. For larger values of the electric force the rate of increase of

---

[1] These ions start as electrons but soon cling to molecules of the gas and become ions except in the case of the inert gases, nitrogen, and hydrogen, when very pure.

the current with the electric force diminishes and the current approaches the saturation value Ie.

Since
$$k = \frac{1}{2}\frac{e}{m}\frac{\lambda}{c},$$

m being the mass of the ion and λ its mean free path, we have

$$i = \frac{IeX}{X + \dfrac{2mc^2}{\lambda e \sqrt{6\pi}}},$$

and when X is small

$$i = \frac{IeX\sqrt{6\pi}\lambda}{2mc^2}.$$

Since mc^2 depends upon the temperature only, the rate of increase of the current with the electric force will be proportional to the mean free path of the ion and so will be greater at low pressures than at high ones. If the pressure of the gas remains constant while the temperature varies, since both λ and mc^2 are proportional to the absolute temperature the ratio of the two will be independent of the temperature, and the variation of i with the temperature will depend solely upon the variations of I.

135. Even when there is no external electric field the diffusion of the ions will produce a small current due to the drift of the negative ions from the illuminated electrode. If n is the density of the negative ions at the illuminated electrode, n' that at the other electrode, the gradient of density is $\dfrac{n-n'}{l}$, where l is the distance between the electrodes; hence if D is the coefficient of diffusion of the negative ions, i the current through unit area,

$$D\frac{(n-n')e}{l} = i.$$

We have also
$$Ie - \frac{cne}{\sqrt{6\pi}} = i,$$

and
$$\frac{cn'e}{\sqrt{6\pi}} = i.$$

Hence
$$Ie - \frac{c}{\sqrt{6\pi}}\frac{li}{D} = 2i,$$

or
$$i = \frac{Ie}{2 + \dfrac{c}{\sqrt{6\pi}} \dfrac{l}{D}}.$$

136. Elster and Geitel[1] and Stoletow[2] found that, with the strength of electric field used by them, the rate of escape of electricity through carbonic acid gas was much greater than that through air or oxygen. Breisig[3] on the contrary found that the

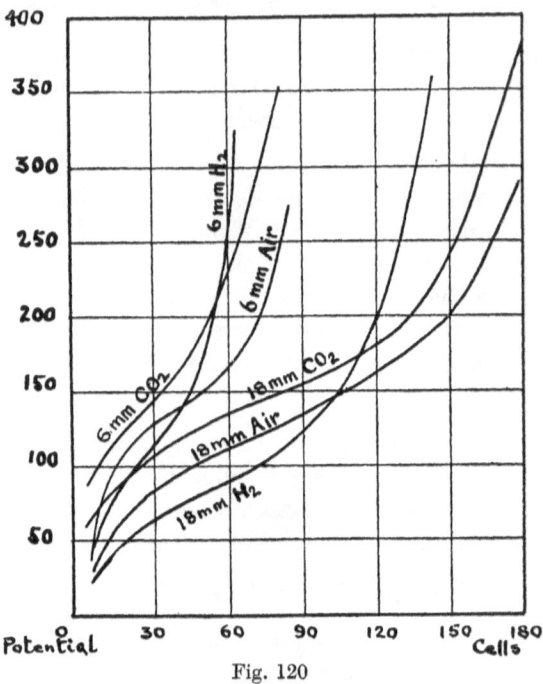

Fig. 120

rate was less through CO_2 than through air: and that it was exceptionally large through the vapours of ether and alcohol. The rate of leak varies so much with the potential difference that a comparison of the rates of leak for the different gases with only one value for the potential difference is not satisfactory and gives little information. What is really wanted is a comparison for the different gases of the curves representing the relation

[1] Elster and Geitel, *Wied. Ann.* xli. p. 161, 1890.
[2] Stoletow, *C. R.* cvii. p. 91, 1888.
[3] Breisig, *Bonn Diss.* 1891; *Wied. Beiblätter*, xvii. p. 60.

between the current and the potential difference. This has been done by Varley[1], one of whose curves is given in Fig. 120. It will be seen that the current due to the photo-electric effect is with weak electric fields greater in air than in hydrogen, with strong fields it is greater in hydrogen than in air. It would also be desirable to have these curves drawn for ultra-violet light of different wave-lengths. The different gases might also cause the currents to differ by altering the surface of the metal either by combining with it or by condensing on its surface.

137. We shall now go on to consider the sudden increase in the current which occurs when the electric field approaches the intensity required to produce a discharge in the dark. We can, I think, explain this by means of some considerations first advanced by the author[2] to explain the ionisation produced when a strong electric field causes a discharge to pass through a gas. When cathode or Lenard rays pass through a gas, the gas becomes a conductor, $i.e.$ it is ionised; hence we see that when very rapidly moving ions pass through a gas and come into collision with its molecules the gas is ionised, the energy required for the ionisation coming from the kinetic energy of the rapidly moving ions. Inasmuch as the ionisation of a molecule of a gas requires the expenditure of a finite amount of work, a moving ion cannot ionise a molecule against which it strikes unless its kinetic energy exceeds a certain critical value, but when its energy does exceed this value then a certain fraction of the number of collisions between the ions and the molecule will result in ionisation. Now when the ions are moving in an electric field, the kinetic energy acquired by the ions will increase as the strength of the field increases, and when the field is strong enough to make the kinetic energy of the ions exceed the critical value, the ions by their collisions will give rise to new ions, and thus there will be an increase both in the number of ions and the current through the gas: it is this increase which is so marked a feature of the currents produced by ultra-violet light when the electric field is strong.

If l is the mean free path of an ion, X the electric force, e the

[1] Varley, *Phil. Trans.* A, ccii. p. 439, 1904.
[2] J. J. Thomson, *Proc. Camb. Phil. Soc.* Feb. 5, 1900; *Phil. Mag.* v. 50, p. 278, 1900.

charge on the ion, then the mean kinetic energy given to the ion by the electric field is Xel; when therefore Xel exceeds a certain critical value, ionisation will take place in a certain fraction of the collisions; let us denote this fraction by $f(Xel)$, $f(x)$ being a function of x which vanishes when x is less than a certain value. If there are n ions per cubic centimetre, then the number of collisions in unit time is equal to nv/l, where v is the average velocity of translation; hence the number of ions produced in unit time per unit volume is $\dfrac{nv}{l} f(Xel)$. A certain number of collisions may result either in the recombination of the ion, or the attachment of the ion to the system against which it collides, so that the ion ceases to be available for carrying the current[1]; let a fraction β of the collisions result in the destruction of the ion as an ionising agent, then the number of these ions which disappear from a cubic centimetre of the gas in unit time is $\beta \dfrac{nv}{l}$[2], hence the excess of the ions produced over those which disappear is equal to

$$\frac{nv}{l} \{f(Xel) - \beta\} \quad \ldots\ldots\ldots\ldots\ldots\ldots(1).$$

We have by the equation of continuity, if u is the average velocity of translation parallel to the axis of x,

$$\frac{dn}{dt} + \frac{d}{dx}(nu) = \frac{nv}{l}\{f(Xel) - \beta\} \quad \ldots\ldots\ldots(2).$$

Now when the ions are moving so rapidly that they have sufficient kinetic energy to act as ionising agents, their velocity must be mainly due to the electric field, since when this field is absent no ionisation is produced. Hence we have approximately

$$v = u.$$

When things are in a steady state we have by (2)

$$\frac{d}{dx}(nu) = \frac{nu}{l}\{f(Xel) - \beta\};$$

---

[1] In the case we are considering, where the ions are initially electrons, this will practically be the case if they become attached to a molecule, owing to the much smaller mobility and ionising power of the system so formed.

[2] We have here neglected the loss of ions due to the recombination of positive and negative ions in comparison with that due to the collision of the ions with the molecules.

integrating we get
$$nu = C\epsilon^{\frac{1}{l}\int_0^x (f(Xel)-\beta)\,dx},$$
or if as a first approximation we regard X as constant we have
$$nu = C\epsilon^{\frac{x}{l}(f(Xel)-\beta)}$$
If the current has reached the saturation stage before ionisation begins, then $nu = I$ when $x = 0$, x being measured from the illuminated plate, hence
$$nu = I\epsilon^{\frac{x}{l}(f(Xel)-\beta)};$$
if d is the distance between the plates, then i the current is the value of nue when $x = d$, since the positive ions carry none of the current at this plate, thus
$$i = eI\epsilon^{\frac{d}{l}(f(Xel)-\beta)} \quad\ldots\ldots\ldots\ldots\ldots(3).$$
When this additional ionisation sets in, the current with a constant value of X increases with the distance between the plates; this effect has been observed by Stoletow[1]. As long as the ionisation is confined to that produced at the metal plate by the ultra-violet light, the current is determined by the electric force, *i.e.* i is a function of X and not of d; when however the secondary ionisation occurs i is a function of both X and d.

The point at which the secondary ionisation begins is when Xel has a certain definite value; as l the mean free path of an ion is inversely proportional to the pressure, the value of X required to start the secondary ionisation will be directly proportional to the pressure.

It is evident that the current cannot go on continually increasing as the pressure diminishes, for, when the free path gets comparable with the distance between the plates, there will be very few collisions, and therefore little if any secondary ionisation; in the limit when the pressure is indefinitely reduced, the number of ions reaching the plate not exposed to the light must equal the number leaving the illuminated plate, hence with our previous notation the limiting current will be equal to Ie.

The value of the free path at the pressure when the current is a maximum is by equation (3) determined by finding the value of l which makes $\{f(Xel) - \beta\}/l$ a maximum; this condition

[1] Stoletow, *Journal de Physique*, ii. 9, p. 468, 1890.

gives $f'(Xel) Xel = f(Xel) - \beta$, an equation to determine Xel; thus when the current is a maximum Xl has a constant value, this coincides with Stoletow's result that if p_m is the pressure at which the current is a maximum, X/p_m is constant.

137·1. Townsend[1] has made a large number of experiments on this effect. He takes $\beta = 0$ and writes α for $\frac{1}{l} f(Xel)$. He finds that the relation between current and voltage is well represented by his formula, and confirms Stoletow's result that X/p_m is constant. He has also shown that his values for α can be represented by the formula $\dfrac{\alpha}{p} = Ne^{-\frac{NVp}{X}}$, where N is the number of collisions made by an ion in passing through 1 cm. of the gas at 1 mm. pressure. This formula can be deduced on theoretical grounds if it is assumed that the collisions of the electrons with the molecules are inelastic, and that if the energy with which the electron strikes the molecule exceeds Ve, ionisation will occur. Partzsch has shown that the agreement between theory and experiment may be somewhat improved by taking account of the fact that ionisation by collision cannot occur till the electron has moved a distance V/X from the plate.

Townsend's results are shown in the following table, compared with the values obtained by the Franck and Hertz method. It

	Ionisation potentials in volts		
Gas	Townsend	Partzsch	Low pressure method
Air	25·0	27·1	—
N_2	27·6	27·9	16·9 (Mohler and Foote)
O_2	—	23·9	15·5 (Mohler and Foote)
H_2	26·0	27·8	{13·6 atom (K. T. Compton) 15·9 molecule (Smyth)
CO_2	23·3	23·5	—
HCl	16·5	—	13·7 (Foote and Mohler)
H_2O	22·4	—	—
A	17·3	—	15·1 (Horton and Davis)
He	14·5	—	25·7 (Horton and Davis)

[1] Townsend, *Electricity in Gases*, chap. viii.

will be seen that, except in the case of the inert gases, they are too large. The discrepancy is perhaps accounted for by the neglect of β.

It must be remembered also that whenever ionisation, followed by recombination, is occurring, there will be radiation emitted, mostly of a very absorbable kind, which will cause the emission of electrons from the metal electrodes and perhaps from the gas. The emission from the cathode will go to increase that due to the outside illumination. This effect is of course familiar in the Franck and Hertz methods of measuring ionisation and resonance potentials. As the radiation is very easily absorbed by the gas it will be most effective at low pressures, but it will always somewhat modify the results of the last four paragraphs.

138·1. The magnifying effect of gas at a suitable pressure on the photo-electric effect, which may easily reach a hundredfold or more, is made use of in the design of photo-electric cells for the measurement of small intensities of light. The alkali metals have chiefly been used for this purpose, and have proved very successful. The sensitivity of the cell can be much increased by passing a slow discharge through it when it contains hydrogen. This causes the formation of a hydride and probably turns the rest of a metal into a very sensitive colloidal form. This appears both to shift the threshold to a larger wave-length and also to increase the general emission. It is then filled with an inert gas at a pressure of about 1 mm., this allows of ionisation by collision and does not cause surface change. Elster and Geitel[1] have prepared cells sensitive to 3×10^{-9} ergs per sq. cm. per sec. for blue light, and 2×10^{-7} ergs per sq. cm. per sec. for orange light, the former being more sensitive than the human eye. For the design and uses of such cells see Hughes' *Report on Photo-electricity*, p. 107. Although each separate cell gives accurate and consistent results, it has not been found possible to make different cells identical in properties, so they need calibrating individually.

Photo-electric Fatigue.

139. The rate of emission of electrons from some metals is much greater when a clean surface is first exposed to ultra-

[1] Elster and Geitel, *Phys. Zeits.* xiv. p. 741, 1913.

violet light than it is after the exposure has lasted for some time. The cause of this effect, which is known as 'photo-electric fatigue,' has not yet been fully made out, and there are some discrepancies in the results of the experiments which have been made upon it. There seems no doubt that it varies greatly (1) for different metals, (2) with the gas in which the metal is placed, and (3) with the quality of the ultra-violet light. Thus Ladenburg[1] found that aluminium in a high vacuum showed no fatigue, whereas the fatigue for silver was very well marked; the surface of the silver was roughened by the ultra-violet light, suggesting that particles of silver had been torn from the surface, while the aluminium surface was not affected. Ladenburg showed that the state of polish of the surface had great influence on the photo-electric effect. Varley found that while zinc became rapidly fatigued in air it did not do so in hydrogen, in this case the fatigue might be ascribed to oxidation of the surface. Fatigue however is by no means always due to oxidation, for Lenard[2] has observed it in the best vacuum he could obtain. Schweidler[3] found that fatigue did not depend upon the electrification of the surface, and that it took place just as rapidly when the surface was positively electrified, when no electricity was escaping from the plate, as when the plate was negatively charged and electricity was streaming out of the plate. Kreusler[4] and Buisson[5] found that fatigue did not take place in the dark and thus was due to the action of light. Hallwachs[6], who worked with surfaces of Cu, CuO and Cu_2O, on the other hand came to the conclusion that fatigue took place just as rapidly in the dark as in the light; he found that it was very much less in small vessels than in large ones or in the open air and he attributed it to the effect of ozone.

139·1. A very large amount of work has been done on photo-electric fatigue, but without leading to any very conclusive results. The success of photo-electric cells as measures of light intensity shows that fatigue is not an essential part of photo-electric phenomena.

[1] Ladenburg, *Ann. der Phys.* xii. p. 558, 1903.
[2] Lenard, *Ann. der Phys.* xii. p. 449, 1903.
[3] Schweidler, *Wien. Ber.* cxii. 11*a*, p. 974, 1903.
[4] Kreusler, *Ann. der Phys.* vi. p. 398, 1901.
[5] Buisson, *Journ. de Phys.* iii. 10, p. 597, 1901.
[6] Hallwachs, *Phys. Zeits.* v. p. 239, 1904.

Hughes points out that in these cells the electrodes are platinum sealed into glass and that there are no taps, wax joints or other sources of contamination. Other experimenters have however found fatigue in the best vacua easily obtainable. In all probability fatigue is due to a variety of causes. One seems to be the formation of a thin film of hydrogen peroxide which has a great absorption for ultra-violet light, but the most common is probably a modification of the surface layer leading to a change in the contact difference of potential and consequent shift of the apparent 'threshold.' Cases are known in which photo-electric effects increase with time[1]. The effect does not seem one of great theoretical importance as regards the action of the light itself. Reference should however be made to the result obtained by several observers that very careful removal of gas films and absorbed gas, as for example by repeated distillation *in vacuo*, renders a metal almost completely insensitive to light, and it has even been suggested that the presence of gas is essential for the existence of a photo-electric effect. The work of Hughes however suggests that what is being observed is a change in the 'threshold' connected with a change in the contact potential difference.

Influence of Temperature on the Photo-electric Effect.

140·1. A number of measurements have been made on the effect of temperature on the photo-electric current. Those made with the active surface in the presence of gas almost all showed a marked effect, generally, but not always, an increase with the temperature. Zeleny[2] found a marked lag, due presumably to the time taken by the surface layer of gas to adjust itself to the changed temperature. Experiments made *in vacuo* however have mostly shown no effect (see for example Nielsen)[3]. This holds both for the total current and for the maximum velocity of emission. The effects observed in the presence of gases must thus be supposed to be secondary ones due to the influence of the gas on the surface layer. The temperature must not be so high that any appreciable thermionic effect is occurring, as this would mask the photo-electric

[1] For an account of the work done up to 1913 see Allen's *Photo-electricity* and Hughes' *Photo-electricity*.
[2] Zeleny, *Phys. Rev.* xii. p. 321, 1901.
[3] Nielsen, *Phys. Rev.* xxv. p. 30, 1925.

effect, which is normally far smaller. Crew[1] has however investigated some cases in which the extra current due to light is comparable with the thermionic current, and much larger than the light causes on the cold surface. He found that oxide-coated filaments of platinum showed the effect, while plain platinum and tungsten did not. It appears to be due to chemical action between the oxides and the platinum, see § 132·2. For a detailed discussion of the effect of temperature the reader is referred to the treatises of Allen and Hughes.

Low Temperatures.

140·2. Ives[2] found a potassium surface in a high vacuum showed a decreased sensitivity on cooling from 20° C. to − 180° C. corresponding to a change in the work function of about ·2 volts. Hornbeck[3] finds a similar result and attributes it to a change in crystal structure. Burt[4] however working on a sodium surface found no change in this range if the surface was kept perfectly clean. Different degrees of contamination produced temperature effects of either sign. Burt formed his surface by driving the sodium electrolytically through the walls of a high vacuum soda glass lamp bulb.

Effect of Ultra-violet Light on the Condensation of Moisture.

141. C. T. R. Wilson[5] has studied the action of ultra-violet light on gases from the point of view of the effect produced by the light on the formation of clouds. His results with intense light have already been described in Chap. VIII; we shall only consider here the effects obtained with very feeble light, as the effects have a direct bearing on the question of the ionisation of air by ultra-violet light, though they do not touch the question as to the effects produced by the extremely absorbable light studied by Lenard. Wilson found that with very feeble ultra-violet light clouds were produced by expansion when this exceeded a definite amount, just as in the case of a gas ionised by X-rays, and that the amount of expansion required was just the same for the ultra-

[1] Crew, *Phys. Rev.* xxviii. p. 1265, 1926.
[2] Ives, *Opt. Soc. Amer. Jour. and Rev. Sci. Inst.* viii. p. 551, 1924.
[3] Hornbeck, *Phys. Rev.* xxiv. p. 631, 1924.
[4] Burt, *Phil. Mag.* xlix. p. 1168, 1925.
[5] C. T. R. Wilson, *Phil. Trans.* A, cxcii. p. 403, 1899.

violet light as for these rays: this at first sight looks as if the ultra-violet light ionised the gas. Wilson, however, found that the clouds produced by ultra-violet light differed from those produced by X-rays, inasmuch as the former were not affected by strong electric fields, whereas the formation of the latter was almost entirely prevented by such fields. If the clouds due to ultra-violet light had been due to the ionisation of the gas, the ions would have been removed by the field and the clouds stopped. At the same time the coincidence between the expansions required for the formation of clouds under ultra-violet light and when ions are present is so remarkable that it makes us very reluctant to believe that the nuclei are different in the two cases; it seems to me that an explanation which is in harmony with the facts is that charged ions do form the nuclei of the drops formed by weak ultra-violet light, but that these ions are produced during the expansion of the gas and are not present when the gas is at rest. These ions might arise in the following way: we have seen in Chap. VIII that under the action of strong ultra-violet light visible clouds are formed without expansion, these clouds being probably due to the formation of hydrogen peroxide, which mixing with the water lowers the vapour pressure; now when the light is very feeble it seems probable that there may still be a formation of drops of water which, however, in consequence of the very small amount of hydrogen peroxide produced by the feeble light, never grow large enough to be visible. Thus we may regard the air exposed to the ultra-violet light as full of exceedingly minute drops of water; when the expansions take place the air will rush violently past the drops and we get a state of things which in many respects is analogous to the bubbling of gas through water; when, however, air bubbles through water there is, as Lord Kelvin[1] has shown, negative electricity in the air and positive in the water; thus when the air rushes past the water drops we should expect the air to contain negative ions, the positive ions being on the drops; the ions once formed would act as nuclei for clouds if the expansion exceeded the value 1·25. If this view is correct, then we should expect the number of ions produced by an expansion greater than 1·25 to increase with the expansion, for in this case

[1] Lord Kelvin, *Proc. Roy. Soc.* A, lvii. p. 335, 1894.

the expansion has to produce the nuclei as well as deposit the clouds, and the more vigorous the expansion the greater would be the number of nuclei produced.

There are some other considerations which may have a bearing on this question; we have seen (p. 334) that the formation of a fresh water surface is accompanied by the liberation of ions; when the cloud of small drops is formed by the weak ultra-violet light there is a creation of new surfaces of water and the probability of the liberation of ions, the positive ions being carried by small water drops would only move very slowly under an electric field and so might not be cleared out by any field it is practicable to apply.

Vincent[1] has made experiments on the clouds formed without expansion by ultra-violet light; he finds that some drops are charged positively, others negatively, while others were without charge. He was not able to detect the presence of hydrogen peroxide.

It is an important meteorological question whether direct sunlight can produce a cloud in the atmosphere without expansion. Wilson was not able to get a cloud in a closed vessel in sunlight with less than the normal expansion 1·25. He points out, however, that the conditions in the open air are more favourable to the production of clouds than those in a closed vessel, for in a closed vessel the drops might diffuse to the sides before they had time to grow to a visible size, while in the atmosphere this way of escape would not be open to them.

Selective Photo-electric Effect.

142·1. Elster and Geitel discovered a very remarkable peculiarity in the photo-electric effect of the liquid sodium-potassium alloy. The photo-electric current showed a marked maximum for a certain angle of incidence, about 60°, but only when the electric vector of the light had a component normal to the surface of the liquid. In one case the ratio of the currents with light polarised with the electric vector in the plane of incidence ($E \parallel$) to that with the electric vector normal to this plane ($E \perp$) was as 117 to 2. The maximum velocity of emission was however the same in the

[1] Vincent, *Proc. Camb. Phil. Soc.* xii. p. 305, 1904.

two cases. Pohl and Pringsheim[1] showed that for $E \parallel$ light the current showed a marked maximum at about λ 3900, while for $E \perp$ light it decreases steadily with increasing wave-length. The effect of course only appears at oblique incidence, since for normal incidence there is no real difference between the directions of polarisation. Pohl and Pringsheim regard the emission for $E \parallel$ light as the sum of two effects, one 'normal' and equal to that for $E \perp$ light, the other a 'selective' effect occurring only over a certain range of wave-lengths. Their results for the selective effect are shown in Fig. 121. Besides this alloy, Na, K, Rb and some of their other

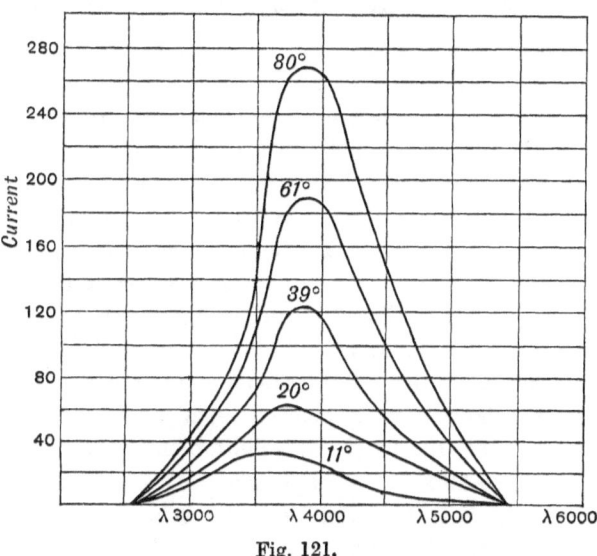

Fig. 121.

alloys show the effect, with varying positions of the maximum, though it is harder to study for solids on account of the difficulty of getting an optically smooth surface. While for a given substance the position of the maximum is constant, the magnitude of the effect varies greatly. Thus for the Na–K alloy the ratio of selective to normal effect at 60° may vary from 300 to 10 or less. Wolf[2], working with the Na–K alloy, finds that the velocity corresponding to the greatest number of electrons is the same for both normal

[1] Pohl and Pringsheim, *Verh. d. Deutsch. Phys. Ges.* xii. p. 215, 1910.
[2] Wolf, *Ann. der Phys.* lxxxiii. p. 1001, 1927.

and selective effects. He used the magnetic deflection method of measuring velocity. It used to be supposed that the existence of a maximum effect at a certain wave-length was proof of a 'selective' effect of this kind, but it now appears that some substances such as Al and Ca show a maximum of a different kind. Thus with Ca the maximum becomes less pronounced as the incidence increases, and there was no abnormal difference between light polarised in different planes. The theory of Richardson's mentioned above leads to the result that all substances should show a maximum effect at two-thirds of the threshold wave length, though this is only partially confirmed by experiments on Al and Na.

The explanation of the selective effect is extremely difficult. In considering it we must first take into account the amount of light absorbed by the surface. Pohl[1] showed from experiments on films of Pt and Cu formed by cathode sputtering, and on Hg, that the ratio of current to light absorbed was the same whatever the plane of polarisation; this is to be considered the normal result. Later Pohl and Pringsheim[2] have shown that in the case of K the maximum is even more marked if the current is reckoned per unit intensity of absorbed light than if it is reckoned per unit intensity of incident light, the maximum being a region of high reflecting power. It has been suggested by Hughes that since strong reflection is associated with strong absorption the maximum may be due to the $E \parallel$ light being absorbed in a very thin layer and so allowing more of the electrons to escape. Some experiments by Miss Frehafer[3] on the reflecting powers of Na and K and on the transparency of thin films, give this theory only partial support. Hughes[4] found that the selective photo-electrons were emitted, on the whole, in directions nearer to the perpendicular to the surface than those of the normal effect.

Gross[5], who found a selective effect in Cu when spluttered in hydrogen, did not obtain it when the metal was spluttered *in vacuo* and considers it to be due to occluded gas in the layer. He used

[1] Pohl, *Verh. d. Deutsch. Phys. Ges.* x. pp. 339, 609, 715, 1909.
[2] Pohl and Pringsheim, *Verh. d. Deutsch. Phys. Ges.* xv. p. 173, 1913.
[3] Frehafer, *Phys. Rev.* xv. p. 110, 1920.
[4] Hughes, *Phys. Rev.* x. p. 490, 1917.
[5] Gross, *Zeits. f. Phys.* xxxi. p. 637, 1925.

light of wave-length 2537 and examined the different behaviour of the surface for light polarised in, and perpendicular to, the plane of incidence. Gold also showed an effect, silver and platinum were doubtful. Admitting air or water vapour destroyed the selective effect, which was always considerably less marked than for the alkalis. Ives[1], and Ives and Johnsrud[2] have obtained some very interesting results with the alkali metals and their alloys. Using thin films deposited *in vacuo*, they found that the thickness of the film was of great importance. Taking the difference in effectiveness of $E \parallel$ and $E \perp$ light as a test of the existence of a selective effect, they found this large for a certain thickness of film, while for very thin films, and for the solid metal, it was hardly more than could be accounted for by differences in absorption of the light. For thin films the nature of the metal on which they were deposited was important. No definite signs of maxima at certain wave-lengths were found, certainly not at the wave-lengths given by Pohl and Pringsheim. Work on liquid surfaces of the metals and alloys showed that the sodium-potassium alloy, originally studied by Elster and Geitel, gave the effect much more than any other. In most other cases the effect was hardly more than might be accounted for by differences in absorption. Maxima were found for certain wave-lengths in most cases, but they were not at the wave-lengths found by Pohl and Pringsheim, nor were they constant for different conditions of the surface. They seemed however to bear some relation to the threshold wave-lengths being generally roughly two-thirds of it (see p. 446). The large difference in the sensitivity of the sodium-potassium alloy to $E \parallel$ and $E \perp$ light was found to disappear on heating. Ives and Stilwell have investigated other alloys of sodium and potassium, besides the one with equal numbers of atoms previously studied. Plotting the ratio of the emission $E \parallel$ to $E \perp$ against composition they find three maxima, one being the original alloy. They observed cases in which crystalline plates were formed on the surface of the molten alloy or metal. These gave abnormally high ratios. Ives considers that part of the difference of his results from those of Pohl and Pringsheim is due to their working with

[1] Ives, *Astrophys. J.* lx. p. 209, 1924.
[2] Ives and Johnsrud, *Astrophys. J.* lx. p. 231, 1924.

rough or colloidal surfaces, except in the case of the Na-K alloy, while his showed specular reflection. He considers that the selective effect requires a peculiar structural arrangement of the atoms which usually occurs only for a certain range of thickness, and in the case of the Na-K alloy is disturbed when the thermal vibration gets too large. This still however leaves unexplained how the difference between the effect of light polarised in the two planes arises, unless we can suppose that the structure forms a system of stationary waves as in Wiener's experiment.

Effects with Cathode Rays.

143·1. A peculiar effect has been observed by Langmuir[1], which he interprets as showing that when light is incident on a caesium surface this acquires the power of reflecting electrons with little loss of energy (*e.g.* electrons of 100 volts energy may retain 80 volts after reflection). The electrons were produced from a hot filament and the 'reflection' was shown by a second caesium surface receiving a negative charge in addition to the photo-electrons from the first surface. In the short account given there is no drawing of the apparatus, and it is not certain how far the effect might be accounted for by a space charge due to photo-electrons from the first surface influencing the motion of the thermo-electrons.

A rather similar effect has been observed by Dember[2] who finds that bombardment of an aluminium surface with electrons of about 100 volts energy, increases the photo-electric current manyfold, even after allowance has been made for the δ-rays emitted. This can be accounted for by a shift of the photo-electric threshold due to the action of the cathode rays on the surface, but the same result would be caused by an effect such as Langmuir claims.

[1] Langmuir, *Science*, lviii. p. 389, 1923.
[2] Dember, *Zeits. f. Phys.* xxxiii. p. 529, 1925.

NAME INDEX

Abraham, 258
Aitken, 312
Akulov, 455
Alberti, 250, 264
Allen, 5, 476
Altberg, 90, 128
Anderson, 450
Andrade, 402, 412, 422, 423
Andrén, 334
Arnold, 364, 455
Arrhenius, 349, 403, 407, 410, 420
Aselmann, 184
Ashley, 173
Aston, 276, 279, 282, 286, 389, 390

Baedeker, 63
Baeyer, 353
Bailey, 61, 143, 147, 148
Bairstow, 97
Baker, 319
Bar, 306
Barnes, 414, 417, 424
Barton, 444
Barus, 310
Bazzoni, 452
Beam, 438
Beattie, 394
Becker, A., 101, 128, 185, 241, 264, 358
Becquerel, 254, 257, 264, 358
Bělǎr, Maria, 132
Bennett, 414, 417, 424
Bennewitz, 442, 444
Bergwitz, 8, 10
Bessel, 211
Bestelmeyer, 252, 253, 260, 264
Bijl, van der, 23, 40
Birckenbach, 286
Blackwood, 49, 186
Blanc, 123, 124, 152, 155, 172
Bloch, 40, 181, 447
Blondlot, 338
Bohr, 257, 456
Bowen, 9–11
Bowman, L. L., 105
Boys, 2
Bragg, W. H., 28
Branly, 341
Braun, 400, 402
Breisig, 468
Bridgman, 357
Broglie, M. de, 184, 185
Bronsted, 285
Brown, F. C., 358, 359, 392
Brown, S. L., 357
Bryan, 414, 415, 426

Bucherer, 260, 263, 264
Buisson, 318, 459, 474
Burt, 476
Burton, 5
Busse, 98, 155, 186

Cavallo, 338
Chattock, 119–124, 133, 135
Chenault, 451, 453
Chien Cha, 441
Child, 115, 190, 371
Classen, 252, 264
Colston, 319
Compton, K. T., 11, 134, 175, 176, 439, 442, 472
Congdon, 362
Cooke, 5, 364, 366
Cornelius, 444
Costa, 283
Coulier, 312
Coulomb, 1, 7
Cravath, 125, 126, 181
Crew, 476
Crookes, 5
Cunningham, 187
Curie, M. et Mme, 253

Davis, 472
Davisson, 355, 356, 364, 381, 388, 444
Dawson, 411, 415
Deininger, 347
Del Rosario, 362
Dember, 482
Dempster, 123, 126, 286, 287, 289, 390, 391
Des Coudres, 242, 290
Dixon, 123, 124
Du Bridge, 444
Du Fay, 339
Dunn, 444
Dushman, 356, 381, 444
Du Tour, 339

Eddington, 11
Edison, 347
Ehrenfert, 180
Ehrenhaft, 180
Einstein, 439
Elster, 3, 5, 324, 339, 340, 341, 348, 435–437, 456, 468, 473, 478, 481
Erikson, 23, 30, 36, 38–40, 98, 114, 128, 133, 149, 151–155, 159, 178, 181
Erman, 402
Evans, 127
Ewald, 381

31–2

NAME INDEX

Fleming, 347–349
Foote, 451–453, 472
Fortrat, 264
Fowler, 416
Franck, 77, 83, 97, 105, 122–124, 133, 134, 153, 155, 163, 447, 472
Frehafer, Miss, 480
Fulcher, 122, 127

Garrett, 115, 388, 395, 397, 398
Gehlhoff, 133
Geiger, 307
Geitel, 3–5, 324, 339–341, 348, 435–437, 456, 468, 473, 478, 481
Gerdien, 101
Gerlach, 306, 307
Germer, 355, 357, 362, 364, 381, 444
Giese, 402
Gilbreath, 450
Goekel, 9
Goetz, 358
Gold, 405–407, 424, 426
Grindley, 111, 114, 123, 149, 152, 155, 181
Gross, 480
Guthrie, 338

Haines, 138, 148, 149
Hallwachs, 435, 445, 454, 474
Hamer, 444, 445
Hankel, 402
Haselfoot, 83
Heaps, 433
Helmholtz, R. von, 310–312, 341
Hendren, 30, 35, 40
Henry, 449
Hensel, 77
Herold, 441, 442, 443
Hertz, 2, 61, 163, 206, 435, 447, 472
Herwig, 402
Hevesy, 286
Hess, 9
Hittorf, 379, 402
Holtz, 401
Hönigschmidt, 286
Hoor, 435
Horton, 346, 472
Hughes, 123, 437, 439, 445, 447, 448, 450, 453, 473, 475, 476, 480
Hulbert, 386
Hupka, 262

Ives, 455, 476, 481

Jaffé, 40
Jeans, 11, 156
Jenkins, 386
Jentzsch, 363
Johannot, 329
Johnsrud, 481
Johnston, 257

Jones, 362

Kadesch, 440
Kalendyk, 393
Kaufmann, 235, 240, 258, 264
Kazda, 444
Kelvin, Lord, 312, 325, 477
Keys, 126, 127
Kidner, 381
Kiessling, 312
Killian, 418
Kingdon, 358, 418, 451
Kleeman, 28, 454
Klemperer, 442
Kluge, 442, 444
Knoblauch, O., 437
Kohlrausch, K. W. F., 101
Kolhörster, 9, 10, 11
Koller, 362, 368
Kondratjeff, 391
Kovarik, 115, 123, 124, 126, 130, 151, 161, 178
Kramers, 453
Kreusler, 460, 461, 474
Kunz, 451

Laby, 234
Ladenburg, 438, 474
Lafay, 117
Langevin, 33, 35, 36, 38, 40, 47, 52, 83, 106, 123, 124, 126, 129, 149, 153, 169, 171, 181, 183, 189
Langmuir, 347, 368, 370, 371, 373, 381, 418, 444, 482
Laporte, 115, 117, 155
Lattey, R. T., 105, 123, 124, 126, 130, 131, 135, 145–147, 162
Lawrence, 438, 452
Lecker, 244
Lenard, 235, 236, 241, 248, 250, 264, 311, 312, 318, 402, 422, 437, 438, 446, 447, 449, 453, 455, 462, 463, 474, 476
Lester, 364
Lichtenecker, 438
Liebrich, 364
Linss, 3
Little, 453
Lockrow, 368
Loeb, 65, 106, 123, 125, 126, 128, 133, 142, 143, 147, 160, 173–176, 180, 181, 426
Lorentz, 258
Ludlam, 449, 453

McCallum, 8
McClelland, 23, 24, 78, 101, 102, 181, 184, 186, 342, 378, 379, 398, 399
McClung, 23
Mache, 101
McKeehan, 189
Mackenzie, 299

NAME INDEX

McLennan, 5, 8, 126, 127
Malassez, 241
Marolli, 397
Marx, 396, 403, 407, 421, 426, 431, 438
Mattauch, 305
Matteucci, 1, 2
Maxwell, 66, 165, 168, 169, 187, 188, 439
Mayer, 131, 133, 155, 174
Mazundar, 418
Meissner, 34
Meyer, 306
Mie, 33, 203, 204
Mietner, 153
Millikan, 9–12, 83, 189, 300, 301, 302, 305, 306, 308, 439–441, 444, 452
Milne, 416, 453
Mohler, 449, 451, 453, 472
Moreau, 403, 421
Moseley, 281
Moulin, 28

Nahrwold, 2
Narr, 2
Netchaewa, 455
Neumann, 261, 263, 264
Neureneuf, 400, 401
Newall, 21
Nielsen, 143, 475
Nolan, 148, 149, 186
Noyes, 417

Owen, 252, 346, 382

Pacini, 8, 12
Palmer, 448, 449
Pannell, 96
Partzsch, 454, 472
Paschen, 264
Pettinelli, 397
Phillips, 38, 39, 52, 110, 123, 132, 149, 151, 177, 178
Pidduck, 148
Piersol, 441
Plimpton, 26, 35
Pohl, 105, 123, 124, 456, 479–481
Pollock, 185
Potter, 362
Pouillet, 401
Predwoditelew, 455
Preece, 347
Priestley, 338
Pringsheim, 208, 319, 456, 479–481
Przibram, K., 125, 126, 132, 151, 172, 174, 334

Ramsauer, 175, 176, 422, 449, 453
Ratner, 122, 123, 128, 134, 135, 145
Rayleigh, Lord, 329
Regener, 305, 307
Reynolds, O., 96
Riccati, 211

Richardson, O. W., 344, 346, 347, 352, 353, 356–365, 367, 369, 381–388, 392, 394, 395, 439, 442, 444–446, 452, 480
Richarz, 310–312
Richtmeyer, 456
Riecke, 33
Rieger, 250
Righi, 435
Ritschinsky, 40
Robb, 204
Robinson, 441
Ross, 442
Rossiger, 362
Rothgeiser, 123, 124, 152
Rowe, 381
Roy, 444, 446
Rücker, 329
Rümelin, 25, 40
Russell, 319, 416
Rutherford, 5, 15, 19, 23, 24, 32, 75, 78, 80, 85, 99, 100, 102, 104, 115, 132, 153, 208, 280–282, 291, 307, 378, 437, 459

Sabine, 441
Sackür, 356
Saha, 415–418
Salles, 77, 83
Samuel, 453
Sault, du, 174
Scheel, 307
Schlicter, 381, 444
Schmidt, G. C., 436
Schmidt, H. W., 153, 394, 395, 436
Schottky, 361, 370, 371, 373
Schuster, 239
Schweidler, von, 18, 459, 462, 464, 474
Seeliger, 204
Seemann, 204
Seitz, 240, 264
Selenyi, 349
Serkof, 450
Sheard, 393, 395
Shimidzu, 336
Simon, 240, 264
Smith, K. K., 355
Smithells, 411, 415
Smyth, 449, 452, 472
Stanton, 96
Stark, 403, 450
Starke, 403
Steubing, 450
Stilwell, 481
Stokes, 187, 188, 291
Stoletow, 435, 437, 454, 457–460, 462, 463, 468, 471, 472
Stoney, 257
Stormer, 222
Strutt, 382
Stuhlman, 441, 454, 455
Suhrmann, 381, 382
Sur, 418

Sutherland, 23, 151, 177

Tetrode, 356
Thirkill, 36, 40, 54
Thomson, G. P., 287, 390
Thomson, J. J., 15, 17, 19, 32, 50, 63, 233, 234, 237, 246, 251, 264, 291, 309, 315, 320, 321, 324, 385, 393, 394, 400, 437, 449, 469
Ting, 362
Tizard, 117, 123, 124, 126, 130, 131, 135, 147
Todd, 123, 124, 126, 127, 398
Townsend, 18, 24, 40, 61, 69, 74, 76, 77, 80–83, 117, 135, 143, 147, 148, 175, 176, 298, 311, 319, 472
Trkal, 180
Tufts, 403, 407
Tyndall, 111, 114, 123, 132, 133, 149, 152, 155
Tyndall, J., 317

Van der Byl, 452
Varley, 464, 469, 474
Vincent, 317, 478
Volmer, 391
Volta, 399, 436

Wahlin, 125, 134, 142, 147, 149, 154, 175, 176
Walker, 123, 124, 204
Walter, 394
Warburg, 2, 133, 242
Waterman, 389
Watson, 211, 338

Watt, 432, 433
Wehnelt, 252, 265, 346, 363, 364, 369
Wellisch, 36, 123–126, 128, 132, 136, 143, 149, 153, 173
Wentzel, 455
West, 449, 453
Westphal, 77, 83
Whiddington, 449
Wiechert, 241, 242, 264
Wiedemann, 13, 402
Wien, W., 265, 266, 274
Wiener, 482
Williams, 451
Williamson, 452
Willows, 395, 397
Wilson, C. T. R., 3–5, 11, 28, 291, 294, 298, 299, 312–320, 476–478
Wilson, H. A., 175, 176, 234, 299, 319, 357, 366–368, 378, 379, 382, 395, 396, 402–409, 411–426, 431–434
Wilson, W., 446
Winkelmann, 78
Wolf, 442, 479
Wolf, M., 311, 312
Wolz, 261, 263, 264
Wright, 8
Wulf, 8

Yen, 124, 125, 133, 149
Young, 357, 358, 381, 444, 445

Zachmann, 412, 414, 415
Zeeman, 257
Zeleny, 75, 80, 86, 87, 90, 91, 98, 123, 124, 130, 131, 186, 189, 190, 475

SUBJECT INDEX

(The references are to the pages)

Absorption, effect on photo-electric emission, 454, 481
Age of ions, effect on mobility, 99, 114, 128, 152, 155, 159
Aggregation of molecules round an ion, see Complex ions
Air flow, turbulent, 96, 99
Alternating currents, conductivity of flames for, 424
Anode rays, 389
Atomic volume, effect on photo-electric emission, 445
Aurora Borealis, 349
Auto-photo-electric effect, 446

Canalstrahlen, see Positive rays
Cathode rays, velocity of, 242
effect on photo-electric emission, 482
e/m for, 229 et seq.
Caves, natural conduction in, 6
Cells, photo-electric, 473
Chemical action, ions by, 185
Clouds, formed on dust, 312
formed on ions, 291, 310 et seq.
in organic vapours, 334
produced by metals, 319
produced by ultra-violet light, 316, 318, 476
relative effect of positive and negative ions in producing, 320, 322
theory of production of, on ions, 325 et seq.
Collisions, inelastic, of electrons, 61, 148, 472
ionisation by, 469
of electrons with gas molecules, 142, 147, 160, 163
of ions with gas molecules, effect on recombination, 44
Columnar ionisation, 28
Comet, effect on natural conduction, 12
Complex ions, 51, 61, 127, 178
Condensation, see Clouds
Contact potential difference, effect on photo-electric emission, 443
influence on thermionic emission, 355, 365

Diffusion of gases, 78

Diffusion of ions, 67 et seq.
coefficient of, 75 et seq.
effect of moisture on, 75, 76
effect on current between parallel plates, 211
effect on photo-electric current, 467
effect on thermionic emission, 374
relation with mobility, 79, 81
Direction of emission of photo-electrons, 441, 454
Distribution of electric intensity between charged plates:
effect of difference of mobility of ions on, 204
effect of ionisation due to salt on, 206, 379
effect of ionisation on, 190, 197 et seq.
Doubly charged ions, 83, 273, 388, 390, 391
Drops, see Clouds
Dust, as carrier of ions, 181, 183, 185
clouds formed on, 312

E, determination of, Chap. VII
oil drops used to determine, 300
e/m, determination of, Chap. VI
for electrons, 229 et seq., 264 (table of values)
for positive ions from hot metals, 385
for positive ions from hot salts, 387
for positive ions from hot wire, 290
for positive rays, 265 et seq.
values of, for electron, 233, 236, 237, 238, 264
Efficiency of photo-electric emission, 453, 456
Electro-negative gases, effect on mobility of negative ions, 133, 155
Electrons, collisions with gas molecules, 142, 147, 160, 163
determination of e/m for, 229 et seq., 264 (table of values)
effect of gases on electron emission from heated solids, 340, 365 et seq.
emission from heated alkali metals, 357, 369, 381
emission from heated carbon, 341, 347, 349, 381

488   SUBJECT INDEX

Electrons (*continued*)
 emission from heated molybdenum, 381
 emission from heated oxides, 368, 381
 emission from heated platinum, 340, 366, 381
 emission from heated tantalum, 381
 emission from heated tungsten, 368, 381
 emission from hot solids, table of constants, 381
 free path of, 57
 from the sun, 349
 heat of emission of, 363
 inelastic collisions of, 61, 148, 472
 mobility of, 130, 133, 148, 158, 160–164, 175, 424–433
 produced from heated solids, 338 et seq.
 recombination with positive ions, 60
 reflection of, 354, 482
 released by light, *see* Photo-electric emission
 thermionic, energy distribution among, 358
 values of e/m for, 233, 236–238, 264 (table)
 variation of mass with speed, 257 et seq.
 variation with temperature of emission from heated compounds, 346
 variation with temperature of emission from heated metals, 344, 350 et seq.
Energy, distribution of, among photo-electric electrons, 441
 distribution of, among positive thermions, 392
 distribution of, among thermionic electrons, 358
 of emission of photo-electrons, 438
 of emission of photo-electrons from thin films, 441

Fatigue, photo-electric, 473
Field strength, effect of, on mobility, 134, 136 et seq., 145, 162, 163
Flames, conduction in, Chap. X
 conductivity of, for alternating currents, 424
 effect of electric force on, 400, 402
 effect of salt on conductivity in, 407 et seq.
 effect of temperature on conductivity in, 415
 electric intensity due to magnetic field, 426
 increased resistance due to magnetic field, 433
 ions drawn from, mobilities of, 101, 184, 378, 379, 400

Flames (*continued*)
 maximum current carried by salt in, 433
 mobility of ions in, 378, 379, 418 et seq.
Free path, of an electron, 57
 of ion, effect of charge on, 48

Height, variation of natural conductivity with, 9
Hot bodies, ions from, *see* Thermionic emission
Hot wire, e/m for positive ions from, 290

Intensity, electric, between charged plates, effect of diffusion on, 204
 effect of ionisation due to salt on, 206, 379, 409
 effect of ionisation on, 190, 197 et seq., 395, 403
Intensity, electric, effect of, on mobility, 134, 136 et seq., 145, 162, 163
Intensity of light, photo-electric emission proportional to, 455
'Intermediate ions,' 185
Ionisation by stages, 452
Ionisation potentials, 472
Isotopes, 276
 separation of, 284
 table of, 288
 use of ions from salts to investigate, 389

Lag, absence of, in photo-electric emission, 438
Large ions, groups of, 186
 mobilities of, 101, 181
 theory of mobility, 187
Light, emission of electrons due to, *see* Photo-electric emission
 influence on thermionic emission, 455, 476
Liquids, recombination in, 23, 38, 40
 photo-electric effects in, 454

Magnetic field, effect on motion of ions, Chap. V
 electric intensity in flames, due to, 426
 increased resistance in flames due to, 433
 in electric wave, effect on motion of ions, 227
 motion of free ions in, 217
 motion of ions in gas in, 215
Mass spectrograph, 276 et seq.
Metals, clouds produced by, 319
Mixed gases, mobilities of ions in, 152, 171
Mobility of electrons, 130, 133–148, 158, 160–164, 175, 424–433

Mobility of normal ions, Chap. III; *see also* 397, 398, and 418–433
  difference between positive and negative ions, 86, 90, 130
  effect of age of ions on, 99, 114, 128, 152, 155
  effect of electro-negative gases on, 133, 155
  effect of field strength on, 134 et seq., 145, 162, 163
  effect of moisture on, 75, 96, 114, 131, 155
  effect of polar molecules, 181
  effect of pressure on, 398
  effect of vapours on, 132
  from fine wire, 97, 134
  from heated metals, 342, 378, 379
  from hot salts, 122, 397
  from sharp points, 119, 122
  in flames, 378, 379, 418 et seq.
  in gases other than their own, 152, 154
  in mixed gases, 152, 171
  indeterminate value of, in certain cases, 133 et seq., 162, 163
  of vapours, 125, 151, 172
  relation with rate of diffusion, 79, 81
  relation with rate of recombination, 55
  suggested continuous variation of, 117
  theory of, 156, 165 et seq.
  unusual mobility, 128, 148, 318
  values of coefficients of, 96, 102, 104, 123–125 (tables)
  variation with pressure, 126, 129, 398
  variation with temperature, 149, 176
Mobility of ions drawn from flames, 101, 184, 378, 379, 399
Mobility of 'intermediate' ions, 185
Mobility of large ions, 181, 318
  groups of, 186
Mobility of recoil atoms, 153
Moisture, effect on conductivity of salt vapour, 393
  effect on mobility, 75, 96, 114, 131, 155
Monomolecular reactions, 145, 160

N, value of, 308
Natural conductivity, at sea, 8
  effect of comet on, 12
  effect of pressure on, 5
  effect of time on, 6
  in caves, 6
  in the open air, 3
  unaffected by temperature, 2
  variation with height, 9
Ne, value of, 80
Negative ions, formation of, 61

Negative ions (*continued*)
  mobility different from that of positive ions, 86, 90, 130
  mobility of, from heated salts, 398
  more efficient in producing clouds, 320, 322

Oil drops, used to determine e, 300
Open air, natural conductivity in, 3

Packing fraction, 283, 284
Penetrating radiation, 9
  absorption of, 10, 11
Photo-electric cells, 473
Photo-electric current in presence of gas, 457, 468
  theory of, 465
Photo-electric emission, Chap. XI
  distribution of energy among electrons, 441
  effect of absorption on, 454, 481
  effect of cathode rays on, 482
  effect of contact potential difference on, 443
  efficiency of, 453, 456
  energy of electrons, 438
  energy of electrons from thin films, 441
  from gases, 446 et seq.
  from liquids, 454
  from vapours, 450
  independent of time, 438
  nature of ions, 437
  proportional to intensity of light, 455
  relation with atomic volume, 445
  selective, 478
  threshold of, 440, 444, 445
  threshold of, effect of gas on, 444
  threshold of, for metallic vapours, 457
  variation with temperature, 475
Photo-electric fatigue, 473
Points, mobility of ions from sharp, 119, 122
Polar molecules, 63
  effect of, on mobility, 181
Positive ions:
  e/m for, from hot metals 286, 385
  e/m for, from hot salts, 387
  emission of, by hot metals, 382 et seq.
  emission of, by hot metals, effect of gas on, 391
  emission of, by hot metals, energy of, 392
  from hot wire, e/m for, 290
  less efficient in producing clouds, 320, 322
  mobility different from that of negative ions, 86, 90, 130, 398
  recombination with electrons, 60
  variation of thermionic current with temperature, 385

Positive rays, 134
  e/m for, 265 et seq.
  secondary, 275
Potential distribution, *see* Distribution of electric intensity
Potentials, ionisation, 472
Pressure, effect on mobility, 126, 129, 398
  effect on natural conduction, 5
  effect on recombination, 35, 47

Radiation, penetrating, 9
  absorption of, 10, 11
Recoil atoms, mobility of, 153
Recombination, 19 et seq.
  coefficient, definition of, 20
  coefficient, values of, 27, 36, 40, 53
  effect of collision of ions with gas molecules on, 44
  in liquids, 23, 38, 40
  of electrons and positive ions, 60
  relation with mobility, 55
  theory of, 40 et seq.
  variation with pressure, 35, 47
  variation with temperature, 38, 51
Reflection of electrons, 354, 482

Salts, e/m for ions from heated, 387
  effect of ionisation from, on distribution of electric intensity, 206, 379, 409
  effect on conductivity of flames, 407 et seq.
  emission of positive ions from heated, 394
  maximum current carried by, in flames, 433
  mobility of ions from, 122, 397, 398
  variation of mobility of ions from, with pressure, 398
Saturation current, 17
  thermionic, 344, 370
Sea, natural conductivity over, 8
Selective photo-electric effect, 478
Sensitivity of photo-electric emission, 456, 473
Size of ion, limit to, 65
Space charge, 370, 375, 451
Spectrograph, mass, 276
Steam jet, effect of electric discharge on, 310
Stokes' Law, 187 et seq., 291, 302, 305
Sub-electron, 305
Sun, electrons from, 349

Temperature, effect of, on conductivity of flame, 415
  effect of, on mobility, 149, 176
  effect of, on recombination, 38, 51
  influence on photo-electric emission, 475

Temperature (*continued*)
  natural conduction not affected by, 2
  variation of thermionic emission, 344, 350 et seq.
Theories, of different efficiency of positive and negative ions in producing clouds, 322
  of mobility, 156, 165 et seq.
  of mobility of large ions, 187
  of photo-electric current in presence of gas, 465
  of production of clouds, 325 et seq.
  of recombination, 40 et seq.
  of thermionic emission, 350 et seq
Thermionic emission:
  e/m for negative ions, 342
  e/m for positive ions from hot metals, 385
  e/m for positive ions from hot salts, 387
  effect of contact potential difference, 355, 365
  effect of diffusion on, 374
  effect of gas on current, 375 et seq.
  energy distribution among electrons, 358
  from alkali metals, 357, 369, 381
  from carbon, 341, 347, 349, 381
  from heated compounds, 346
  from heated metals, 338 et seq.
  from heated metals, effect of gases on, 340, 365 et seq.
  from heated metals, mobility of ions, 342, 378, 379
  from heated salts, 374
  from molybdenum, 381
  from oxides, 368, 381
  from platinum, 340, 366, 381
  from tantalum, 381
  from tungsten, 368, 381
  heat absorbed in, 363
  influenced by light, 455, 476
  of positive ions, 382 et seq.
  saturation of, 344, 370
  table of constants, 381
  theory of, 350 et seq.
  variation with temperature, 344, 350 et seq.
Thermo-luminescence, 383
Threshold, photo-electric, 440, 444, 445
  effect of gas on, 444
  of metallic vapours, 451
Time, effect of, on natural conduction, 6
  photo-electric emission independent of, 438
Tracks, Wilson, 334
Turbulence of air flow, 96, 99

Ultra-violet light:
  abnormal mobility of ions produced by, 318
  clouds produced by, 316–318, 476

SUBJECT INDEX

Ultra-violet light (*continued*)
  ions from, *see* Photo-electric emission

Vapours, effect on mobility, 132
  hot, conduction in, 392, 418
  mobility of ions of, 125, 151, 172
  organic, clouds in, 334
  photo-electric emission from, 450
  water, effect on mobility, 75, 96, 114, 131, 155

Velocity of cathode rays, 242

Water vapour, effect of, on conductivity of heated salts, 393
  effect of, on mobility, 75, 96, 114, 131, 155

Wave, electric, effect of, on motion of ions, 227

Wire, mobility of ions from fine, 97, 134

For EU product safety concerns, contact us at Calle de José Abascal, 56–1°, 28003 Madrid, Spain or eugpsr@cambridge.org.

www.ingramcontent.com/pod-product-compliance
Lightning Source LLC
LaVergne TN
LVHW091526060526
838200LV00036B/507